PROTEIN INTERACTIONS
Computational Methods, Analysis and Applications

PROTEIN INTERACTIONS
Computational Methods, Analysis and Applications

Editor

M Michael Gromiha
Indian Institute of Technology Madras, India

World Scientific

NEW JERSEY · LONDON · SINGAPORE · BEIJING · SHANGHAI · HONG KONG · TAIPEI · CHENNAI · TOKYO

Published by

World Scientific Publishing Co. Pte. Ltd.

5 Toh Tuck Link, Singapore 596224

USA office: 27 Warren Street, Suite 401-402, Hackensack, NJ 07601

UK office: 57 Shelton Street, Covent Garden, London WC2H 9HE

Library of Congress Control Number: 2020931570

British Library Cataloguing-in-Publication Data
A catalogue record for this book is available from the British Library.

PROTEIN INTERACTIONS
Computational Methods, Analysis and Applications

Copyright © 2020 by World Scientific Publishing Co. Pte. Ltd.

All rights reserved. This book, or parts thereof, may not be reproduced in any form or by any means, electronic or mechanical, including photocopying, recording or any information storage and retrieval system now known or to be invented, without written permission from the publisher.

For photocopying of material in this volume, please pay a copying fee through the Copyright Clearance Center, Inc., 222 Rosewood Drive, Danvers, MA 01923, USA. In this case permission to photocopy is not required from the publisher.

ISBN 978-981-121-186-7 (hardcover)
ISBN 978-981-121-187-4 (ebook for institutions)
ISBN 978-981-121-188-1 (ebook for individuals)

For any available supplementary material, please visit
https://www.worldscientific.com/worldscibooks/10.1142/11596#t=suppl

Typeset by Stallion Press
Email: enquiries@stallionpress.com

Dedicated to the memory of my beloved father

Contents

Chapter 9   Predicting protein-binding sites in nucleic acids   243

Kyungsook Han

Chapter 10   Docking algorithms and scoring functions   257

*Arina Afanasyeva, Chioko Nagao and Kenji
Mizuguchi*

Preface

The interactions of proteins with other molecules such as proteins, nucleic acids, carbohydrates and small molecules are vital for many cellular functions. The importance of these interactions to folding, stability, binding and function of protein complexes has been extensively studied using experimental and computational approaches. The availability of high-speed computers with the extreme storage capacity encouraged computational biologists to develop efficient algorithms and resources. Further, recent advances in big data analytics and artificial intelligence have narrowed down the gap between computational predictions and experimental observations. This book is devoted to providing a forum for both fundamental aspects and applications of computational methods on different perspectives: databases, sequence and structural analysis, basic concepts to study the interface residues, statistical methods and machine learning techniques, assessment procedures and computer-aided drug design.

The highlight of the book is the coverage of all types of protein complexes (protein–protein, protein–DNA, protein–RNA, protein–carbohydrate and protein–ligand) and wide aspects of investigations ranging from database development to drug design. The web addresses of the available databases and online tools are provided in appropriate chapters. In addition, the literature has been thoroughly surveyed and important methods have been highlighted in detail.

This book will be of immense use and a valuable guide to the students and researchers working on proteins and their interactions as well as to those who have the interest to study protein interactions using computational approaches.

The book is broadly classified into three categories: protein–protein, protein–nucleic acid and protein–ligand interactions. Part I deals with protein–protein interactions with a systematic analysis of structural and dynamical aspects of conserved protein–protein complexes, prediction of binding site residues in structured and disordered regions of proteins from amino acid sequence, prediction of protein–protein complexes using docking methodologies, factors influencing the binding affinity of protein–protein complexes, computational tools for predicting the binding affinity and changes in binding affinity upon mutations as well as the consequence of mutations in functions of protein–protein complexes.

Part II is devoted to protein–DNA and protein–RNA interactions focusing on analysis and prediction. This part includes various databases available in the literature on protein–nucleic acid interactions, structural analysis of protein–nucleic acid complexes, computational approaches for understanding the recognition mechanism, prediction of DNA and RNA binding proteins and their binding sites, prediction of protein binding sites in nucleic acids, docking algorithms and scoring functions as well as the latest developments in protein–RNA docking.

Part III is focused on protein–ligand interactions covering both protein–carbohydrate and protein–ligand interactions. The chapter on protein–carbohydrate interactions includes databases for carbohydrates and protein–carbohydrate complexes, identification and analysis of binding sites, prediction of binding sites, binding affinity and molecular dynamics simulations for understanding the recognition mechanism of protein–carbohydrate complexes. In protein–ligand interactions, the concept of quantitative structure–activity relationship has been highlighted with details on goals, methodology and development of models, dimensionality of properties, correlation and classification problems as well as applications of QSAR to identify lead compounds in drug discovery. Further, the role of protein–ligand interactions has been illustrated with the flexibility of macromolecular targets, sampling of ligands, screening and visualization of protein–ligand interactions. Moreover, docking algorithms and scoring functions for binding free energy estimations have been illustrated along with the advantages and pitfalls of machine learning methods.

In essence, this book would be a valuable resource to students and researchers to strengthen the knowledge about the studies on proteins and their interactions.

M. Michael Gromiha

Acknowledgments

I am deeply indebted to Professor P.K. Ponnuswamy, who introduced me to the field of proteins, and encouraged me to strengthen the knowledge in protein research.

I am grateful to Prof. S. Pongor, Prof. Y. Akiyama, Prof. Y-h. Taguchi and Prof. I. Simon for their continuous support and encouragements.

I sincerely thank Prof. N. Srinivasan, Prof. L. Kurgan, Prof. M. Zacharias, Prof. O. Keskin, Prof. J.M. Shifman, Prof. K. Han, Prof. K. Mizuguchi, Prof. H-B. Shen, Prof. K. Veluraja, Prof. D. Velmurugan and Prof. B. Jayaram for accepting my invitation and contributing a chapter to the book.

I also extend my thanks to Dr. S.K. Rayala, Dr. Y. Kumar, Dr. R. Nagarajan, Dr. P. Anoosha, Dr. Dhanusha, Dr. Jino Blessy, Dr. C. Ramakrishnan, Dr. R. Sakthivel, K. Vishnupriya, S. Jemimah, Ambuj, R. Prabakaran, A. Kulandaisamy, S. Shanmugam, Nisha, R.A. Jayaram and B. Lalithamaheswari for their efforts to prepare necessary sections, figures and valuable suggestions.

It is my pleasure to acknowledge my wife A. Mary Thangakani and children Michael Mozim, Angela Shalom and Michael Abejo for their support to edit the book. I also immensely acknowledge my mother for her valuable guidance.

I warmly thank Ms. Xiao Ling for efficiently managing the production of the book and World Scientific for publishing the book.

Finally, I thank all my well wishers and friends who encouraged me to edit this book.

About the Editor

M. Michael Gromiha received his PhD in physics from Bharathidasan University, India, and served as a post-doctoral fellow, STA fellow, RIKEN researcher, Research scientist, and Senior scientist at the International Center for Genetic Engineering and Biotechnology (ICGEB), Italy, The Institute of Physical and Chemical Research (RIKEN), Japan and the Institute of Advanced Industrial Science and Technology (AIST), Japan till 2010. Currently, he is working as a Professor at the Indian Institute of Technology (IIT) Madras, India. His main research interests include sequence and structural analysis, prediction, folding, and stability of globular and membrane proteins, protein–protein, protein–nucleic acid and protein–ligand interactions, development of bioinformatics databases and tools, structure-based drug design and next-generation sequence analysis. He has published more than 220 research articles, 40 reviews, seven editorials, and a book entitled *Protein Bioinformatics: From Sequence to Function* by Elsevier/Academic Press. His papers received more than 10,000 citations and his h-index is 55. He is an Associate Editor of *BMC Bioinformatics*, Section Editor of *Current Protein and Peptide Science* as well as an Editorial Board member of *Scientific Reports, Biology Direct, Journal of Bioinformatics and Computational Biology, Genes* and *Current Computer Aided Drug Design*.

He has received several awards including the Oxford University Press Bioinformatics prize, Okawa Science Foundation Research Grant, Young Scientist Travel awards from ISCB, JSPS, AMBO and ICTP, Best Paper Award at ICIC2011, ICTP Associateship Award, ICMR International Fellowship for Senior Biomedical Scientists,

INSA Senior Scientist Award, Best Paper Award in Bioinformatics by the Department of Biotechnology, India, Institute Research and Development Award from IIT Madras, Outstanding Performance Award from Initiative for Parallel Bioinformatics (IPAB), Tokyo Institute of Technology, Japan and Tamil Nadu Scientist Award (TANSA) from Tamil Nadu State Council for Science and Technology, India.

Part I
Protein–Protein Interactions

Chapter 1

Structural and dynamical aspects of evolutionarily conserved protein–protein complexes

Himani Tandon, Sneha Vishwanath and
Narayanaswamy Srinivasan*

*Molecular Biophysics Unit, Indian Institute of Science,
Bangalore 560012, India*

Proteins interact with other proteins to perform vital cellular functions. Understanding various features of protein–protein interfaces is important for their prediction as well as designing inhibitors/ activators of protein–protein interactions (PPIs). In this article, the structural and dynamical characteristics of PPIs and their evolutionary conservation have been reviewed. The extent of partner retention and similarity in the quaternary structure and quaternary states of protein–protein complexes (PPCs) depends on the evolutionary divergence among related proteins. Similar interacting modes are observed in protein complexes if the related proteins share high sequence identity, and at low sequence identities, interface residues may differ. Presence of additional structural elements also brings diversity among the evolutionarily related PPCs. These observations suggest caution when predicting interfaces based on homologous complexes with low sequence identities. Binding of two proteins can

*ns@iisc.ac.in

also lead to change in structure and/or dynamics at the interfaces as well as distant sites in PPCs. These changes often have functional relevance and regulate protein (dis)assembly through allosteric communication. Various examples have been discussed in this article to understand how sequence identities between evolutionarily conserved PPCs affect their interfacial characteristics and how the binding of partner proteins affects their structure and dynamics.

Abbreviations: PPCs: protein–protein complexes, PPIs: protein–protein interactions, PDB: protein data bank, OMP: orotidine 5′-phosphate, PLP: pyridoxal 5′-phosphate, Epha2: ephrin type-A receptor 2, SHIP2: phosphatidylinositol-3, 4, 5-trisphosphate 5-phosphatase 2, SAM: sterile alpha motif, Tcf4: transcription factor 4, KID: kinase-inducible domain, KIX: KID-interacting domain, cAMP: cyclic adenosine monophosphate, CREB: cAMP response element-binding protein, CBP: CREB-binding protein, BFM: bacterial flagellar motor

1.1. Introduction

Proteins often interact with other molecular entities in the cell to perform various biological functions. Such interactions can be with other proteins, small molecules, nucleic acids, lipids, or carbohydrates. In this article, we focus on protein–protein interactions (PPIs) caused by physical binding between proteins. These interactions are important contributors to the processes such as replication, transcription, translation, and signal transduction (Figure 1.1) (Levy and Pereira-Leal, 2008; Reichmann *et al.*, 2007; Schreiber and Keating, 2011; Vidal *et al.*, 2011). Some proteins can perform their functions without getting associated with another protein, but majority of the proteins interact with other protein/s to carry out their biological functions. The complex network of interactions that a protein makes with other proteins provides an efficient way of recycling the limited repertoire of these molecules to accomplish functional diversity (Janin and Wodak, 2002).

Stability and specificity of protein–protein complexes (PPCs) are of utmost importance to maintain the cellular machinery. Spatial and

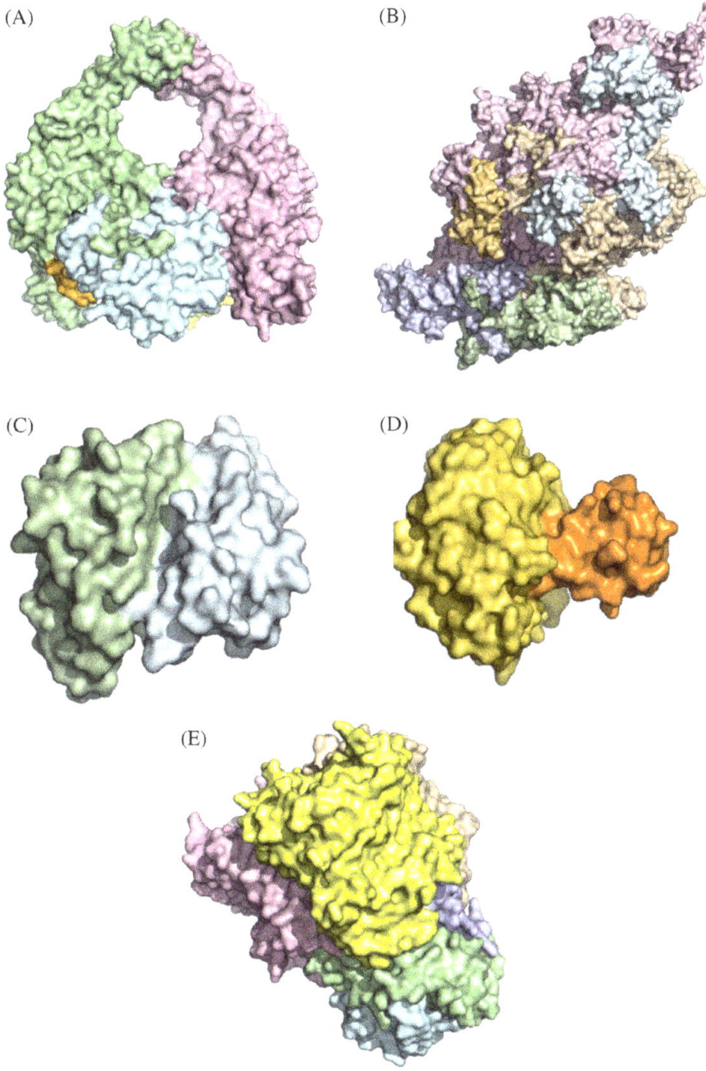

Figure 1.1. Examples of protein complexes involved in processes such as replication, transcription, and signal transduction. (A) Complex of DNA gyrase subunit A, and subunit B. (B) Multi-protein complex of RNA polymerase. (C) Homodimeric complex of T-cell receptor. (D) Heterodimeric complex of trypsin and APPI molecule. (E) Multi-protein symmetric complex of ammonia channel and nitrogen regulatory protein P-II. Figures have been generated using PyMOL v1.7 (Delano, 2002).

temporal controls of PPIs are one of the ways to achieve specific and functionally relevant interactions (Vishwanath *et al.*, 2017). Perturbations in PPI networks within a cell have debilitating effects (Ryan and Matthews, 2005). Despite extensive work in the field of PPIs, it is still non-trivial to predict PPIs in the endogenous context (Ezkurdia *et al.*, 2009; Sorret *et al.*, 2016). Further understanding of features of PPIs is important for several reasons. A better understanding of protein–protein interfaces can help in deciphering the mechanisms of formation of large macromolecular assemblies and can also aid in fitting of atomic-level models in medium-resolution cryo-EM density maps (Joseph *et al.*, 2016; Levy *et al.*, 2008; Marsh and Teichmann, 2015). Targeting PPIs is desirable in managing certain disease conditions (Sheng and Georg, 2014; Skwarczynska and Ottmann, 2015).

1.1.1. *Classification of protein–protein complexes*

Protein–protein complexes (PPCs) can be categorized into various kinds depending on different features (Nooren and Thornton, 2003). Associated polypeptide chains of identical sequence are termed as *homomeric* complexes, whereas PPCs between proteins of non-identical sequences are termed as *heteromeric* complexes (Jones and Thornton, 1996). Detailed analyses of homomeric and heteromeric protein complexes have revealed that former complexes are more stable and abundant in the cell than the latter (Lukatsky *et al.*, 2007). Homomeric complexes are grossly symmetric (Levy *et al.*, 2006). However, a study by Swapna *et al.* (2012c) has suggested global and local asymmetry in many of the homomeric complexes. PPCs can be further classified as obligate and non-obligate depending on the stability of the component structures in the unbound form. If the structures of the components are stable *in vivo* in the unbound form, then such PPCs are termed as non-obligate, and if they are unstable *in vivo* in the absence of the interacting partner, then such PPCs are termed as obligate (Jones and Thornton, 1996). Majority of the homomeric complexes are obligate complexes (Dey *et al.*, 2010).

Depending upon the *in vivo* lifetime of the PPCs, the protein complexes are described as either *permanent* or *transient*. Most of the obligate complexes are permanent complexes and non-obligate complexes are transient complexes (Jones and Thornton, 1996). The permanent complexes have higher interaction affinities (K_d values are often in lower nM range) than transient complexes (K_d values are often in the range of µM to nM) (Perkins *et al.*, 2010). Depending on the interaction affinities, the transient complexes can be further classified as "weak" or "strong." Transient complexes are often part of regulatory and signaling process in the cell (Acuner Ozbabacan *et al.*, 2011; Perkins *et al.*, 2010).

1.1.2. *Characteristics of protein–protein interfaces*

Two proteins interact with each other through patches of residues termed as interface residues. (Chakrabarti and Janin, 2002). Availability of 3D structures for a plethora of PPCs in Protein Data Bank (PDB) (Berman *et al.*, 2000) has led to a good understanding of the various features of PPIs. Different studies have explored their shape (Jones and Thornton, 1996), size (Mintseris and Weng, 2003), architecture (Chakrabarti and Janin, 2002), residue composition (De *et al.*, 2005), hydrophobicity (Conte *et al.*, 1999; Jones and Thornton, 1996), planarity (Jones *et al.*, 2002), circularity (Jones and Thornton, 1996), residue conservation (Choi *et al.*, 2009), and packing efficiency (Conte *et al.*, 1999). A protein in a PPC may have multiple interface regions depending on the number of proteins it interacts within the complex (Choi *et al.*, 2009; Kim *et al.*, 2006). Given the number of possible interface regions for a protein, prediction of interface region is non-trivial for many proteins. But some distinguishing features of interface residues such as enhanced residue conservation as compared to rest of the surface residues (De *et al.*, 2005; Mintseris and Weng, 2005), reduced flexibility (Jones *et al.*, 2002; Swapna *et al.*, 2012a), and modular architecture (Keskin *et al.*, 2008; Reichmann *et al.*, 2007) can help in predicting the interface regions to an extent (Janin *et al.*, 2008). Protein–protein interface residues can be further divided into core and rim residues (Chakrabarti and Janin, 2002; Levy, 2010). Interface residues that are completely buried on formation of the complex are termed as core residues, and those interface residues that are partially buried are termed as rim residues (Chakrabarti and Janin, 2002; Levy, 2010).

Not all the interface residues contribute equally to the stability of PPCs. Certain residues contribute significantly more to the binding energy and are termed as "hot-spot residues" (Keskin *et al.*, 2005; Moreira *et al.*, 2007). Mutation of such hot-spot residues has been reported to change the binding energy of the complex by at least 2 kcal/mol (Bogan and Thorn, 1998). It has been estimated that about 9.5% of the interfacial residues are hot-spot residues (Moreira *et al.*, 2007). The amino acid composition of these hot-spot residues is distinctive with tryptophan (21%), arginine (13.3%), and tyrosine (12.3%) occurring frequently (Bogan and Thorn, 1998). The bulky nature of these residues contributes to large change in energies when mutated. Hot-spot residues are known to cluster together at the interface (Keskin *et al.*, 2005). Stability of PPCs has been reported to be influenced (either stabilization of the PPCs or destabilization of PPCs) by alterations in the interface residues. These alterations can be post-translation modification of interface residues, binding of small molecules at the interface region, and mutation at the interface region (David *et al.*, 2012; Gao and Skolnick, 2012; Nishi *et al.*, 2011; Sheng and Georg, 2014).

Here the structural and dynamical features of PPCs and their variation in evolution have been discussed. This includes reviewing the extent of partner retention during evolution, the structural changes in the subunits on complexation, and how the dynamics associated with multi-protein complexes can be related to function.

1.2. Evolutionary perspective on protein–protein complexes (PPCs)

Studies on 3D structures of proteins have revealed that the homologous proteins have similar 3D structures and functions (Sadowski and Jones, 2009). Similar trends have been observed for the structures of homologous PPCs (Aloy *et al.*, 2003; Sudha *et al.*, 2015; Teichmann, 2002). Two PPCs (say AB and A′B′) are said to be homologous if A is evolutionarily related to A′ and B is evolutionarily related to B′.

Such pairs of PPCs are termed as "interologs" (Walhout *et al.*, 2000) (Figure 1.2A). If the homologous subunits in interologs share high sequence identity, the arrangement of the subunits is similar in interologs (Aloy *et al.*, 2003), but if the sequence identities are low, the arrangement of subunits can differ in interologs (Faure *et al.*, 2012).

Also for closely related interologs, the interface residues show significant physical, chemical, and structural similarities, whereas the interface residues of distantly related interologs show differences in many cases (Rekha *et al.*, 2005). Differences in the geometry of interologs and interface residues properties can arise because of distinct interfaces in a pair of interologs. Some interologs have been reported to interact using alternate binding modes (Hamp and Rost, 2012). Such alternate binding modes introduce promiscuity in the homologous PPCs due to varying selection pressures (Andreani *et al.*, 2012; Gandhi *et al.*, 2006; Suthram *et al.*, 2005; Yu *et al.*, 2004). In the following sections, we briefly discuss different evolutionary perspectives on PPCs.

1.2.1. *Sequence and functional similarities dictate the conservation of PPIs*

Identification of interactions between proteins in cellular milieu is a non-trivial problem. Despite improvements in the technologies to detect PPIs, the space of known PPIs is far from completion (Aloy and Russell, 2004; Vidal, 2016). One of the most widely used computational approaches is "interolog annotation" (Yu *et al.*, 2004). This approach is based on the observation that if two proteins are known to interact in an organism, then proteins of another organism that are evolutionarily related to interacting proteins are also predicted to interact. For example, if protein A interacts with protein B in *Homo sapiens*, then the evolutionarily related protein of A in *Bos taurus* say A′ and evolutionary related protein of B in *Bos taurus* say B′ are predicted to interact. Though this approach has been successful in many cases, it is associated with high false-positive rates, especially when the proteins are distantly related (Mika and Rost, 2006; Yu *et al.*, 2004). A detailed analyses of annotation transfer between four

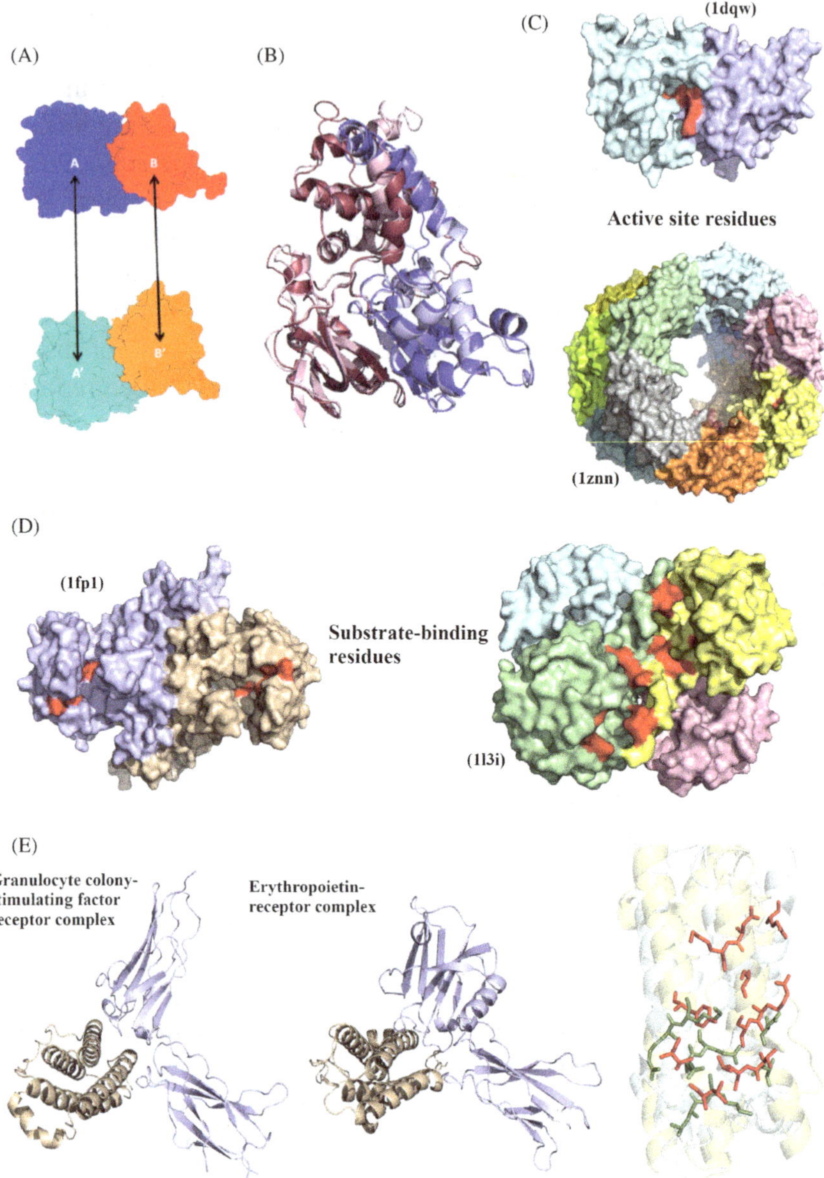

Figure 1.2. Features of evolutionarily conserved PPCs. (A) Cartoon representation of interologs (A and A′ are related to each other and B and B′ are related to each other). (B) An example of distantly related interologs — nitrile hydratase alpha subunit (dark blue) and nitrile hydratase beta subunit (dark pink) from *Pseudomonas putida*

genomes of model organisms showed that PPIs are well conserved between protein pairs with at least 80% sequence identity. At sequence identities <40%, the confidence of annotating a protein pair as interacting goes low (Yu *et al.*, 2004). It has to be noted that there are examples in literature where the PPIs are conserved at low sequence identity, for example, nitrile hydratase alpha subunit and nitrile hydratase beta subunit from *Pseudomonas putida* and nitrile hydratase alpha subunit and nitrile hydratase beta subunit from *Pseudonocardia thermophila* share low sequence identities – 49.3% and 38.5%, respectively, but the interactions between the proteins as well as the geometry of the complex are conserved (Figure 1.2B). Many PPCs may be specific to certain stages of the life cycle of an organism or specific conditions in the cell. Interologs of such PPCs may not be present in other organisms, for example, only a handful of interacting pairs in *Plasmodium falciparam* are conserved in other organisms (Suthram *et al.*, 2005; Zinman *et al.*, 2011).

Extensive analysis of the high coverage genomic data, expression data, PPI data, and genetic interaction data for yeast strains, nematode, and fruit-fly has suggested that interactions are usually conserved for proteins that share similar functions. These sets of highly

---

Figure 1.2. (*Continued*)   and nitrile hydratase alpha subunit (light blue) and nitrile hydratase beta subunit (light pink) from *Pseudonocardia thermophila* share sequence identity –49.3% and 38.5%, respectively. (C) Active site residues (colored in red) located at subunit interface in OMP decarboxylase (PDB code 1dqw) and at the internal surface of PLP synthase (PDB code 1znn). (D) Substrate-binding residues (colored in red) located near the dimer interface of chalcone-O-methyl transferase (PDB code 1fp1) and at the tetramer interface of Precorrin-6Y methyl transferase (PDB code 1l3i). (E) Example of homologous proteins with alternate binding modes. Homologous cytokine–receptor complexes — Granulocyte colony-stimulating factor receptor complex (PDB code 1cd9) and erythropoietin–receptor complex (PDB code 1eer) are shown. Cytokines in the respective structures are shown in wheat color, and the receptors are shown in light purple color. On the right of the panel, the interface residues of the cytokines are shown in red and green for PDB codes 1cd9 and 1eer, respectively. Many of the interface residues are not topologically equivalent. Examples of structures shown in Panels (C) and (D) are taken from Sudha and Srinivasan (2016) and panel (E) shows structures of examples of PPCs taken from Sudha *et al.* (2014). The figures have been generated using PyMOL ver 1.7 (Delano, 2002).

conserved interacting proteins that share similar functions are termed as "functional modules" (Zinman *et al.*, 2011). It has also been observed that PPIs within a functional module are conserved better than PPIs between modules. For example, 46.5% of within-module PPIs were observed to be conserved between *Saccharomyces cerevisiae* and *Schizosaccharomyces pombe*. Moreover, it has been observed that proteins interacting within large multi-protein complexes are conserved better in evolution (Rajagopala *et al.*, 2014). For example, protein pairs are conserved in bacteria closely related to *Escherichia coli*, and the conservation goes down as the evolutionary distance of the bacteria increases from *E. coli*. However, interactions between proteins encoded by essential genes in *E. coli* were found conserved in other bacteria irrespective of their evolutionary distances (Rajagopala *et al.*, 2014). Interactions involving proteins that interact with many proteins during its lifetime, termed as hub proteins, are conserved better than non-hub proteins (Fox *et al.*, 2009). The non-retention of the interactions across organisms is not due to rewiring of protein interaction network by partner switching in organisms but mostly due to loss of the genes coding the subunits (van Dam and Snel, 2008). Out of 5,960 protein pairs present in the same complex in *Homo sapiens*, 2,216 protein pairs are missing in *S. cerevisiae*, because both proteins lack orthologs. Absence of 1,828 protein pairs was reported because one of the proteins does not have orthologs. For the remaining 1,916, about ~10% could never be co-purified (van Dam and Snel, 2008).

1.2.2. *Variations in quaternary structure and quaternary state of homologous homomeric complexes during evolution*

Once the interologs have been identified, it is intuitive to extrapolate the 3D arrangement of PPCs, whose structures have been determined, to all its interologs whose structures have not been determined yet. It is one of the underlying principles of an approach called "template-based docking" (Baspinar *et al.*, 2014; Kundrotas *et al.*, 2012; Mosca *et al.*, 2013). Here we consider extension of such an approach to oligomeric proteins that form quaternary structures. In this approach, subunits of an oligomeric protein are docked by using the known quaternary structure of its interologs as template. Like the preservation of interactions between proteins, conservation of quaternary structure depends on the sequence identity (Aloy *et al.*, 2003; Levy *et al.*, 2008).

However, unfortunately, quaternary structures and quaternary state (the number of subunits in oligomeric proteins) are not always conserved during evolution. Interestingly, even for a few closely related homomeric PPCs, starting from the text book examples of hemoglobin and myoglobin, the quaternary states are different (Hashimoto *et al.*, 2011; Perica *et al.*, 2012). In some cases of homologous oligomeric proteins, despite conserved quaternary states, their quaternary structures are radically different. These differences have been attributed to amino acid substitutions at the interfaces, subunit geometry, insertions or deletions at the interface, and the subunit flexibility (Marsh and Teichmann, 2014). A detailed analysis of distantly related homomers has revealed that only 23% of the analyzed homomeric complexes have same quaternary state (Sudha and Srinivasan, 2016). Similar analysis of a data set of closely related homomeric proteins revealed that only 31% of the pairs have identical quaternary state (Sudha and Srinivasan, 2016). However, in 22% of the pairs from the data set of closely related homomers, though the quaternary states are different, structural superposition of the pairs suggested close substructural similarity. Differences between the quaternary states and quaternary structures of homologous oligomers seem to translate into evolution of non-equivalent functional sites. Two examples of this kind are discussed here.

a. Ribulose-phosphate-binding barrel

Orotidine 5′-phosphate (OMP) decarboxylase and pyridoxal 5′-phosphate (PLP) synthase enzymes share the same tertiary fold and share 14% sequence identity. OMP decarboxylase is a homodimer whereas PLP synthase is a homo-dodecamer. Although they share the gross function of nucleotide transport and metabolism, their active site residues occupy different spatial location in the proteins (Figure 1.2C).

b. S-adenosyl-1-methionine-dependent methyl transferase

Isoliquiritigenin 2′-O-methyl transferase and precorrin-6y methyl transferase share the same tertiary fold and share 11% sequence identity. The former is a dimer and the latter is a tetramer. In this case also, both the enzymes retain the general transferase function; however, the substrate binds at the dimer interface of the former and tetramer interface of the latter (Figure 1.2D).

Observation of alternate binding mode suggests that homologous oligomeric proteins and PPCs can utilize different interfaces for binding (Hamp and Rost, 2012). The extensive sequence divergence and additional secondary structural elements at the interface have been attributed as the reasons behind these observations. Such observations further strengthen the idea that interfaces are plastic, and this evolutionary plasticity tunes the proteins for specific interactions. It can be thought of as nature's way of generating new functional features for the existing limited repertoire of tertiary folds such as homologous proteins binding with alternate modes (Figure 1.2E).

1.3. Conservation of structural features of interface residues of interologs

Do the interologs with similar quaternary structure have similar interface regions? Similarity of the interface regions of interologs can be compared at the levels of amino acid residues and structure. Several previous studies have shown a moderate level of conservation of the interface residues as compared to the other surface residues (Choi *et al.*, 2009; Guharoy and Chakrabarti, 2005; Keskin *et al.*, 2005; Mintseris and Weng, 2005). Among the interface residues, core residues are conserved better than the rim residues (Guharoy and Chakrabarti, 2005). Different levels of conservation have been observed for permanent and transient PPCs. Studies have shown that interface residues are conserved better in permanent complexes than the transient complexes (De *et al.*, 2005; Mintseris and Weng, 2005; Sukhwal and Sowdhamini, 2013). Permanent complexes have higher proportion of hydrophobic residues at the interface than transient complexes (Sukhwal and Sowdhamini, 2013). This feature can be attributed to the complementarity in the interaction interfaces as well as correlated substitutions. The hot-spot residues have also been shown to be conserved in homologous transient PPCs and can be employed as an effective way of conferring specificity in PPCs (Keskin *et al.*, 2005).

During evolution, structures are conserved better than the sequence. Similarly, it has been observed that the structure of interface regions is conserved better than the amino acid residues that constitute the interface region. Conservation of structure of interface residues of interologs can be measured by enumerating the number of "topologically equivalent" interface residues. Two residues from related protein structures that occupy equivalent position on superposing the structures optimally are termed as "topologically equivalent" (Figure 1.3A). In the following sections, conservation of structure and its attributes have been discussed in detail.

1.3.1. *Similarity in the structural features of interface regions in interologs*

Detailed analyses of structures of interologs have revealed that majority of the interface residues are not topologically equivalent in the interologs, suggesting that spatial locations of interface residues are not highly conserved (Andreani *et al.*, 2012; Hamp and Rost, 2012; Sudha *et al.*, 2015). About a quarter of the interface residues were reported to be switched out of the interfaces (Andreani *et al.*, 2012). This observation has an important implication that even when interologs bind in similar geometry, the structural features of interface residues may not be similar. An analysis of the amino acid conservation of these interface residues revealed that the conservation of interface residues in interologs does not always suggest that the conformation of the interface residues is same in the interologs. Further analyses of the pairs of topologically equivalent interacting residues (an example is shown in Figure 1.3B) revealed that only 18% of the studied interologs show high conservation. This abysmally low value

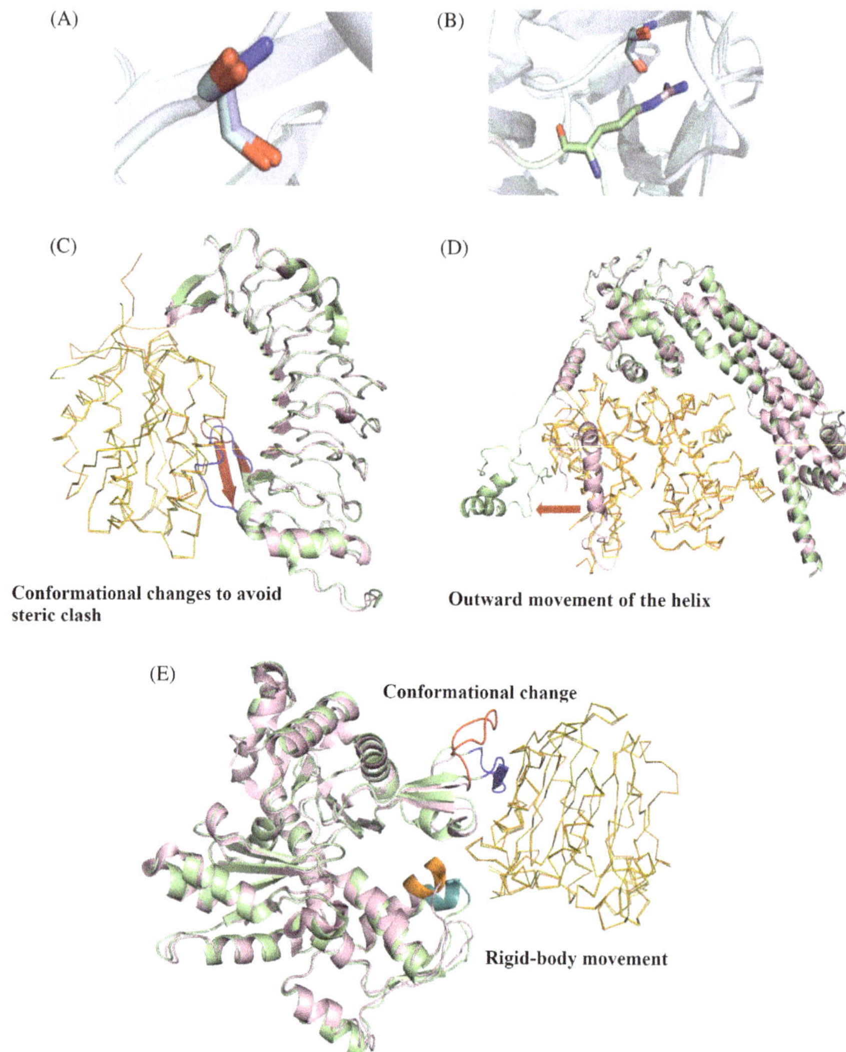

Figure 1.3.   Changes observed at the interfaces. (A) Topologically equivalent residues are shown in sticks. (B) Topologically equivalent interacting residues. (C–E) Examples of interface regions undergoing structural changes upon binding. (C) An example where the conformational changes occur at the interface to avoid steric clashes (PDB code 1M10; Glycoprotein Ib alpha & von Willebrand factor). The interface regions that show conformational changes are colored in red and blue. (D) An example of movement of the interface to avoid steric clashes (PDB code 1Y64; BNI1 protein and alpha actin). The large outward movement is depicted with red arrow. (E) An example of interface region showing rigid-body movement on

suggests that interologs are stabilized by different interactions. This suggests that the popular approach of directly extrapolating conserved residues as interface residues during the comparative modelling of PPCs is not always valid (Sudha *et al.*, 2015). Structural features of core interface residues are conserved better than the rim residues. Furthermore, additional structural elements at the protein interfaces can generate diversity in PPCs during evolution (Plach *et al.*, 2017). These elements can be extra loops or entire secondary structures at the interfaces that help in maintaining the specificity of interactions.

Another important aspect of PPCs is the occurrence of correlated amino acid substitutions at the interfaces. Correlated amino acid substitutions imply that a residue substitution at the interface during the course of evolution is compensated by a mutation of its interacting residue partner in such a way that the interaction between the binding partners is not affected (Andreani *et al.*, 2012). Such correlated mutations at the interface have been suggested as an indicator of structural similarity at the interface (Rodriguez-Rivas *et al.*, 2016). Though the interaction pattern and interfacial nature of the residues are not highly conserved, it has been observed that the conservation rate of interface residues increases up to 91%, if only pairs of coevolving residues are taken into consideration (Rodriguez-Rivas *et al.*, 2016). This feature has been explored to identify interface residues in many PPCs, for example, in human pyruvate dehydrogenase complex (Rodriguez-Rivas *et al.*, 2016). Moreover, the conservation of the structure of the apolar interacting residues has been observed to be better than the polar interacting residues (Andreani *et al.*, 2012).

Figure 1.3. (*Continued*) complexation (PDB code 1ATN; deoxyribonuclease I and actin). The rigid-body movement is shown in orange and teal colors, whereas the region of interface undergoing conformational change is shown in red and blue colors. In (C–E), the bound form is colored in pale green and unbound in light pink. The binding partner undergoing structural changes is represented as cartoon. Figures have been generated using PyMOL v1.7 (Delano, 2002). Example shown in panel (C–E) has been picked from a previous study (Swapna *et al.*, 2012b).

1.3.2. *Structural features of interfacial residues in the bound and unbound forms of proteins*

How far is the conformation of the interface residues similar in the bound and unbound forms? Some residues (~20%) have been observed to have nearly same conformation in the bound and unbound forms, whereas a majority of the interface residues have markedly different conformations (Swapna *et al.*, 2012a; Yogurtcu *et al.*, 2008). Interface residues undergo conformational changes on binding its interacting partner to avoid steric clashes (Figure 1.3C–E). It must be noted that the changes in the conformation can occur in the side chains, backbone, or both. These residues, which undergo change, usually populate the core interface region and form spatial clusters (Swapna *et al.*, 2012a). It has been observed in many complexes that the interface region of one of the proteins is nearly premade and the other interface regions undergo conformational changes on complexation. About 65% of the conserved interfacial residues have been observed to be conformationally invariant and rigid in their unbound forms. Conformational variability of the core interfacial residues was found to be comparable to many other buried residues. Some of these residues contribute mainly to the stability of interface structure in the unbound forms and hence are hot-spots as well (Swapna *et al.*, 2012a; Vishwanath *et al.*, 2017). Importance of these conformationally invariant residues at the interface has not been completely understood yet. It has been suggested that these residues decrease the entropy cost during the complex formation and may as well direct the transition of the protein from the unbound to the bound form (Vishwanath *et al.*, 2017; Yogurtcu *et al.*, 2008).

1.4. Protein–protein complexes (PPCs) are fuzzy entities

Proteins are not static in the cellular milieu, as represented by the structural models proposed using X-ray diffraction (Marsh *et al.*, 2012). Dynamics of proteins, especially at the functional and regulatory sites, is important for its biological function. The interface residues in PPCs are also constantly in motion, where the interactions between the residues both within and across interface regions are constantly strengthened and weakened. As the interface residues are solvent exposed in the unbound form, they are generally more dynamic than in the bound form, where they are either partially or fully buried. In this section, we would first discuss about different types of "fuzziness" observed in PPCs followed by discussion on the influence/changes in the dynamics at regions other than interface in PPCs and its implication in the (dis)assembly of proteins in PPCs.

1.4.1. *Protein interfaces show variation in dynamics in different conditions*

Comparisons of multiple crystal structures of identical PPCs have provided undisputed evidence of dynamic plasticity of the PPIs (Hamp and Rost, 2012). This dynamic plasticity could be because of flexibility of side chains, backbone, or both in the bound form, thus having different conformation in different crystals. Further studies of PPCs using nuclear magnetic resonance and long molecular dynamic simulations have provided support for the flexibility of the interface residues in PPCs (Marsh and Teichmann, 2015). Flexibility of the interface residues results in constant breaking and making of interactions between interface residues both within and across interface regions. This constant process of making and breaking of interactions helps in transition from association to dissociation and vice versa in transient PPCs. For example, NMR studies and microsecond scale MD simulations of EphA2:SHIP2 complex have shown that SAM domains of these two proteins change their orientation with respect to each other through rotation and slight shifting of the interfaces (Lee *et al.*, 2012; Zhang and Buck, 2013). Shifting of hydrogen bonds and appearance of intermediary salt bridges have been observed during the simulations, suggesting that a higher density of salt bridges near the interface enables the observed transitions (Zhang and Buck, 2013).

Another interesting manifestation of dynamics of interface residues is observed in the case of intrinsically disordered proteins/regions (IDP/IDR). IDP/IDR do not have an ordered structure in

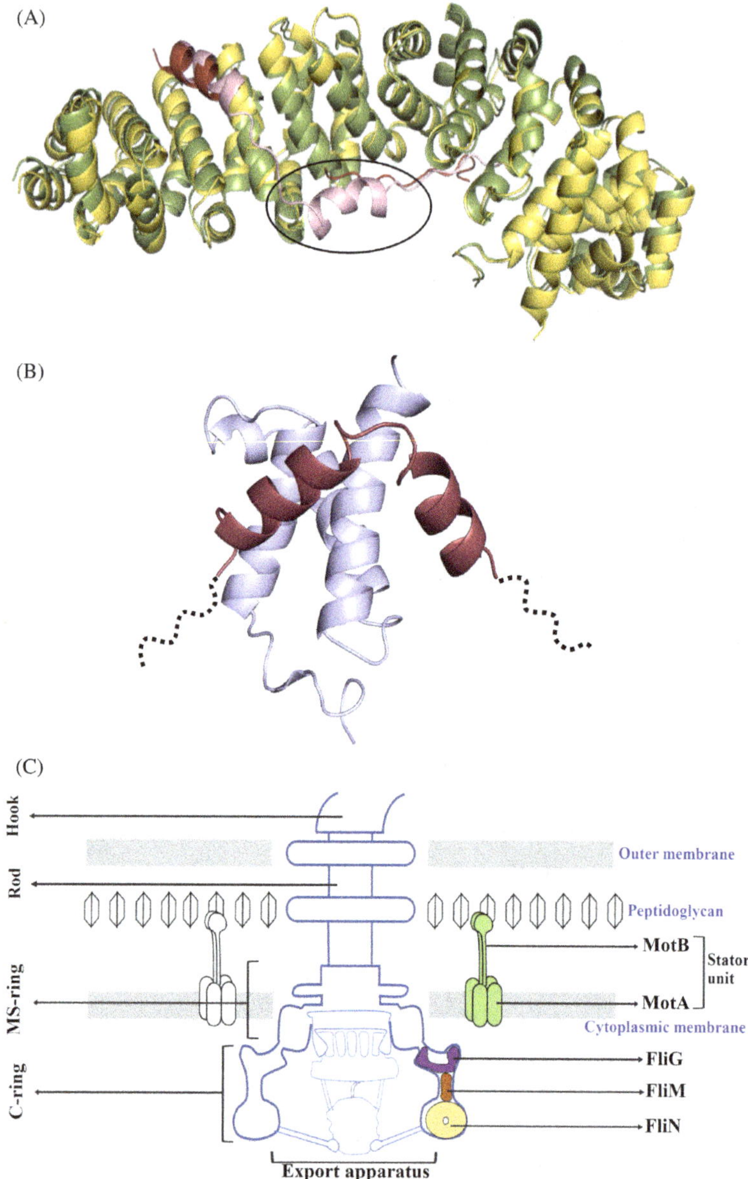

Figure 1.4.   Dynamic properties of PPCs. (A) An example of static disorder in protein complexes. Superposition of two structures of Tcf4:β-catenin complexes are shown. In one structure (PDB code 1jdh), the middle segment of Tcf4 (encircled in black) is alpha helical, whereas in the other structure (PDB code 1jpw), this region

the unbound form and acquire an ordered or near to ordered structure on binding (Oldfield and Dunker, 2014; Wright and Dyson, 1999). It has to be noted that in few cases, the interface region retains the disordered nature (Fuxreiter, 2018; Tompa and Fuxreiter, 2008). This prevalence of disordered regions in PPCs makes them fuzzy in nature and this feature of disorder in the bound form is termed as "fuzziness" (Fuxreiter, 2018; Tompa and Fuxreiter, 2008). Fuzziness has been observed to be an intrinsic feature of all the higher order protein organizations, ranging from static amyloids/prions, prion-like, and multivalent signaling complexes as well as cytoplasmic and nuclear granules (Henzler-Wildman and Kern, 2007). Two types of fuzziness in PPCs exist – static and dynamic. If the protein or its segment is completely ordered in the bound state but adopts multiple and few stable conformations, it is considered as static fuzziness. Whereas if a part or whole of the protein does not undergo disorder-to-order transition upon binding and remains in an ensemble of conformations, which is in rapid equilibrium, it is referred to as dynamic fuzziness (Tompa and Fuxreiter, 2008). An example of static fuzziness is the interaction of Tcf4 with β-catenin. The disordered catenin-binding domain of Tcf4 binds to β-catenin in an extended conformation (Graham *et al.*, 2001), whereas its middle segment adopts several distinct conformations (Figure 1.4A) and establishes alternative salt bridges that are crucial for Tcf4–β-catenin binding. An example of dynamic disorder is binding of the disordered kinase-inducible domain (KID) of the cAMP response element-binding protein (CREB) to the KIX domain of CREB-binding protein (CBP) (Figure 1.4B). KID adopts a helix-turn-helix conformation in the

---

Figure 1.4. (*Continued*)   is disordered. This segment is suggested to adopt several conformations (Graham *et al.*, 2001). (B) Binding of KID domain of CREB to CBP is an example of dynamic disorder. The flanking regions of KID domain (raspberry color) remain disordered (denoted as dotted region) in the bound state (PDB code 1kdx). The figures have been generated using PyMOL v1.7 (Delano, 2002). (C) Diagrammatic representation of the bacterial flagellar motor (BFM) showing the locations of C-ring and stator subunits. Example shown in panel (C) has been picked from Sowa and Berry (2008).

bound state for 29 residues, and the remaining flanking region remains disordered. It has been shown that the flanking region is important for binding and its deletion lowers the affinity of the complex (Zor *et al.*, 2002). The prevalence of fuzziness in PPCs has been attributed to the context-dependent functional state of the proteins (Miskei *et al.*, 2017). It enables rewiring of the interactions within or between the proteins via alternate motifs and generates new functions via allosteric motifs (Fuxreiter, 2018; Tompa and Fuxreiter, 2008).

1.4.2. *Dynamics may alter in regions other than interface in PPCs on complexation*

When two proteins interact with each other, the interface region undergoes conformational as well as dynamic changes to facilitate formation of PPCs (Gunasekaran *et al.*, 2004; Nussinov, 2012; Swapna *et al.*, 2012b; Tsai *et al.*, 2009). The intrinsic flexibility of the unbound form strongly affects the extent of conformational change. In general, larger is the flexibility associated with the protein, higher is the conformational differences observed in the bound form (Marsh and Teichmann, 2015). Earlier studies have shown that these differences in conformation and dynamics need not be confined to interface regions but may extend to non-interface regions (Grünberg *et al.*, 2006; Swapna *et al.*, 2012b). These differences have been observed to have functional implications like binding of another biomolecule at the region of altered dynamics and are well conserved during the evolution (Swapna *et al.*, 2012b).

Similar observations have been reported in the context of domain–domain interactions within a protein (Vishwanath *et al.*, 2018). As domain–domain interaction within a protein and interactions between proteins have similar features (Jones *et al.*, 2002), it can be concluded that dynamics and structural differences at a region away from interface are a general feature than exception. Because majority of the interface residues become rigid on forming PPCs, these altered dynamics have been proposed to redistribute the entropy of the residues in such a way that the entropy of the protein remains similar in the bound and unbound forms (Grünberg *et al.*, 2006).

1.4.3. *Role of protein dynamics in assemblies*

Dynamic nature of protein–protein interfaces facilitates protein re-organization in macromolecular assemblies (Tusk *et al.*, 2018). Molecular machines such as bacterial flagellar motor (BFM), ribosomes, nuclear pore complex, and spliceosome once formed were believed to be permanent, but structural and dynamics studies revealed that BFM undergoes subunit exchange during its functional lifetime (Blair and Berg, 1988; Leake *et al.*, 2006; Tusk *et al.*, 2018). BFM is an energy-driven rotary motor embedded in bacterial membrane, which assists in bacterial swimming motility (Sowa and Berry, 2008). This flagellar rotor comprises a shaft coupled to an inner-membrane spanning in the structural core, and the cytoplasmic C-ring forms the surface that is pushed by the stator units (Figure 1.4C). The stator unit exchange is exploited to allow for rapid adaptation of motor function to changes in energy availability and output requirements (Tusk *et al.*, 2018). Ribosomes, nuclear pore complexes, and spliceosomes have been reported to show subunit exchange. In the case of bacterial ribosomes, the subunit exchange is proposed to play a role in ribosome damage repair (Pulk *et al.*, 2010). For nuclear pore complexes, subunit dynamics and exchange enable compositional flexibility (Knockenhauer and Schwartz, 2016), and in the case of spliceosomes, reversibility of subcomplex-association steps during assembly increases efficiency and fidelity of target recognition (Larson and Hoskins, 2017).

1.5. Conclusion

PPIs are essential for the proper functioning of biological processes. In this article, we have discussed the evolutionary, structural, and dynamical aspects of PPCs. Conservation of structural features of interfaces is dependent on the sequence identity between the interologs. Different evolutionary pressures operate on transient and permanent PPIs, hence different conservation trends and interface properties have been observed. Although at high sequence identities (>80%) between the complexes, the interaction partners as well as the geometrical features of the interfaces are conserved, lower sequence identities call for an eye of caution while transferring annotations. Moreover, topologically equivalent interface residues of a complex A–B may not always be present at the interface in the homologous complex A′–B′. Hence, care must be taken during comparative modeling and template-based docking. Protein–protein interfaces can be multifaceted, and the binding modes can vary among interologs. Differences in the attributes of interfacial residues in the unbound and bound forms of a protein have also been discussed. The flexibility of interfacial residues decreases upon complexation, and redistribution in the flexibility of non-interfacial regions has been reported. Some of the core and conserved interfacial residues do not show significant changes in the structural conformation between the bound and unbound forms, suggesting their role as anchor residues in the formation of PPCs. Dynamics associated with PPIs have an important role to play in eliciting function. These dynamics are responsible for the allosteric communication within or between the protein subunits and may result in subunit exchange, which in turn seems important for protein (dis)assembly. Understanding of protein–protein interfaces from structural, dynamics, and evolutionary point-of-view can further help in designing better interfaces or inhibitors/activators of certain PPIs that can prove advantageous under certain disease conditions.

Acknowledgments

This project is supported by an Indo-French Centre for the Promotion of Advanced Research/CEFIPRA collaborative grant (number 5302 – 2). This research is also supported by Indian Institute of Science — Department of Biotechnology partnership program and Mathematical Biology program of Department of Science and Technology to NS. NS is a J.C. Bose National fellow. Support for infrastructural facilities from Fund for Improvement of Science and Technology infrastructure (FIST), DST and Centre for Advanced Studies (CAS), University Grants Commission (UGC) is acknowledged. HT is a DST-INSPIRE fellow.

References

Acuner Ozbabacan, S.E., Engin, H.B., Gursoy, A., and Keskin, O. (2011). Transient protein–protein interactions. *Protein Eng. Des. Sel.* 24, 635–648.

Aloy, P., Ceulemans, H., Stark, A., and Russell, R.B. (2003). The relationship between sequence and interaction divergence in proteins. J. Mol. Biol. 332, 989–998.

Aloy, P., and Russell, R.B. (2004). Ten thousand interactions for the molecular biologist. *Nat. Biotechnol.* 22, 1317–1321.

Andreani, J., Faure, G., and Guerois, R. (2012). Versatility and invariance in the evolution of homologous heteromeric interfaces. *PLoS Comput. Biol.* 8, e1002677.

Baspinar, A., Cukuroglu, E., Nussinov, *et al.* (2014). PRISM: A web server and repository for prediction of protein–protein interactions and modeling their 3D complexes. *Nucleic Acids Res.* 42, W285–W289.

Berman, H.M., Westbrook, J., Feng, Z., *et al.* (2000). The protein data bank www.rcsb.org. *Nucleic Acids Res.* 28, 235–242.

Blair, D.F., and Berg, H.C. (1988). Restoration of torque in defective flagellar motors. *Science.* 242, 1678–1681.

Bogan, A.A., and Thorn, K.S. (1998). Anatomy of hot spots in protein interfaces. *J. Mol. Biol.* 280, 1–9.

Chakrabarti, P., and Janin, J. (2002). Dissecting protein–protein recognition sites. *Proteins Struct. Funct. Genet.* 47, 334–343.

Choi, Y.S., Yang, J.S., Choi, Y., *et al.* (2009). Evolutionary conservation in multiple faces of protein interaction. *Proteins Struct. Funct. Bioinforma.* 77, 14–25.

Conte, L. Lo, Chothia, C., and Janin, J. (1999). The atomic structure of protein–protein recognition sites. *J. Mol. Biol.* 285, 2177–2198.

David, A., Razali, R., Wass, M.N., *et al.* (2012). Protein–protein interaction sites are hot spots for disease-associated nonsynonymous SNPs. *Hum. Mutat.* 33, 359–363.

De, S., Krishnadev, O., Srinivasan, N., and Rekha, N. (2005). Interaction preferences across protein–protein interfaces of obligatory and non-obligatory components are different. *BMC Struct. Biol.* 5, 15.

Delano, W.L. (2002). *The PyMOL Molecular Graphics System.* Schrodinger.

Dey, S., Pal, A., Chakrabarti, P., *et al.* (2010). The subunit interfaces of weakly associated homodimeric proteins. *J. Mol. Biol.* 398, 146–160.

Ezkurdia, I., Bartoli, L., Fariselli, P., *et al.* (2009). Progress and challenges in predicting protein–protein interaction sites. *Brief. Bioinform.* 10, 233–246.

Faure, G., Andreani, J., and Guerois, R. (2012). InterEvol database: Exploring the structure and evolution of protein complex interfaces. *Nucleic Acids Res.* 40, D847–D856.

Fox, A., Taylor, D., and Slonim, D.K. (2009). High throughput interaction data reveals degree conservation of hub proteins. *Pacific Symp. Biocomput.* 391–402.

Fuxreiter, M. (2018). Fuzziness in protein interactions — A historical perspective. *J. Mol. Biol.* 430, 2278–2287.

Gandhi, T.K.B., Zhong, J., Mathivanan, S., *et al.* (2006). Analysis of the human protein interactome and comparison with yeast, worm and fly interaction datasets. *Nat. Genet.* 38, 285–293.

Gao, M., and Skolnick, J. (2012). The distribution of ligand-binding pockets around protein–protein interfaces suggests a general mechanism for pocket formation. *Proc. Natl. Acad. Sci.* 109, 3784–3789.

Graham, T.A., Ferkey, D.M., Mao, F., *et al.* (2001). Tcf4 can specifically recognize β-catenin using alternative conformations. *Nat. Struct. Biol.* 8, 1048–1052.

Grünberg, R., Nilges, M., and Leckner, J. (2006). Flexibility and conformational entropy in protein–protein binding. *Structure* 14, 683–693.

Guharoy, M., and Chakrabarti, P. (2005). Conservation and relative importance of residues across protein–protein interfaces. *Proc. Natl. Acad. Sci.* 102, 15447–15452.

Gunasekaran, K., Ma, B., and Nussinov, R. (2004). Is allostery an intrinsic property of all dynamic proteins? *Proteins Struct. Funct. Genet.* 57, 433–443.

Hamp, T., and Rost, B. (2012). Alternative protein–protein interfaces are frequent exceptions. *PLoS Comput. Biol.* 8, e1002623.

Hashimoto, K., Nishi, H., Bryant, S., *et al.* (2011). Caught in self-interaction: Evolutionary and functional mechanisms of protein homooligomerization. *Phys. Biol.* 8, 035007.

Henzler-Wildman, K., and Kern, D. (2007). Dynamic personalities of proteins. *Nature* 450, 964–972.

Janin, J., Bahadur, R.P., and Chakrabarti, P. (2008). Protein–protein interaction and quaternary structure. *Q. Rev. Biophys.* 41, 133–180.

Janin, J., and Wodak, S.J. (2002). Protein modules and protein–protein interaction: Introduction. *Adv. Protein Chem.* 61, 1–8.

Jones, S., Marin, A., and Thornton, J.M. (2002). Protein domain interfaces: Characterization and comparison with oligomeric protein interfaces. *Protein Eng. Des. Sel.* 13, 77–82.

Jones, S., and Thornton, J.M. (1996). Principles of protein–protein interactions. Proc. Natl. Acad. Sci. 93, 13–20.

Joseph, A.P., Swapna, L.S., Rakesh, R., *et al.* (2016). Use of evolutionary information in the fitting of atomic level protein models in low resolution cryo-EM map of a protein assembly improves the accuracy of the fitting. *J. Struct. Biol.* 195, 294–305.

Keskin, O., Gursoy, A., Ma, B., and Nussinov, R. (2008). Principles of protein–protein interactions: What are the preferred ways for proteins to interact? *Chem. Rev.* 108, 1225–1244.

Keskin, O., Ma, B., and Nussinov, R. (2005). Hot regions in protein–protein interactions: The organization and contribution of structurally conserved hot spot residues. *J. Mol. Biol.* 345, 1281–1294.

Kim, W.K., Henschel, A., Winter, C., *et al.* (2006). The many faces of protein–protein interactions: A compendium of interface geometry. *PLoS Comput. Biol.* 2, 1151–1164.

Knockenhauer, K.E., and Schwartz, T.U. (2016). The nuclear pore complex as a flexible and dynamic gate. *Cell* 164, 1162–1171.

Kundrotas, P.J., Zhu, Z., Janin, J., *et al.* (2012). Templates are available to model nearly all complexes of structurally characterized proteins. *Proc. Natl. Acad. Sci.* 109, 9438–9441.

Larson, J.D., and Hoskins, A.A. (2017). Dynamics and consequences of spliceosome E complex formation. Elife 6, e27592.

Leake, M.C., Chandler, J.H., Wadhams, G.H., *et al.* (2006). Stoichiometry and turnover in single, functioning membrane protein complexes. *Nature* 443, 355–358.

Lee, H.J., Hota, P.K., Chugha, P., *et al.* (2012). NMR structure of a heterodimeric SAM: SAM complex: Characterization and manipulation of EphA2 binding reveal new cellular functions of SHIP2. *Structure* 20, 41–55.

Levy, E.D. (2010). A simple definition of structural regions in proteins and its use in analyzing interface evolution. *J. Mol. Biol.* 403, 660–670.

Levy, E.D., Erba, E.B., Robinson, C. V., *et al.* (2008). Assembly reflects evolution of protein complexes. *Nature* 453, 1262–1265.

Levy, E.D., and Pereira-Leal, J.B. (2008). Evolution and dynamics of protein interactions and networks. *Curr. Opin. Struct. Biol.* 18, 349–357.

28   *H. Tandon et al.*

Levy, E.D., Pereira-Leal, J.B., Chothia, C., *et al.* (2006). 3D complex: A structural classification of protein complexes. *PLoS Comput. Biol.* 2, 1395–1406.

Lukatsky, D.B., Shakhnovich, B.E., Mintseris, J., *et al.* (2007). Structural similarity enhances interaction propensity of proteins. *J. Mol. Biol.* 365, 1596–1606.

Marsh, J.A., and Teichmann, S.A. (2014). Protein flexibility facilitates quaternary structure assembly and evolution. *PLoS Biol.* 12, e1001870.

Marsh, J.A., and Teichmann, S.A. (2015). Structure, dynamics, assembly, and evolution of protein complexes. *Annu. Rev. Biochem.* 84, 551–575.

Marsh, J.A., Teichmann, S.A., and Forman-Kay, J.D. (2012). Probing the diverse landscape of protein flexibility and binding. *Curr. Opin. Struct. Biol.* 22, 643–650.

Mika, S., and Rost, B. (2006). Protein–protein interactions more conserved within species than across species. *PLoS Comput. Biol.* 2, 0698–0709.

Mintseris, J., and Weng, Z. (2003). Atomic contact vectors in protein–protein recognition. *Proteins Struct. Funct. Genet.* 53, 629–639.

Mintseris, J., and Weng, Z. (2005). Structure, function, and evolution of transient and obligate protein–protein interactions. *Proc. Natl. Acad. Sci. U. S. A.* 102, 10930–10935.

Miskei, M., Gregus, A., Sharma, R., *et al.* (2017). Fuzziness enables context dependence of protein interactions. *FEBS Lett.* 591, 2682–2695.

Moreira, I.S., Fernandes, P.A., and Ramos, M.J. (2007). Hot spots — A review of the protein–protein interface determinant amino-acid residues. *Proteins Struct. Funct. Genet.* 68, 803–812.

Mosca, R., Céol, A., and Aloy, P. (2013). Interactome 3D: Adding structural details to protein networks. *Nat. Methods.* 10, 47–53.

Nishi, H., Hashimoto, K., and Panchenko, A.R. (2011). Phosphorylation in protein–protein binding: Effect on stability and function. *Structure.* 19, 1807–1815.

Nooren, I.M.A., and Thornton, J.M. (2003). Diversity of protein–protein interactions. *EMBO J.* 22, 3486–3492.

Nussinov, R. (2012). How do dynamic cellular signals travel long distances? *Mol. Biosyst.* 8, 22–26.

Oldfield, C.J., and Dunker, A.K. (2014). Intrinsically disordered proteins and intrinsically disordered protein regions. *Annu. Rev. Biochem.* 83, 553–584.

Perica, T., Chothia, C., and Teichmann, S.A. (2012). Evolution of oligomeric state through geometric coupling of protein interfaces. *Proc. Natl. Acad. Sci.* 109, 8127–8132.

Perkins, J.R., Diboun, I., Dessailly, B.H., *et al.* (2010). Transient protein–protein interactions: Structural, functional, and network properties. *Structure.* 18, 1233–1243.

Plach, M.G., Semmelmann, F., Busch, F., *et al.* (2017). Evolutionary diversification of protein–protein interactions by interface add-ons. *Proc. Natl. Acad. Sci.* 114, E8333–E8342.

Pulk, A., Liiv, A., Peil, L., *et al.* (2010). Ribosome reactivation by replacement of damaged proteins. *Mol. Microbiol.* 75, 801–814.

Rajagopala, S. V., Sikorski, P., Kumar, A., *et al.* (2014). The binary protein–protein interaction landscape of *Escherichia coli. Nat. Biotechnol.* 32, 285–290.

Reichmann, D., Rahat, O., Cohen, M., *et al.* (2007). The molecular architecture of protein–protein binding sites. *Curr. Opin. Struct. Biol.* 17, 67–76.

Rekha, N., Machado, S.M., Narayanan, C., *et al.* (2005). Interaction interfaces of protein domains are not topologically equivalent across families within superfamilies: Implications for metabolic and signaling pathways. *Proteins Struct. Funct. Genet.* 58, 339–353.

Rodriguez-Rivas, J., Marsili, S., Juan, D., *et al.* (2016). Conservation of coevolving protein interfaces bridges prokaryote–eukaryote homologies in the twilight zone. *Proc. Natl. Acad. Sci.* 113, 15018–15023.

Ryan, D.P., and Matthews, J.M. (2005). Protein–protein interactions in human disease. *Curr. Opin. Struct. Biol.* 15, 441–446.

Sadowski, M.I., and Jones, D.T. (2009). The sequence–structure relationship and protein function prediction. *Curr. Opin. Struct. Biol.* 19, 357–362.

Schreiber, G., and Keating, A.E. (2011). Protein binding specificity versus promiscuity. *Curr. Opin. Struct. Biol.* 21, 50–61.

Sheng, C., and Georg, G.I. (2014). Targeting protein–protein interactions by small molecules. *Annu. Rev. Pharmacol. Toxicol.* 54, 435–456.

Skwarczynska, M., and Ottmann, C. (2015). Protein–protein interactions as drug targets. *Future Med. Chem.* 7, 2195–2219.

Sorret, L.L., DeWinter, M.A., Schwartz, D.K., *et al.* (2016). Challenges in predicting protein–protein interactions from measurements of molecular diffusivity. *Biophys. J.* 111, 1831–1842.

Sowa, Y., and Berry, R.M. (2008). Bacterial flagellar motor. *Q. Rev. Biophys.* 41, 103–132.

Sudha, G., Singh, P., Swapna, L.S., *et al.* (2015). Weak conservation of structural features in the interfaces of homologous transient protein–protein complexes. *Protein Sci.* 24, 1856–1873.

Sudha, G., and Srinivasan, N. (2016). Comparative analyses of quaternary arrangements in homo-oligomeric proteins in superfamilies: Functional implications. *Proteins Struct. Funct. Bioinforma.* 84, 1190–1202.

Sukhwal, A., and Sowdhamini, R. (2013). Oligomerisation status and evolutionary conservation of interfaces of protein structural domain superfamilies. *Mol. Biosyst.* 9, 1652–1661.

Suthram, S., Sittler, T., and Ideker, T. (2005). The Plasmodium protein network diverges from those of other eukaryotes. *Nature* 438, 108–112.

Swapna, L.S., Bhaskara, R.M., Sharma, J., *et al.* (2012a). Roles of residues in the interface of transient protein–protein complexes before complexation. *Sci. Rep.* 2, 334.

Swapna, L.S., Mahajan, S., de Brevern, A.G., *et al.* (2012b). Comparison of tertiary structures of proteins in protein–protein complexes with unbound forms suggests prevalence of allostery in signalling proteins. *BMC Struct. Biol.* 12, 6.

Swapna, L.S., Srikeerthana, K., and Srinivasan, N. (2012c). Extent of structural asymmetry in homodimeric proteins: Prevalence and relevance. *PLOS ONE* 7, e36688.

Teichmann, S.A. (2002). The constraints protein–protein interactions place on sequence divergence. *J. Mol. Biol.* 324, 399–407.

Tompa, P., and Fuxreiter, M. (2008). Fuzzy complexes: Polymorphism and structural disorder in protein–protein interactions. *Trends Biochem. Sci.* 33, 2–8.

Tsai, C.J., Del Sol, A., and Nussinov, R. (2009). Protein allostery, signal transmission and dynamics: A classification scheme of allosteric mechanisms. *Mol. Biosyst.* 5, 207–216.

Tusk, S.E., Delalez, N.J., and Berry, R.M. (2018). Subunit exchange in protein complexes. *J. Mol. Biol.* 430, 4557–4579.

van Dam, T.J.P., and Snel, B. (2008). Protein complex evolution does not involve extensive network rewiring. *PLoS Comput. Biol.* 4, e1000132.

Vidal, M. (2016). How much of the human protein interactome remains to be mapped? *Sci. Signal.* 9, eg7–eg7.

Vidal, M., Cusick, M.E., and Barabási, A.-L. (2011). Interactome networks and human disease. *Cell* 144, 986–998.

Vishwanath, S., de Brevern, A.G., and Srinivasan, N. (2018). Same but not alike: Structure, flexibility and energetics of domains in multi-domain proteins are influenced by the presence of other domains. *PLOS Comput. Biol.* 14, e1006008.

Vishwanath, S., Sukhwal, A., Sowdhamini, R., *et al.* (2017). Specificity and stability of transient protein–protein interactions. *Curr. Opin. Struct. Biol.* 44, 77–86.

Walhout, A.J., Sordella, R., Lu, X., *et al.* (2000). Protein interaction mapping in *C. elegans* using proteins involved in vulval development. *Science* 287, 116–122.

Wright, P.E., and Dyson, H.J. (1999). Intrinsically unstructured proteins: Re-assessing the protein structure-function paradigm. *J. Mol. Biol.* 293, 321–331.

Yogurtcu, O.N., Erdemli, S.B., Nussinov, R., *et al.* (2008). Restricted mobility of conserved residues in protein–protein interfaces in molecular simulations. *Biophys. J.* 94, 3475–3485.

Yu, H., Luscombe, N.M., Lu, H.X., *et al.* (2004). Annotation transfer between genomes: Protein–protein interologs and protein-DNA regulogs. *Genome Res.* 14, 1107–1118.

Zhang, L., and Buck, M. (2013). Molecular simulations of a dynamic protein complex: Role of salt-bridges and polar interactions in configurational transitions. *Biophys. J.* 105, 2412–2417.

Zinman, G.E., Zhong, S., and Bar-Joseph, Z. (2011). Biological interaction networks are conserved at the module level. *BMC Syst. Biol.* 5, 134.

Zor, T., Mayr, B.M., Jane Dyson, H., *et al* (2002). Roles of phosphorylation and helix propensity in the binding of the KIX domain of CREB-binding protein by constitutive (c-Myb) and inducible (CREB) activators. *J. Biol. Chem.* 277, 42241–42248.

Chapter 2

A comprehensive overview of sequence-based protein-binding residue predictions for structured and disordered regions

Amita Barik and Lukasz Kurgan*

Department of Computer Science, Virginia Commonwealth University, Richmond, VA 23284, USA

Knowledge of protein–protein interactions (PPIs) facilitates annotation of protein functions and drug development efforts. Computational prediction of PPIs is motivated by a growing gap between the number of known and the number of functionally annotated protein sequences. This chapter focuses on sequence-based predictors of protein-binding residues (PBRs). These methods rely on predictive models that were developed from training data where the native annotations of PBRs were collected either from structures of protein–protein complexes (structure-annotated predictors) or from the intrinsically disordered/unstructured regions (disorder-annotated predictors). This is the first overview that considers both groups of methods. A comprehensive set of 36 predictors including 19 structure-annotated and 17 disorder-annotated tools have been surveyed. The fact that six new methods were published in 2018 alone suggests that this is still an active research area. Their availability, impact, predictive architectures, and outputs have been discussed.

*lkurgan@vcu.edu

These predictors rely on different combinations of five main types of inputs and typically utilize machine learning-derived predictive models. Methods with Web servers are the most highly cited, with the average citation count of 104, compared to the overall average of 74. The lack of solutions that combine structure-annotated and disorder-annotated data to accurately predict PBRs in both structured and disordered regions has been noted.

2.1. Introduction

Proteins interact with nucleic acids, lipids, a variety of small ligands, and proteins, prompting development of computational predictors of these interactions [1–10]. Understanding protein–protein interactions (PPIs) is important for a variety of applications, such as annotation of protein functions [11], drug discovery [12–14], study of disease mechanisms [15, 16], and the development of PPI networks [17, 18]. Several publicly available data repositories and resources that archive PPI data at the molecule (protein) and molecular (residue or atomic) levels have been developed. Examples include the Mentha resource that integrates data about PPIs at the protein level [19], BioLip that annotates protein-binding residues (PBRs) [20], and Protein Data Bank (PDB) that provides access to detailed atomic-level structures of protein–protein complexes [21–23]. However, the scope of these resources is very limited given the large number of already sequenced proteins, which has recently reached over 137 million (source: UniProt [24, 25] as of Jan 14, 2019). To compare, Mentha offers access to only 741 thousand interactions (as of Jan 13, 2014), BioLip to 21.5 thousand residue-level interaction sites (as of Jan 4, 2019), and PDB to close to 27 thousand complexes (as of Jan 14, 2019). The very wide gap between the expected and annotated number of interactions motivates the development of fast and cost-effective computational predictors of PPIs [8, 26–32], which can be used to support more tedious, labor-intensive, and relatively expensive experimental techniques [33–35].

Numerous computational methods for the prediction of PPIs have been developed [8, 26–32]. They can be broadly classified into two categories based on the input: structure based and sequence based [8, 31]. The structure-based methods are limited in scope to a relatively small set of proteins that have structure and proteins for which structure can be accurately predicted. The sequence-based methods require only protein sequences as input and thus they can be applied to make predictions for any of the millions of the currently sequenced proteins. They predict PPIs at either the whole protein level (whether or not a given protein interacts with another protein) or the residue level (whether particular amino acids in the protein sequence interact with proteins). The prediction of interactions at the residue level arguably provides more insight/more detailed information and hence we focus on these methods. The residue-level methods predict PBRs using either a single protein sequence or a pair of sequences. This chapter surveys and describes a large collection of predictors that require a single protein sequence as input. These methods predict PBRs for every residue in the input protein chain. Discussion concerning a relatively small set of protein-pair-based methods [36–40] can be found in [8]. These methods find PBRs for a pair of protein sequences that are presumably interacting with each other.

The single-sequence predictors of PBRs take a protein sequence as their input and provide output in either binary format (each residue in the input protein sequence is classified as either PBR or non-PBR) or both numeric (propensity score that quantifies likelihood that a given residue is protein binding) and binary formats. Figure 2.1 shows example predictions of PBRs for the nucleoplasmin protein from *Xenopus* (UniProt ID: P05221) that were generated with the DisoRDPbind predictor [41, 42]. A thin blue line at the top of the figure shows the putative propensity scores that are generated for each residue in this protein sequence. Higher values of these scores correspond to a higher likelihood that a given residue binds with proteins. These propensities are also used to generate the binary predictions. Residues with scores > threshold (0.807) are assumed to interact with proteins, whereas the remaining residues are assumed not to interact. This threshold was pre-optimized on a benchmark data set to ensure that DisoRDPbind generates a low false-positive rate of 10% (rate of

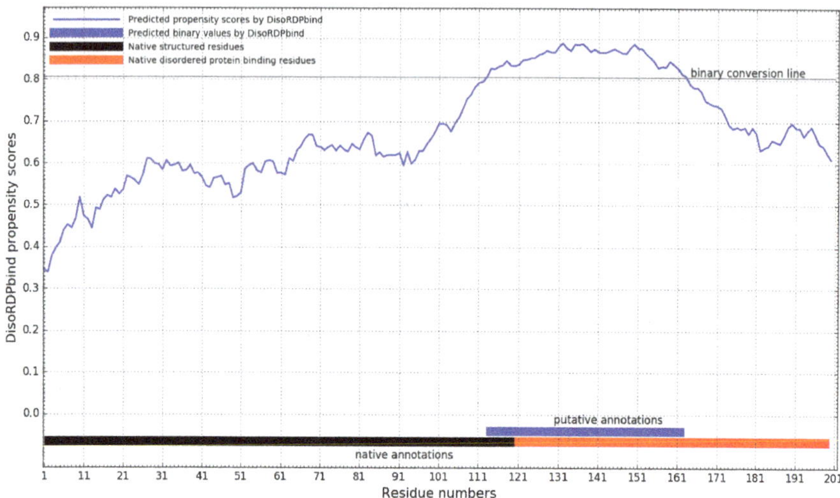

Figure 2.1. Prediction of PBRs for the nucleoplasmin protein from *Xenopus* (UniProt ID: P05221). Native annotations of PBRs that were obtained from DisProt database (DisProt ID: DP00217) are shown with black and red horizontal bars at the bottom of the figure. The putative annotations generated with the DisoRDPbind method are shown with a thin blue line (for the putative numeric propensities) and with blue horizontal bars (for the putative binary predictions).

incorrectly predicted PBRs). The binary predictions that stretch between positions 113 and 163 are represented by a thick blue horizontal line at the bottom of Figure 2.1. The native annotations of PBRs were obtained from the DisProt database [43, 44] and include intrinsically disordered protein-binding regions between positions 153 and 171 [45] and between positions 121 and 200 [46]. The predictions are in general in good agreement with the native PBRs. Although DisoRDPbind fails to generate correct binary predictions at the C-terminus that is annotated as protein binding (positions 164 to 200), it still provides relatively high propensity scores for this region that suggests high likelihood of protein binding. This example not only explains format and interpretation of a sample prediction but also shows that the putative propensities can be used to effectively supplement the binary predictions.

The single-sequence-based predictors of PBRs are being continually developed over the last couple of decades [8, 26–32]. We provide a comprehensive overview of these predictors focusing on their availability, impact, predictive architectures, and outputs.

2.2. Computational prediction of protein-binding residues from sequence

The single-sequence-based predictors of PBRs are typically derived with machine learning (ML) algorithms [47, 48]. These algorithms compute predictive models from a training data set that is annotated with native PBRs. The ML algorithms optimize the architecture and parameters of the models such that the disagreements between their outputs and the native annotations in the training data set are minimized. After the training is completed, the resulting models can be used to accurately predict PBRs in sequences of proteins that are not included in the training data set [8].

The single-sequence-based predictors of PBRs are divided into two types based on the training data sets that they utilize: structure annotated and disorder annotated. The structure-annotated methods rely on training data sets where the annotations of PBRs are derived from structures of protein–protein complexes, typically collected from PDB [21–23]. The disorder-annotated predictors are optimized based on training data sets with intrinsically disordered protein-binding regions. The intrinsically disordered regions (IDRs) lack a stable three-dimensional structure, and they typically materialize as ensembles of multiple conformational states [49–56]. They are highly abundant in nature, particularly among eukaryotic organisms [57–60]. The structural plasticity of IDRs facilitates efficient interactions with a variety of different molecules [61–71], including proteins [72–76]. Many IDRs undergo disorder-to-structure transitions concomitant with their protein-binding activity [72,77–80], which means that some of these interactions are covered by the protein–protein complexes in PDB. Annotations of disordered PBRs can be collected from a variety of sources including PDB (based on regions with missing three-dimensional structure) [81], DisProt [43, 44], and MobiDB [82, 83]. Moreover, both structure-annotated and disorder-annotated methods can be broadly classified into two

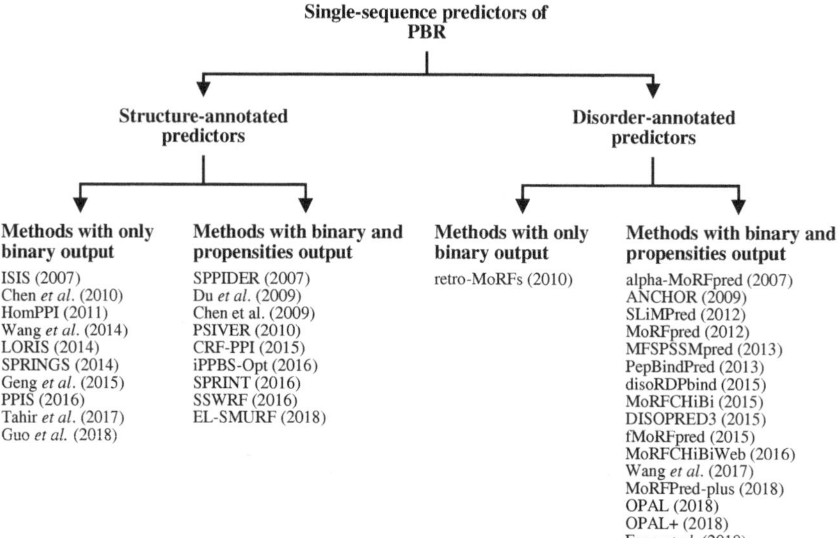

Figure 2.2.   Classification of single-sequence-based predictors of PBRs based on the type of annotations used to generate the underlying predictive models and the outputs that they produce.

categories based on the format of outputs that they generate. As discussed in Figure 2.1, a given predictor can produce either binary scores or both binary and propensity scores. Figure 2.2 illustrates a classification of current single-sequence-based predictors of PBRs on the basis of the annotations and the outputs. The efforts directed toward development of these predictors are well balanced and include 19 structure-annotated methods and 17 disorder-annotated methods. However, although most of the disorder-annotated predictors provide arguably more useful set of both output types, about half of the structure-annotated methods provide only the binary outputs.

Several surveys of the single-sequence-based predictors of PBRs have been published [8, 26–32]. The main drawback of these articles is that they focus specifically on only the disorder-annotated [32] or structure-annotated [8, 26–31] methods. This is the first article that bridges that divide and discusses both types of computational predictors.

2.2.1. *Overview of sequence-based predictors of PBRs*

Well over 30 single-sequence predictors of PBRs have been developed. Figure 2.3 depicts a historical timeline of these efforts. The structure-annotated methods are represented using black bars (total of 19), whereas disordered-annotated methods are shown with white bars (total of 17). The first two methods, the structure-annotated ISIS [84] and the disorder-annotated alpha-MoRFpred [85, 86], were released in 2007. We observe that although initially the structure-annotated predictors were dominant (six structure-annotated vs. three disorder-annotated methods were published between 2007 and 2010), recent years have observed a substantial shift toward the development of the disorder-annotated methods (10 disorder-annotated vs. 9 structure-annotated methods were released between 2015 and 2018). The fact that 2018 alone has seen six new methods [87–92] suggests that this is still an active research area.

Table 2.1 provides references and information about the year of publication, availability, and citation counts for the 36 predictors. We provide Uniform Resource Locators (URLs) for the implementations that are currently (as of Jan 7, 2019) available for 24 of the 36 predictors. The implementations can be provided in two ways: as Web

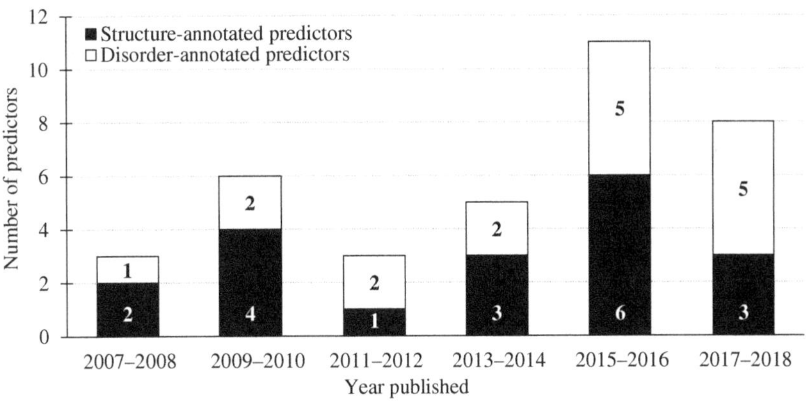

Figure 2.3. Historical timeline of the development of the single-sequence predictors of PBRs. The numbers inside the bars represent the number of predictors developed in the corresponding years.

Table 2.1.   Overview of the single-sequence-based predictors of PBRs.

Type	Method	Ref.	Year Published	Availability Type	Availability URL	Citations Total	Citations Annual
Structure-annotated predictors	ISIS	[84]	2007	NA	NA	244	22
	SPPIDER	[105]	2007	WS	http://sppider.cchmc.org/	272	25
	Du et al.	[106]	2009	NA	NA	11	1
	Chen et al.	[100]	2009	SC	NLA	126	14
	PSIVER	[107]	2010	WS	http://mizuguchilab.org/PSIVER/	114	14
	Chen et al.	[108]	2010	SC	http://mail.ustc.edu.cn/~bigeagle/BMCBioinfo2010/index.htm	39	5
	HomPPI	[109]	2011	WS	http://ailab1.ist.psu.edu/PSHOMPPIv1.2/index.html	55	8
	Wang et al.	[110]	2014	NA	NA	50	13
	LORIS	[111]	2014	SC	https://sites.google.com/site/sukantamondal/software	24	6
	SPRINGS	[112]	2014	SC	https://sites.google.com/site/predppis/	12	3
	CRF-PPI	[113]	2015	SC	http://csbio.njust.edu.cn/bioinf/CRF-PPI	6	2
	Geng et al.	[114]	2015	NA	NA	10	3
	iPPBS-Opt	[102]	2016	WS	http://www.jci-bioinfo.cn/iPPBS-Opt	94	47
	PPIS	[115]	2016	SC	http://csbio.njust.edu.cn/bioinf/PPIS	13	7
	SPRINT	[116]	2016	SC	http://sparks-lab.org/server/SPRINT/	32	16
	SSWRF	[101]	2016	SC	NLA	30	15
	Tahir et al.	[117]	2017	NA	NA	4	4
	Guo et al.	[90]	2018	NA	NA	6	6
	EL-SMURF	[91]	2018	SC	http://github.com/QUST-AIBBDRC/EL-SMURF/	0	0

	Method	Ref	Year	Type	URL	Total citations	Annual citations
Disorder-annotated predictors	alpha-MoRFpred	[85, 86]	2007	NA	NA	445	40
	ANCHOR	[93–95]	2009	WS + SC	http://anchor.enzim.hu	386	43
	retro-MoRFs	[118]	2010	NA	NA	27	3
	SLiMPred	[119]	2012	WS	http://bioware.ucd.ie/~compass/biowareweb//Server_pages/slimpred.php	54	9
	MoRFpred	[103,104]	2012	WS	http://biomine.cs.vcu.edu/servers/MoRFpred/	192	32
	MFSPSSMpred	[96]	2013	WS + SC	http://webapp.yama.info.waseda.ac.jp/fang/MoRFs.php	33	7
	PepBindPred	[120]	2013	WS	http://bioware.ucd.ie/~compass/biowareweb/Server_pages/pepbindpred.php	17	3
	DisoRDPbind	[41, 42]	2015	WS	http://biomine.cs.vcu.edu/servers/DisoRDPbind/	44	15
	MoRFCHiBi	[97]	2015	WS + SC	https://morf.msl.ubc.ca/index.xhtml	35	12
	DISOPRED3	[98]	2015	WS + SC	http://bioinf.cs.ucl.ac.uk/disopred	199	66
	fMoRFpred	[80]	2015	WS	http://biomine.cs.vcu.edu/servers/fMoRFpred/	35	12
	MoRFCHiBiWeb	[99]	2016	WS + SC	http://morf.chibi.ubc.ca:8080/mcw/index.xhtml	22	11
	Wang *et al.*	[121]	2017	NA	NA	2	2
	MoRFPred-plus	[92]	2018	SC	https://github.com/roneshsharma/MoRFpred-plus/wiki/MoRFpred-plus	8	8
	OPAL	[87]	2018	WS	http://www.alok-ai-lab.com/tools/opal/	8	8
	OPAL+	[88]	2018	SC	https://github.com/roneshsharma/OPAL-plus/wiki/OPAL-plus-Download	0	0
	Fang *et al.*	[89]	2018	NA	NA	0	0

The methods are divided into two groups: 19 structure annotated and 17 disorder annotated, and they are sorted by the publication year in the ascending order within each group. The "Type" column indicates whether a given method is available as the online webserver (WS) and/or standalone source code (SC); NA means that neither WS nor SC is available. The "URL" column gives the page where the method can be found as of Jan 7, 2019, where NLA means that the method is "no longer available", whereas the published article claims that it was originally available. The "Total citations" column gives the number of citations collected from Google Scholar on Jan 7, 2019. To avoid duplicate counting of citations for methods that are published in multiple articles, we use the one with the highest number of citations. The "Annual citations" column gives an average number of citations per year since a given method was published.

servers (WSs) and/or as source code (SC). The WSs are comparatively easier to use and primarily target less computer savvy users who want to perform *ad hoc* predictions. The computations are performed on the server side, and the end user only needs access to Internet and a Web browser to run the predictions. The predicted results are returned to the users typically via email or/and the Web browser. Only 4 (21%) out of the 19 structure-annotated predictors provide WSs, whereas 11 (65%) out of 17 of the disorder-annotated methods have WS facilities. The source code is helpful for users who want to run the predictions on their hardware, which could be because they need to predict a large data set of proteins or embed a given predictor into a larger bioinformatics pipeline. About 37% (7 out of 19) of the structure-annotated methods and 41% (7 out of 17) of the disorder-annotated methods provide this option. Finally, five disorder-annotated predictors including ANCHOR [93–95], MFSPSSM [96], MoRFChiBi [97], DISOPRED3 [98], and MoRFCHiBiWeb [99] are available as both WS and SC. Moreover, 12 predictors were either never made available to the community (10 methods) or the support was discontinued after they were made available at the time of publication. The latter includes the predictor by Chen *et al.* [100] and SSWRF [101].

Table 2.1 also quantifies citations, arguably one of the most important aspects of impact generated by these predictors. We note that we use only one reference, the one with the highest citation counts, for the few methods that were published in multiple articles to avoid duplicate counting when measuring these values. We quantify the total and the annual number of citations collected from Google Scholar. Overall, the 36 predictors were cited 2649 times, with an impressive average of 74 citations per method and median of 31 citations per method. The average and median increase to 104 and 54 for the 15 methods that offer WSs. The annual citation counts are more suitable for direct comparisons between methods, and they reveal that the most-cited methods include DISOPRED3 [98], iPPBS-Opt [102], ANCHOR [93–95], alpha-MoRFpred [85, 86], and MoRFpred [103,104], all of which secure over 30 citations annually. Interestingly, the availability of these predictors is directly correlated with their citations. Predictors that are not available and those that are available as only standalone code have the average annual citations of 10.3 and 5.2, respectively. The methods that are available as only WSs secure on average 17.3 citations per year, whereas the five predictors that are provided as both standalone code and WS have accumulated an average of 27.8 citations annually. This agrees with common sense because computational methods that are not made available or that have to be installed locally are less likely to be utilized by the end users, and thus less likely to be cited.

2.2.2. *Architectures of the predictors of PBRs*

Table 2.2 provides details about the architectures of the 36 single-sequence predictors of PBRs. We deconstruct the architectures into three major parts: inputs, predictive models, and outputs, and we discuss each of these individually.

The five commonly used elements of the input are computed from the amino acid sequence (AAS), evolutionary information (EVO), and from three relevant predicted structural properties: putative relative solvent accessibility (pRSA), putative secondary structure (pSS) and putative intrinsic disorder (pDIS). AAS-based input typically quantifies amino acid composition, residue-level physiochemical properties, and/or position of amino acids in the sequence. EVO is usually calculated from the position-specific scoring matrix generated from the input protein chain with the PSI-BLAST algorithm [122]. While AA and EVO are calculated directly from the protein sequence, the other three input elements (pRSA, pSS, and pDIS) are predicted from the sequence with bioinformatics tools. RSA is a measure of residue-level solvent exposure, which is calculated by dividing the predicted solvent accessible surface of a given residue in the input protein sequence by the maximum possible solvent accessible surface area of the same amino acid type. Virtually all structure-annotated methods, except for [117], use pRSA, compared to only 5 out of 17 disorder-annotated tools. This is expected because disordered protein-binding regions do not have a well-defined surface when

Table 2.2.   Predictive architecture and outputs generated by the single-sequence-based predictors of PBRs.

Type	Method	Ref.	AAS	EVO	pRSA	pSS	pDIS	Predictive Model	Binary	Propensity
	ISIS	[84]		√	√	√		ML (NN)	√	
	SPPIDER	[105]	√	√	√			ML (KNN)	√	√
	Du et al.	[106]	√	√	√			ML (SVM)	√	√
	Chen et al.	[100]	√	√	√			ML (RF)	√	√
	PSIVER	[107]		√	√			ML (NB)	√	√
	Chen et al.	[108]	√	√	√			ML (SVM)	√	
	HomPPI	[109]		√	√		√	SF	√	
	Wang et al.	[110]		√	√			ML (SVM)	√	
	LORIS	[111]	√	√	√	√		ML (RLF)	√	
	SPRINGS	[112]	√	√	√	√		ML (NN)	√	
	CRF-PPI	[113]	√	√	√			ML (RF)	√	√
	Geng et al.	[114]		√	√			ML (NB)	√	
	iPPBS-Opt	[102]	√		√			ML (KNN)	√	√
	PPIS	[115]	√	√	√			ML (RF)	√	
	SPRINT	[116]	√	√	√	√		ML (SVM)	√	√
	SSWRF	[101]	√	√	√			ML (SVM, RF)	√	√
	Tahir et al.	[117]	√	√				ML (KNN, PNN, SVM)	√	
	Guo et al.	[90]	√	√	√			ML (SVM)	√	
	EL-SMURF	[91]	√	√	√			ML (RF)	√	√

Type column spanning label: Structure-annotated predictors

Inputs (spanning header over AAS, EVO, pRSA, pSS, pDIS). Outputs (spanning header over Binary, Propensity).

Disorder-annotated predictors	References	AAS	EVO	pRSA	pSS	pDIS	Predictive model	Outputs	
alpha-MoRFpred	[85, 86]	√			√	√	ML (NN)	√	√
ANCHOR	[93–95]	√					SF	√	√
retro-MoRFs	[118]	√				√	SF	√	
SLiMPred	[119]	√		√	√	√	ML (NN)	√	√
MoRFpred	[103,104]	√	√	√		√	ML (SVM)	√	√
MFSPSSMpred	[96]		√				ML (SVM)	√	√
PepBindPred	[120]				√	√	ML (NN)	√	√
DisoRDPbind	[41, 42]	√			√	√	ML (LR)	√	√
MoRFCHiBi	[97]	√					ML (SVM)	√	√
DISOPRED3	[98]	√	√				ML (SVM)	√	√
fMoRFpred	[80]	√			√	√	ML (SVM)	√	√
MoRFCHiBiWeb	[99]	√	√			√	ML (NB)	√	√
Wang et al.	[121]	√	√	√	√		ML (SVM)	√	√
MoRFPred-plus	[92]	√	√				ML (HMM)	√	√
OPAL	[87]	√		√		√	ML (SVM)	√	√
OPAL+	[88]	√		√	√		ML (SVM)	√	√
Fang et al.	[89]		√				ML (SVM)	√	√

The methods are divided into two groups: 19 structure annotated and 17 disorder annotated, and they are sorted by the publication year in the ascending order within each group. The "Input" subcolumns include information extracted directly from the amino acid sequence (AAS), evolutionary information (EVO), and structural properties predicted from the sequence that include putative relative solvent accessibility (pRSA), putative secondary structure (pSS), and putative intrinsic disorder (pDIS). The √ means that a given input type is utilized as one of the inputs, whereas blank cell indicates that it is not considered. The "Predictive model" column categorizes the models into two groups: those generated with ML algorithms and those that rely on a scoring function (SF) generated either by an empirical formula or using an alignment score. The ML models include neural network (NN), K-nearest neighbor (KNN), probabilistic neural network (PNN), support vector machine (SVM), random forest (RF), naïve Bayes (NB), regularized logistic function (RLF), logistic regression (LR), and hidden Markov model (HMM). The √ in the "Outputs" column means that a given type of output is generated by the predictor.

compared to the structured protein–protein interfaces. The pSS is generated with one of the popular tools [123–126] that include PSIPRED [127], PROFphd [128], and Distill [129]. The pDIS is clearly relevant for the disorder-annotated predictors, and there are many accurate algorithms that can be used to provide this input [130–135]. Correspondingly, majority of the disorder-annotated predictors (9 out of 17) use this input, compared to only one structure-annotated method. Overall, we observe that none of the methods uses all five types of inputs, whereas majority of the methods (22 out of 36) use between three and four input types.

Over 90% of methods (33 out of 36), which exclude just one structure-annotated and two disorder-annotated predictors, apply ML algorithms to generate predictive models. Some of the frequently used ML models are SVMs (used by 16 methods), NNs (6 predictors), RF (5 methods), KNNs (3 methods), and NB (3 methods). The three non-ML predictors rely on relatively simple predictive models in the form of an empirical formula or sequence alignment [93–95, 109, 118].

The outputs of the predictors, which we explain in the "Introduction" section, take two forms: binary values and propensity scores. Although all predictors discussed here generate binary outputs, some of them do not provide the propensities. To be more precise, all but one disorder-annotated predictor provide both binary and propensity scores, whereas only 47% of the structure-annotated methods (9 out of 19 methods) generate both outputs. This lack of propensities is a drawback because these values can be used to provide useful context for the binary predictions. For instance, propensities can be used to find missing putative PBRs (i.e., residue that are not predicted as PBRs in binary but which have relatively high propensities, which is illustrated in Figure 2.1) or to find lower quality predictions of PBRs (i.e., residues predicted as PBRs in binary but with relatively low propensity scores).

Overall, our analysis reveals that the single-sequence predictors of PBRs are characterized by a wide range of architectures that rely on five major types of inputs and that use a variety of ML-derived predictive models. The fact that none of the current predictors applies all five input types opens an opportunity to develop more accurate models that would take advantage of all types of inputs.

2.3. Summary and recommendations

The low coverage and importance of the current annotations of PPIs have stimulated the development of numerous single-sequence predictors of PBRs. We discuss two categories of these computational predictors that were trained using structure-annotated vs. disorder-annotated data sets. We show that recent development efforts have shifted toward the disorder-annotated predictors, with four methods that were released in 2018 alone. Most of these methods are made available to the community as either SC or WS, with only a few that offer both options. Empirical analysis reveals that methods that include WSs are cited at much higher rates, with an average of 104 citations. We also summarize the predictive architecture of the 36 predictors of PBRs. They rely on various combinations of five main types of inputs and use a wide range of primarily ML-derived predictive models.

Although most of the predictors are made available to the community as WSs and/or standalone programs, one-third (12 out of 36) were not released or their availability was discontinued. They are unlikely to be ever used and consequently they are cited substantially less often. We believe that methods that are not made available should not be published and that the subsequent support should be guaranteed by the authors as part of the publication process. These standards are already imposed in some of the lead publication venues, such as *Nucleic Acids Research* where WS articles are expected to be functional and maintained for at least 5 years after publication. We also stress the importance of the propensity outputs that provide a useful and effective context to the binary outputs, which we demonstrate in Figure 2.1. The designers of these methods should strive to provide the propensities, which unfortunately is not the case for about half of the structure-annotated methods.

A perhaps surprising observation is the separation between the disorder-annotated and the structure-annotated methods. There is not a single method that combines both types of training data to provide a more complete solution capable of predicting PBRs in both structured and disordered regions. Such cross-over solutions should be developed in the near future.

A recent comparative review reveals that the structure-annotated predictors generate moderately accurate predictions of PBRs [8]. The binary outputs of the best structure-annotated predictors offer accuracies at about 80% (the high value is driven by the large majority of easy to predict non-binding residues) and modest levels of correlation between the predicted and native PBRs, with the Matthews correlation coefficient of 0.21. The putative propensities that they produce are characterized by moderate AUC values in the 0.65–0.69 range, where the overall AUC values range between 0.5 and 1. We also emphasize a recent empirical observation that these methods heavily cross-predict residues that bind other ligands (RNA, DNA, and a variety of small ligands) as PBRs. As many as 20–40% of the residues that interact with these other ligands are predicted as PBRs, when compared to similar levels of sensitivity (rate of correct predictions for native PBRs) [8]. This means that the current methods predict PBRs at similar rates among the native PBRs as among the residues that bind the other ligands. This likely stems from the fact that training data sets of these methods are solely use proteins with PBRs, lacking a sufficient population of proteins that interact with other ligands. This results in an inability to train the predictive models that could differentiate between PBRs and other types of ligand-binding residues. Thus, we propose that more accurate solutions that use better training data sets and that can more specifically target PBRs should be developed.

Acknowledgments

This research was supported in part by the National Science Foundation (grant 1617369) and the Robert J. Mattauch Endowment funds to LK.

References

1. Ding, X.M., *et al.* Computational prediction of DNA-protein interactions: A review. *Curr Comput Aided Drug Des.* 2010;6(3): 197–206.
2. Chen, K. and Kurgan, L. Investigation of atomic level patterns in protein — small ligand interactions. *PLOS ONE.* 2009;4(2):e4473.
3. Sudha, G., Nussinov, R., and Srinivasan, N. An overview of recent advances in structural bioinformatics of protein–protein interactions and a guide to their principles. *Prog Biophys Mol Biol.* 2014;116(2–3): 141–150.
4. Fornes, O., *et al.* On the use of knowledge-based potentials for the evaluation of models of protein–protein, protein–DNA, and protein–RNA interactions. *Adv Protein Chem Struct Biol.* 2014;94:77–120.
5. Barik, A., *et al.* Molecular architecture of protein–RNA recognition sites. *J Biomol Struct Dyn.* 2015;33(12):2738–2751.
6. Chowdhury, S., Zhang, J., and Kurgan, L. In silico prediction and validation of novel RNA binding proteins and residues in the human proteome. *Proteomics.* 2018;18:e1800064.
7. Zhao, H., Yang, Y., and Zhou, Y. Prediction of RNA binding proteins comes of age from low resolution to high resolution. *Mol Biosyst.* 2013;9(10):2417–2425.
8. Zhang, J. and Kurgan, L. Review and comparative assessment of sequence-based predictors of protein-binding residues. *Brief Bioinform.* 2018;19(5):821–837.
9. Hu, G., *et al.* Finding protein targets for small biologically relevant ligands across fold space using inverse ligand binding predictions. *Structure.* 2012;20(11):1815–1822.
10. Chen, K., *et al.* A critical comparative assessment of predictions of protein-binding sites for biologically relevant organic compounds. *Structure.* 2011;19(5):613–621.
11. Orii, N. and Ganapathiraju, M.K. Wiki-pi: A web-server of annotated human protein–protein interactions to aid in discovery of protein function. *PLOS ONE.* 2012;7(11):e49029.
12. Sperandio, O. Editorial: Toward the design of drugs on protein–protein interactions. *Curr Pharm Des.* 2012;18(30):4585.
13. Petta, I., *et al.* Modulation of protein–protein interactions for the development of novel therapeutics. *Mol Ther.* 2016;24(4):707–718.

14. Athanasios, A., *et al.* Protein–protein interaction (PPI) network: Recent advances in drug discovery. *Curr Drug Metab.* 2017;18(1): 5–10.

15. Kuzmanov, U. and Emili, A. Protein–protein interaction networks: Probing disease mechanisms using model systems. *Genome Med.* 2013;5(4):37.

16. Nibbe, R.K., *et al.* Protein–protein interaction networks and subnetworks in the biology of disease. *Wiley Interdiscip Rev Syst Biol Med.* 2011;3(3):357–367.

17. De Las Rivas, J. and Fontanillo, C. Protein–protein interaction networks: Unraveling the wiring of molecular machines within the cell. *Brief Funct Genomics.* 2012;11(6):489–496.

18. Hao, T., *et al.* Reconstruction and application of protein–protein interaction network. *Int J Mol Sci.* 2016;17(6):907.

19. Calderone, A., Castagnoli, L., and Cesareni, G. Mentha: A resource for browsing integrated protein-interaction networks. *Nat Methods.* 2013;10(8):690–691.

20. Yang, J., Roy, A., and Zhang, Y. BioLiP: A semi-manually curated database for biologically relevant ligand-protein interactions. *Nucleic Acids Res.* 2013;41(Database issue):D1096–D1103.

21. Burley, S.K. PDB40: The Protein Data Bank celebrates its 40th birthday. *Biopolymers.* 2013;99(3):165–169.

22. Burley, S.K., *et al.* Protein Data Bank (PDB): The single global macromolecular structure archive. *Methods Mol Biol.* 2017;1607:627–641.

23. Berman, H.M., *et al.* The Protein Data Bank. *Nucleic Acids Res.* 2000;28(1):235–242.

24. The UniProt, C. UniProt: The universal protein knowledgebase. *Nucleic Acids Res.* 2017;45(D1):D158–D169.

25. UniProt, C. UniProt: A hub for protein information. *Nucleic Acids Res.* 2015;43(Database issue):D204–D212.

26. Ezkurdia, I., *et al.* Progress and challenges in predicting protein–protein interaction sites. *Brief Bioinform.* 2009;10(3):233–246.

27. Fernandez-Recio, J. Prediction of protein binding sites and hot spots. *WIRES Comput Mol Sci.* 2011;1(5):680–698.

28. Aumentado-Armstrong, T.T., Istrate, B., and Murgita, R.A. Algorithmic approaches to protein–protein interaction site prediction. *Algorithms Mol Biol.* 2015;10:7.

29. Xue, L.C., *et al.* Computational prediction of protein interfaces: A review of data driven methods. *FEBS Lett.* 2015;589(23):3516–3526.

30. Esmaielbeiki, R., *et al.* Progress and challenges in predicting protein interfaces. *Brief Bioinform.* 2016;17(1):117–131.

31. Maheshwari, S. and Brylinski, M. Predicting protein interface residues using easily accessible on-line resources. *Brief Bioinform.* 2015;16(6): 1025–1034.

32. Meng, F., Uversky, V.N., and Kurgan, L. Comprehensive review of methods for prediction of intrinsic disorder and its molecular functions. *Cell Mol Life Sci.* 2017;74(17):3069–3090.

33. Rigaut, G., *et al.* A generic protein purification method for protein complex characterization and proteome exploration. *Nat Biotechnol.* 1999;17(10):1030–1032.

34. Sobott, F. and Robinson, C.V. Protein complexes gain momentum. *Curr Opin Struct Biol.* 2002;12(6):729–734.

35. Yates, J.R., 3rd., Mass spectrometry. From genomics to proteomics. *Trends Genet.* 2000;16(1):5–8.

36. Pitre, S., *et al.* PIPE: A protein–protein interaction prediction engine based on the re-occurring short polypeptide sequences between known interacting protein pairs. *BMC Bioinformatics.* 2006;7:365.

37. Shi, M.G., *et al.* Predicting protein–protein interactions from sequence using correlation coefficient and high-quality interaction dataset. *Amino Acids.* 2010;38(3):891–899.

38. Chang, D.T.H., Syu, Y.T., and Lin, P.C. Predicting the protein–protein interactions using primary structures with predicted protein surface. *BMC Bioinformatics.* 2010;11:S3.

39. Amos-Binks, A., *et al.* Binding site prediction for protein–protein interactions and novel motif discovery using re-occurring polypeptide sequences. *BMC Bioinformatics.* 2011;12:225.

40. Xia, B., *et al.* PETs: A stable and accurate predictor of protein–protein interacting sites based on extremely-randomized trees. *IEEE Trans Nanobioscience.* 2015;14(8):882–893.

41. Peng, Z., *et al.* Prediction of disordered RNA, DNA, and protein binding regions using disoRDPbind. *Methods Mol Biol.* 2017;1484: 187–203.

42. Peng, Z. and Kurgan, L. High-throughput prediction of RNA, DNA and protein binding regions mediated by intrinsic disorder. *Nucleic Acids Res.* 2015;43(18):e121.

43. Vucetic, S., *et al.* DisProt: A database of protein disorder. *Bioinformatics.* 2005;21(1):137–140.

44. Piovesan, D., *et al.* DisProt 7.0: A major update of the database of disordered proteins. *Nucleic Acids Res.* 2016;D1:D219–D227.

45. Conti, E. and Kuriyan, J., Crystallographic analysis of the specific yet versatile recognition of distinct nuclear localization signals by karyopherin alpha. *Structure.* 2000;8(3):329–338.

46. Hierro, A., *et al.* Structural and functional properties of *Escherichia coli*-derived nucleoplasmin. A comparative study of recombinant and natural proteins. *Eur J Biochem.* 2001;268(6):1739–1748.

47. Larranaga, P., *et al.* Machine learning in bioinformatics. *Brief Bioinform.* 2006;7(1):86–112.

48. Kurgan, L. and Zhou, Y. Machine learning models in protein bioinformatics. *Curr Protein Pept Sci.* 2011;12(6):455.

49. Wright, P.E. and Dyson, H.J. Intrinsically unstructured proteins: Re-assessing the protein structure-function paradigm. *J Mol Biol.* 1999;293(2):321–331.

50. Uversky, V.N., Gillespie, J.R., and Fink, A.L. Why are "natively unfolded" proteins unstructured under physiologic conditions? *Proteins.* 2000;41(3):415–427.

51. Dunker, A.K., *et al.* Intrinsically disordered protein. *J Mol Graph Model.* 2001;19(1):26–59.

52. Uversky, V.N. and Dunker, A.K. Understanding protein non-folding. *Biochim Biophys Acta.* 2010;1804(6):1231–1264.

53. Habchi, J., *et al.* Introducing protein intrinsic disorder. *Chem Rev.* 2014;114(13):6561–6588.

54. Uversky, V.N. Introduction to intrinsically disordered proteins (IDPs). *Chem Rev.* 2014;114(13):6557–6560.

55. van der Lee, R., *et al.* Classification of intrinsically disordered regions and proteins. *Chem Rev.* 2014;114(13):6589–6631.

56. Lieutaud, P., *et al.* How disordered is my protein and what is its disorder for? A guide through the "dark side" of the protein universe. *Intrinsically Disord Proteins.* 2016;4(1):e1259708.

57. Xue, B., Dunker, A.K., and Uversky, V.N. Orderly order in protein intrinsic disorder distribution: Disorder in 3500 proteomes from viruses and the three domains of life. *J Biomol Struct Dyn.* 2012;30(2):137–149.

58. Ward, J.J., *et al.* Prediction and functional analysis of native disorder in proteins from the three kingdoms of life. *J Mol Biol.* 2004;337(3):635–645.

59. Peng, Z., *et al.* Exceptionally abundant exceptions: Comprehensive characterization of intrinsic disorder in all domains of life. *Cell Mol Life Sci.* 2015;72(1):137–151.

60. Peng, Z., Mizianty, M.J., and Kurgan, L. Genome-scale prediction of proteins with long intrinsically disordered regions. *Proteins.* 2014; 82(1):145–158.

61. Dyson, H.J. Roles of intrinsic disorder in protein-nucleic acid interactions. *Mol Biosyst.* 2012;8(1):97–104.

62. Varadi, M., *et al.* Functional advantages of conserved intrinsic disorder in RNA-binding proteins. *PLOS ONE.* 2015;10(10):e0139731.

63. Meng, F., *et al.* Compartmentalization and functionality of nuclear disorder: Intrinsic disorder and protein–protein interactions in intra-nuclear compartments. *Int J Mol Sci.* 2015;17(1):pii.

64. Peng, Z., *et al.* More than just tails: Intrinsic disorder in histone proteins. *Mol Biosyst.* 2012;8(7):1886–1901.

65. Xue, B., *et al.* Protein intrinsic disorder as a flexible armor and a weapon of HIV-1. *Cell Mol Life Sci.* 2012;69(8):1211–1259.

66. Fan, X., *et al.* The intrinsic disorder status of the human hepatitis C virus proteome. *Mol Biosyst.* 2014;10(6):1345–1363.

67. Xue, B., *et al.* Structural disorder in viral proteins. *Chem Rev.* 2014;114(13):6880–6911.

68. Peng, Z., *et al.* A creature with a hundred waggly tails: Intrinsically disordered proteins in the ribosome. *Cell Mol Life Sci.* 2014;71(8): 1477–1504.

69. Wu, Z., *et al.* In various protein complexes, disordered protomers have large per-residue surface areas and area of protein-, DNA- and RNA-binding interfaces. *FEBS Lett.* 2015;589(19 Pt A):2561–2569.

70. Na, I., *et al.* Autophagy-related intrinsically disordered proteins in intra-nuclear compartments. *Mol Biosyst.* 2016;12(9):2798–2817.

71. Peng, Z., *et al.* Resilience of death: Intrinsic disorder in proteins involved in the programmed cell death. *Cell Death Differ.* 2013;20(9):1257–1267.

72. Fuxreiter, M., *et al.* Disordered proteinaceous machines. *Chem Rev.* 2014;114(13):6806–6843.

73. Tompa, P., *et al.* Close encounters of the third kind: Disordered domains and the interactions of proteins. *Bioessays.* 2009;31(3): 328–335.

74. Patil, A. and Nakamura, H. Disordered domains and high surface charge confer hubs with the ability to interact with multiple proteins in interaction networks. *FEBS Lett.* 2006;580(8):2041–2045.

75. Hu, G., *et al.* Functional analysis of human hub proteins and their interactors involved in the intrinsic disorder-enriched interactions. *Int J Mol Sci.* 2017;18(12):pii.

76. Xue, B., *et al.* Stochastic machines as a colocalization mechanism for scaffold protein function. *FEBS Lett.* 2013;587(11):1587–1591.

77. Dyson, H.J. and Wright, P.E. Coupling of folding and binding for unstructured proteins. *Curr Opin Struct Biol.* 2002;12(1): 54–60.

78. Mohan, A., *et al.* Analysis of molecular recognition features (MoRFs). *J Mol Biol.* 2006;362(5):1043–1059.

79. Vacic, V., *et al.* Characterization of molecular recognition features, MoRFs, and their binding partners. *J Proteome Res.* 2007;6(6): 2351–2366.

80. Yan, J., *et al.* Molecular recognition features (MoRFs) in three domains of life. *Mol Biosyst.* 2016;12(3):697–710.

81. DeForte, S. and Uversky, V.N. Resolving the ambiguity: Making sense of intrinsic disorder when PDB structures disagree. *Protein Sci.* 2016;25(3):676–688.

82. Piovesan, D., *et al.* MobiDB 3.0: More annotations for intrinsic disorder, conformational diversity and interactions in proteins. *Nucleic Acids Res.* 2018;46(D1):D471–D476.

83. Di Domenico, T., *et al.* MobiDB: A comprehensive database of intrinsic protein disorder annotations. *Bioinformatics.* 2012;28(15):2080–2081.

84. Ofran, Y. and Rost, B. ISIS: Interaction sites identified from sequence. *Bioinformatics.* 2007;23(2):e13–e16.

85. Oldfield, C.J., *et al.* Comparing and combining predictors of mostly disordered proteins. *Biochemistry.* 2005;44(6):1989–2000.

86. Cheng, Y., *et al.* Mining alpha-helix-forming molecular recognition features with cross species sequence alignments. *Biochemistry.* 2007;46(47):13468–13477.

87. Sharma, R., *et al.* OPAL: Prediction of MoRF regions in intrinsically disordered protein sequences. *Bioinformatics.* 2018;34(11): 1850–1858.

88. Sharma, R., *et al.* OPAL+: Length-specific moRF prediction in intrinsically disordered protein sequences. *Proteomics.* 2018;19: e1800058.

89. Fang, C., *et al. Identifying MoRFs in disordered proteins using enlarged conserved features.* Proceedings of the 2018 6th international conference on bioinformatics and computational biology. Chengdu, China: ACM; 2018:50–54.

90. Guo, H., *et al.* Predicting protein–protein interaction sites using modified support vector machine. *Int J Mach Learn Cybern.* 2018; 9(3):393–398.

91. Wang, X., *et al.* Protein–protein interaction sites prediction by ensemble random forests with synthetic minority oversampling technique. *Bioinformatics.* 2018.

92. Sharma, R., *et al.* MoRFPred-plus: Computational identification of MoRFs in protein sequences using physicochemical properties and HMM profiles. *J Theor Biol.* 2018;437:9–16.

93. Meszaros, B., Simon, I., and Dosztanyi, Z. Prediction of protein binding regions in disordered proteins. *PLOS Comput Biol.* 2009;5(5): e1000376.

94. Dosztanyi, Z., Meszaros, B., and Simon, I. ANCHOR: Web server for predicting protein binding regions in disordered proteins. *Bioinformatics.* 2009;25(20):2745–2746.

95. Meszaros, B., Erdos, G., and Dosztanyi, Z. IUPred2A: Context-dependent prediction of protein disorder as a function of redox state and protein binding. *Nucleic Acids Res.* 2018;46(W1):W329–W337.

96. Fang, C., *et al.* MFSPSSMpred: Identifying short disorder-to-order binding regions in disordered proteins based on contextual local evolutionary conservation. *BMC Bioinformatics.* 2013;14:300.

97. Malhis, N., and Gsponer, J. Computational identification of MoRFs in protein sequences. *Bioinformatics.* 2015;31(11):1738–1744.

98. Jones, D.T. and Cozzetto, D. DISOPRED3: Precise disordered region predictions with annotated protein-binding activity. *Bioinformatics.* 2015;31(6):857–863.

99. Malhis, N., Jacobson, M., and Gsponer, J. MoRFchibi SYSTEM: Software tools for the identification of MoRFs in protein sequences. *Nucleic Acids Res.* 2016;44:W488–W493.

100. Chen, X.W. and Jeong, J.C. Sequence-based prediction of protein interaction sites with an integrative method. *Bioinformatics.* 2009; 25(5):585–591.

101. Wei, Z.S., *et al.* Protein–protein interaction sites prediction by ensembling SVM and sample-weighted random forests. *Neurocomputing.* 2016;193:201–212.

102. Jia, J.H., *et al.* iPPBS-Opt: A sequence-based ensemble classifier for identifying protein–protein binding sites by optimizing imbalanced training datasets. *Molecules.* 2016;21(1):E95.

103. Disfani, F.M., *et al.* MoRFpred, a computational tool for sequence-based prediction and characterization of short disorder-to-order transitioning binding regions in proteins. *Bioinformatics.* 2012;28(12): i75–i83.

104. Oldfield, C.J., Uversky, V.N., and Kurgan, L. Predicting functions of disordered proteins with MoRFpred. *Methods Mol Biol.* 2018;1851: 337–352.

105. Porollo, A. and Meller, J. Prediction-based fingerprints of protein–protein interactions. *Proteins.* 2007;66(3):630–645.

106. Du, X.Q., Cheng, J.X., and Song, J. Improved prediction of protein binding sites from sequences using genetic algorithm. *Protein J.* 2009;28(6):273–280.

107. Murakami, Y. and Mizuguchi, K. Applying the naive Bayes classifier with kernel density estimation to the prediction of protein–protein interaction sites. *Bioinformatics.* 2010;26(15):1841–1848.

108. Chen, P. and Li, J. Sequence-based identification of interface residues by an integrative profile combining hydrophobic and evolutionary information. *BMC Bioinformatics.* 2010;11:402.

109. Xue, L.C., Dobbs, D., and Honavar, V. HomPPI: A class of sequence homology based protein–protein interface prediction methods. *BMC Bioinformatics.* 2011;12:244.

110. Wang, D.A., Wang, R., and Yan, H. Fast prediction of protein–protein interaction sites based on extreme learning machines. *Neurocomputing.* 2014;128:258–266.

111. Dhole, K., *et al.* Sequence-based prediction of protein–protein interaction sites with L1-logreg classifier. *J Theor Biol.* 2014;348:47–54.

112. Singh, G., *et al.* SPRINGS: Prediction of protein–protein interaction sites using artificial neural networks. *J Proteomics Computational Biol.* 2014;1.

113. Wei, Z.S., *et al.* A cascade random forests algorithm for predicting protein–protein interaction sites. *IEEE Trans Nanobioscience.* 2015;14(7):746–760.

114. Geng, H.J., *et al.* Prediction of protein–protein interaction sites based on naive Bayes classifier. *Biochem Res Int.* 2015.

115. Liu, G.H., Shen, H.B., and Yu, D.J. Prediction of protein–protein interaction sites with machine-learning-based data-cleaning and post-filtering procedures. *J Membr Biol.* 2016;249(1–2):141–153.

116. Taherzadeh, G., *et al.* Sequence-based prediction of protein-peptide binding sites using support vector machine. *J Comput Chem.* 2016;37(13):1223–1229.

117. Tahir, M. and Hayat, M. Machine learning based identification of protein–protein interactions using derived features of physiochemical properties and evolutionary profiles. *Artif Intell Med.* 2017;78:61–71.

118. Xue, B., Dunker, A.K., and Uversky, V.N. Retro-MoRFs: Identifying protein binding sites by normal and reverse alignment and intrinsic disorder prediction. *Int J Mol Sci.* 2010;11(10):3725–3747.

119. Mooney, C., *et al.* Prediction of short linear protein binding regions. *J Mol Biol.* 2012;415(1):193–204.

120. Khan, W., *et al.* Predicting binding within disordered protein regions to structurally characterised peptide-binding domains. *PLOS ONE.* 2013;8(9):e72838.

121. Wang, Y., *et al.* A sequence-based computational method for prediction of MoRFs. *RSC Adv.* 2017;7(31):18937–18945.

122. Altschul, S.F., *et al.* Gapped BLAST and PSI-BLAST: A new generation of protein database search programs. *Nucleic Acids Res.* 1997;25(17):3389–3402.

123. Zhang, H., *et al.* Critical assessment of high-throughput standalone methods for secondary structure prediction. *Brief Bioinform.* 2011;12(6):672–688.

124. Meng, F. and Kurgan, L. Computational prediction of protein secondary structure from sequence. *Curr Protoc Protein Sci.* 2016; 86:2.3.1–2.3.10.

125. Chen, K. and Kurgan, L. Computational prediction of secondary and supersecondary structures. *Methods Mol Biol.* 2013;932: 63–86.

126. Meng, F., Uversky, V., and Kurgan, L. Computational prediction of intrinsic disorder in proteins. *Curr Protoc Protein Sci.* 2017;88: 2.16.1–2.16.14.

127. Buchan, D.W., *et al.* Scalable web services for the PSIPRED protein analysis workbench. *Nucleic Acids Res.* 2013;41(Web Server issue):W349–W357.

128. Rost, B. Prediction in 1D: Secondary structure, membrane helices, and accessibility in structural bioinformatics. In: Bourne, P., and Weissig, H., eds. *Structural Bioinformatics.* New Jersey: Wiley; 2002:559–588.

129. Bau, D., *et al.* Distill: A suite of web servers for the prediction of one-, two- and three-dimensional structural features of proteins. *BMC Bioinformatics.* 2006;7:402.

130. Hu, G., *et al.* Quality assessment for the putative intrinsic disorder in proteins. *Bioinformatics.* 2018;35:1692–1700.

131. Walsh, I., *et al.* Comprehensive large-scale assessment of intrinsic protein disorder. *Bioinformatics.* 2015;31(2):201–208.

132. Peng, Z.L. and Kurgan, L. Comprehensive comparative assessment of in-silico predictors of disordered regions. *Curr Protein Pept Sci.* 2012;13(1):6–18.

133. Peng, Z. and Kurgan, L. On the complementarity of the consensus-based disorder prediction. *Pac Symp Biocomput.* 2012:176–187.

134. Deng, X., Eickholt, J., and Cheng, J. A comprehensive overview of computational protein disorder prediction methods. *Mol Biosyst.* 2012;8(1):114–1121.

135. Dosztanyi, Z., Meszaros, B., and Simon, I. Bioinformatical approaches to characterize intrinsically disordered/unstructured proteins. *Brief Bioinform.* 2010;11(2):225–243.

Chapter 3
Prediction of protein–protein complex structures by docking

Danial Pourjafar-Dehkordi and Martin Zacharias*

Technical University of Munich, Physics Department, James Franck Str. 1, Garching D-85748, Germany

Most cellular functions are mediated by the interaction of proteins to form dimeric or multimeric complexes as functional elements. Protein–protein complex structure can be predicted by computational protein–protein docking methods. A variety of docking methodologies are available and can be accessed as Web servers or stand-alone applications. The chapter is intended to give an overview on available approaches, on the refinement and scoring of predicted structures and discusses recent progress and future challenges of the protein–protein docking field.

3.1. Introduction

Protein–protein complex formation is an essential element of the majority of cellular processes. DNA replication, transcription and translation of genetic information but also many metabolic and transport processes require the assembly of complexes consisting of multiple proteins and other biological molecules [1, 2]. A prerequisite for understanding the function of such biological complexes is

*martin.zacharias@ph.tum.de

the knowledge of the structure of the corresponding protein–protein interactions [3–5]. It is possible to determine the experimental structure of protein–protein complexes, for example, using X-ray crystallography. However, this method requires purification of large amounts of partner proteins and formation of relatively stable complexes that can be crystallized to form regular crystals. Especially in the case of transient complexes or for multi-protein complexes that are in a dynamic equilibrium with the isolated components or sub-complexes, this may not be feasible [5, 6]. Nevertheless, an impressive number of several thousand protein–protein complexes have been determined using X-ray crystallography (~5000) [7]. In recent years, new developments in the area of CryoEM (Electron Microscopy under cryogenic conditions) have revolutionized the structural biology especially of large biological assemblies [8]. In contrast to X-ray crystallography, no crystals of the assembly are necessary but only a sufficient number of object images from different viewpoints that are combined and averaged to obtain a high-resolution view of the structure [9]. The technique has, however, a lower limit for the system size because of the limited contrast of small particles relative to the noisy background. A third experimental technique to elucidate the structure of mostly dimeric complexes is nuclear magnetic resonance (NMR) [10, 11]. Although for high-resolution structure determination limited to complexes of small size (~20 kD), it can be used to assist in structural modelling and affinity determination for larger complexes if the structure of the partner proteins is known [11]. Despite the great progress in recent years, the experimental determination of protein–protein complexes remains a challenging task and it will be impossible to determine experimentally all putative and transient protein–protein complexes of a cell in the foreseeable future [5].

In particular, complexes of weakly or transiently interacting protein partners are often not stable enough to allow experimental structure determination at high (atomic) resolution. The number of different proteins in a cell is in the order of thousands or tens of thousands [3, 12]. However, the number of relevant complexes and assemblies is at least in the hundreds of thousands [7]. There are experimental approaches to detect such interactions but for a detailed understanding of the interactions knowledge of the structure of all these complexes is desirable. In order to obtain realistic structural models of complexes, accurate structural modeling and structure prediction are of increasing importance [4, 12, 13].

Due to the growing number of solved protein–protein complex structures, it is often possible to generate a structure model of a complex by using a known complex structure as a template [14]. In fact, it has been found that the majority of stable natural protein–protein interactions can in principle be modeled by a template-based approach [15]. The main task is then to find an appropriate and realistic alignment of the target protein sequence to the template sequence. Template-based protein–protein complex modeling has been reviewed [16] and will not be part of the present chapter. However, even template-based models require often further refinement of the model structure that will be discussed in the final section of the chapter.

In case of no appropriate target-template similarity and if the structures of the isolated protein partners or of closely related proteins are known, it is possible to use computational docking methods to generate putative complex structures [4, 17]. Protein–protein docking is defined as a computational technique to predict the structure of a complex consisting of two or more partners starting from the structure of isolated protein monomers.

Protein-binding processes are driven by associated changes in free energy that correspond to structural and physicochemical properties of the binding partners. The "lock and key" concept of binding proposed by E. Fischer [18] emphasizes the importance of optimal complementarity at binding interfaces as a decisive factor to achieve high binding affinity and specificity. The great majority of protein–protein docking algorithms use the optimal complementarity as the main target criteria for predicting interactions. Typically, the partner structures are considered as rigid bodies and the associated docking methods are termed rigid body protein–protein docking [4, 13]. The principles of these methods will be described in the first section of the chapter.

Nevertheless, biological macromolecules such as proteins (but also RNA and DNA) are usually not rigid and can undergo various types of motion at physiological temperatures. Based on the frequent experimental observation of significant conformational changes of the binding partners upon binding, the induced fit concept of partner proteins has evolved [18, 19]. The protein partners appear to induce conformational changes during binding that are a prerequisite for specific complex formation. Note that in principle all protein binding processes require a certain degree of conformational adaptation, but for some cases these changes are on average smaller than the size of a hydrogen atom. Besides of the induced-fit concept the idea of a pre-existing ensemble of several inter-convertible conformational states of proteins at equilibrium has been postulated [19, 20]. Within this ensemble there are structures close to the bound and unbound forms and binding to partner molecules shifts the equilibrium toward the bound form. However, every conformation is, in principle, accessible even in the unbound state albeit with a potentially low statistical weight. Hence, the original induced fit concept is a special case of ensemble selection where only the presence of a ligand gives rise to an appreciable concentration of the bound partner structure. In many cases for achieving realistic predictions protein–protein docking approaches need to account for such conformational changes preferably already during the early systematic docking stages. The efficient inclusion of conformational changes either directly or in the form of pre-calculated ensembles in docking approaches will be at the focus of the second section of the chapter.

For many protein–protein interactions biochemical and biophysical experiments can help to identify protein residues that participate in the interaction with a partner or can set upper and lower bounds for residues in the vicinity of the binding site [21–23]. In addition, bioinformatics data, e.g., conservation of surface residues or co-evolution of residues on partners of complexes that are found in many organisms, are available [5]. This kind of data can be included in protein–protein

Table 3.1. Critical assessment of predicted interaction (CAPRI) protein–protein docking criteria.

Quality	% Native contacts	Ligand RMSD	Interface RMSD
High	≥50	RMSD ≤ 1 Å	or RMSD ≤ 1 Å
Medium	≥30	1 Å < RMSD ≤ 5 Å	or 1 Å < RMSD ≤ 2 Å
Acceptable	≥10	5 Å < RMSD ≤ 10 Å	or 2 Å < RMSD ≤ 4 Å
Incorrect	<10		

docking and an overview of the various available methods will be covered in the third section.

Frequently, complex structures obtained from an initial rigid or semi-flexible protein–protein docking or comparative template-based modeling programs are of limited accuracy. To generate an accurate realistic structural model, further structural refinement is required. Indeed, the majority of current protein–protein docking protocols are split into a preliminary exhaustive systematic docking search that is supplemented by a refinement step for a subset of putative complexes from the initial search. Occasionally, several refinement steps are involved [4, 26].

In the last section of this chapter such flexible refinement steps and options for scoring predicted complex geometries will be discussed and possible directions of improvement will be outlined. Finally, in the conclusion section, recent developments of the prediction of proteins interaction geometries using brute-force molecular dynamics (MD) simulations including full flexibility of binding partners and explicit inclusion of solvent molecules will also be discussed (Table 3.1).

3.2. Rigid body protein–protein docking approaches

In protein–protein docking using rigid partner structures, the task is to rapidly and exhaustively generate a variety of possible interaction geometries with significant complementarity but no or little overlap of partner structures. The degrees of freedom are limited to three translational and three rotational degrees of one partner with respect to the other. A variety of computational methods have been developed in recent years to efficiently generate a large number of putative binding geometries. In order to implicitly account for conformational adjustment of binding partners, non-specific sterical overlaps between docking partners are typically tolerated up to a certain limit. Among the most common methods are fast Fourier transform (FFT) correlation techniques [16, 24, 27, 28] to efficiently locate overlaps between complementary protein surfaces and geometric hashing methods to rapidly match geometric surface descriptors of proteins (Figure 3.1) [25, 26]. In the FFT-docking approach (pioneered by Katchalski-Katzir *et al.* [24]), the two protein partners are represented by cubic grids; the grid points are assigned discrete values for inside, outside and on the surface of the protein. A geometric complementarity score can be calculated for the two binding partners by computing the correlation of the two grids representing each protein. Instead of summing up all the pair products of the grid entries, one can make use of the Fourier correlation theorem. The corresponding correlation integral can easily be computed in Fourier space. The discrete Fourier transform for the receptor grid needs to be calculated only once. Due to the special shifting properties of Fourier transforms the different translations of the ligand grid with respect to the receptor grid can be computed by a simple multiplication in Fourier space. This process is repeated for various relative orientations of the two proteins. Several available computer programs for protein–protein docking use the Cartesian FFT algorithm (Table 3.2).

A disadvantage of standard Cartesian FFT-based correlation methods is the need to perform FFTs for each relative orientation of one protein molecule with respect to the partner. Typically the relative orientation of partners is modified in steps of $10–15°$, but this can vary between available programs [27]. The discretization of the relative orientation can be avoided by correlating spherical polar basis functions that represent, for example, the surface shape of protein

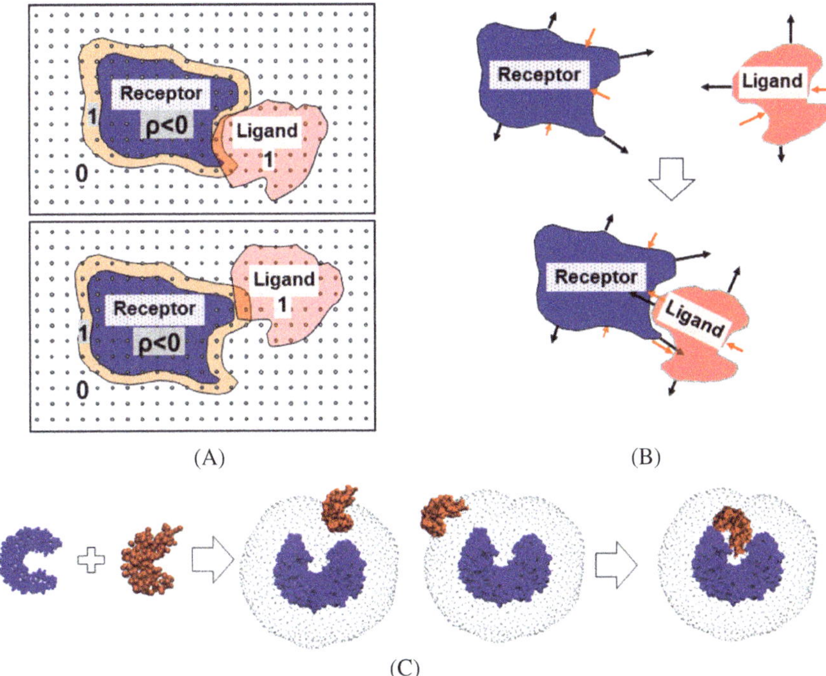

Figure 3.1.   Illustration of the most common types of systematic docking methodologies. (A) Docking based on solving a correlation task: the overlap of a ligand protein (Ligand) with the surface of a receptor protein (receptor) can be calculated by solving a correlation integral [24]. The correlation is positive as long as the ligand overlaps with the surface region (in orange) but becomes unfavorable with increasing overlap with the interior of the receptor (in blue). Using a grid to discretize the protein, the correlation can be solved rapidly by fast Fourier transformation (FFT) and the best possible overlap for a given relative protein orientation can be extracted (illustrated here for two examples, upper and lower panel in A). The resolution of the indicated FFT grid to discretize the protein determines how accurate the convolution can be determined. (B) Geometric hashing to match surface descriptors on both protein partners [25]. In this case concave (red arrows) and convex areas (black) on the protein surface are illustrated. Docking is performed by finding best matches of concave and convex regions (for this particular example, other matching characteristics can also be used). (C) Docking by systematic energy minimization or dynamics simulations starting from thousands of different relative placements of the partner proteins (in red and blue van der Waals representation) to obtain locally stable docking geometries (on the right of the panel in C).

Table 3.2. Protein–protein docking programs and associated Websites or Web servers.

Program	Search method	Protein representation	Website
3D Dock [42]	Correlation FFT	Discrete	http://www.sbg.bio.ic.ac.uk/docking/
ATTRACT [35]	Guided: mult. Minimization	Coarse grain	www.attract.ph.tum.de/
ClusPro [28]	Correlation FFT	discrete	https://cluspro.bu.edu/login.php
DOT [43]	Correlation FFT	Discrete	https://www.sdsc.edu/CCMS/DOT/
GRAMM-X [44]	Correlation FFT	Discrete	http://vakser.compbio.ku.edu/resources/gramm/grammx/
HADDOCK [41]	Guided: data-driven, MD	Atomic	http://haddock.chem.uu.nl/
Hex [45]	Correlation polar FFT	Discrete	http://hexserver.loria.fr/
ICM-Disco [46]	Guided: MC minimization	Atomic	http://www.molsoft.com/icm_pro.html
MolFit [47]	Correlation FFT	Discrete	http://www.weizmann.ac.il/Chemical_Research_Support/molfit/
PatchDock [32]	Geometric	Surface	https://bioinfo3d.cs.tau.ac.il/PatchDock/
RosettaDock [37]	Monte Carlo + minimization	Coarse grain to atomic	https://www.rosettacommons.org/software
ZDock [27]	Correlation FFT	Discrete	http://zdock.umassmed.edu/
MEGADOCK [48]	Correlation FFT	Discrete	http://www.bi.cs.titech.ac.jp/megadock/
SwarmDock [39]	Monte Carlo on a Swarm	Discrete	https://bmm.crick.ac.uk/~svc-bmm-swarmdock/
F2Dock [49]	Correlation FFT	Discrete	http://www.cs.utexas.edu/~bajaj/cvc/software/f2dock.shtml
FroDock [50]	Guided: distance restraints	Discrete	http://frodock.chaconlab.org/
pyDock [51]	Correlation FFT	Discrete	https://life.bsc.es/pid/pydockweb
PROBE [52]	Geometric, MC minimization	Discrete	http://pallab.serc.iisc.ernet.in/probe/

molecules [29]. With recent approaches it is then also possible to solve the whole multi-dimensional search process (orientation and translation) in Fourier space instead of solving only the translation correlation task for discrete relative orientations of the proteins in Cartesian space [30]. Using FFT correlation techniques, it is possible to solve a rigid protein–protein docking problem for typically sized protein partners in a matter of minutes on standard computer workstation systems. It has been successfully applied in the field of protein–protein docking and is part of most protein–protein docking Web servers (Table 3.2). Recently, new multidimensional correlation methods have been developed that allow the correlation of multiterm potentials. Each function needs to be expressed in terms of spherical basis functions characterizing the surface properties of the protein partners [31]. With the help of such rapid scoring methods, it is possible to calculate the full partition function of the rigid docking problem and within this limitation can extract thermodynamic and structural properties [30]. One significant draw-back of FFT correlation methods is that it is only directly applicable to protein dimer prediction and requires sequential application when it comes to molecular assemblies consisting of many protein partners.

Besides the FFT correlation method, geometric hashing is also used to rapidly identify possible protein–protein binding arrangements (Figure 3.1). It has been developed originally to match complementary subsections of several data sets. The protein surface is typically represented as a set of triangles, which are stored in a hash table. By means of a hash key, similar matching triangles on the surface of protein partners can be quickly identified. During docking, these triangles comprise points on a molecular surface, having a certain geometrical (concave, convex) or physico-chemical (polar, hydrophobic) character. By matching triangles belonging to different molecules and being of complementary character, putative complex geometries can be identified. PATCHDOCK is an example program employing geometric hashing [32].

A third type of approaches uses multistart energy minimization, Brownian Dynamics or Monte Carlo simulations on rigid body degrees of freedom to generate protein–protein docking complexes [4]. A great advantage of these methods is that it is possible to include different types of conformational flexibility already at the initial systematic search step. However, the computational demand is higher compared to FFT-based correlation methods or geometric hashing. Often the search needs to be limited to predefined regions of the binding partners. A second significant advantage is the straightforward application of the method to simultaneous docking of multiple protein partners. To reduce the computational costs, coarse-grained (reduced) protein models instead of atomistic models can be employed [33, 34]. This allows to energy minimize many thousands of start configurations. Available programs that belong to this class are ATTRACT [35, 36], RosettaDock [37, 38], SwarmDock [39] and HADDOCK [40, 41]. A number of docking programs are also available as Web servers. A list of protein–protein docking programs and Web servers is given in Table 3.2.

With the help of the community-wide CAPRI experiment [53–55], the progress in protein–protein docking prediction methods has been extensively monitored over more than the last 15 years. In this challenge, participating groups test the performance of docking methods in blind predictions of protein–protein complex structures. For protein partners with small differences between unbound and bound conformation and some experimental hints on the interaction region, accurate predictions of complex structures are possible [55, 56]. However, when protein partners undergo significant conformational changes upon association or for protein homology models, the predictions are often incorrect or of limited accuracy. Computational approaches to realistically predict protein–protein binding geometries need to account for such conformational changes.

3.3. Inclusion of protein flexibility during systematic docking searches

The performance of protein–protein docking approaches on bound partner structures (meaning the protein structures have been extracted from the known complex structure) is typically much better than using unbound partner structures (structures that have been determined in the absence of the partner). Better performance indicates in this case both the deviation of the predicted complex structure closest to the native complex and the ranking or scoring of this near-native predicted complex. The reason is conformational changes in the binding partners that accompany the complex formation process. Depending on the target difficulty, i.e., the conformational difference between unbound and bound conformation, none of the predicted complexes from rigid docking approaches come sufficiently close to the native complex to detect this complex as the most realistic prediction [55, 56]. Hence, especially in these cases, it is desirable to efficiently account for conformational changes already during the systematic docking search. For many protein–protein complex cases, the associated conformational changes can involve local side chains and loop transitions of the protein surface but also global motions of large protein domains.

Often in protein–protein docking, the experimental partner structures are not available but only of close homologs. In such cases, structural models of one or both partners need to be generated using comparative modeling based on a known template structure with sufficient sequence similarity. These homology-modeled structures are typically of lower accuracy than experimentally determined structures and can contain side chain or loop misplacements that can interfere with rigid docking of the partner proteins. A significant fraction of known protein–protein interactions can be modeled based on homology to known protein–protein complexes which in many cases also requires efficient flexible refinement to provide accurate structural models. Several approaches of flexible refinement, ensemble docking and explicit inclusion of flexibility during the entire docking process have been developed to include possible conformational changes during docking (Figure 3.2) [4].

(A)

(B)

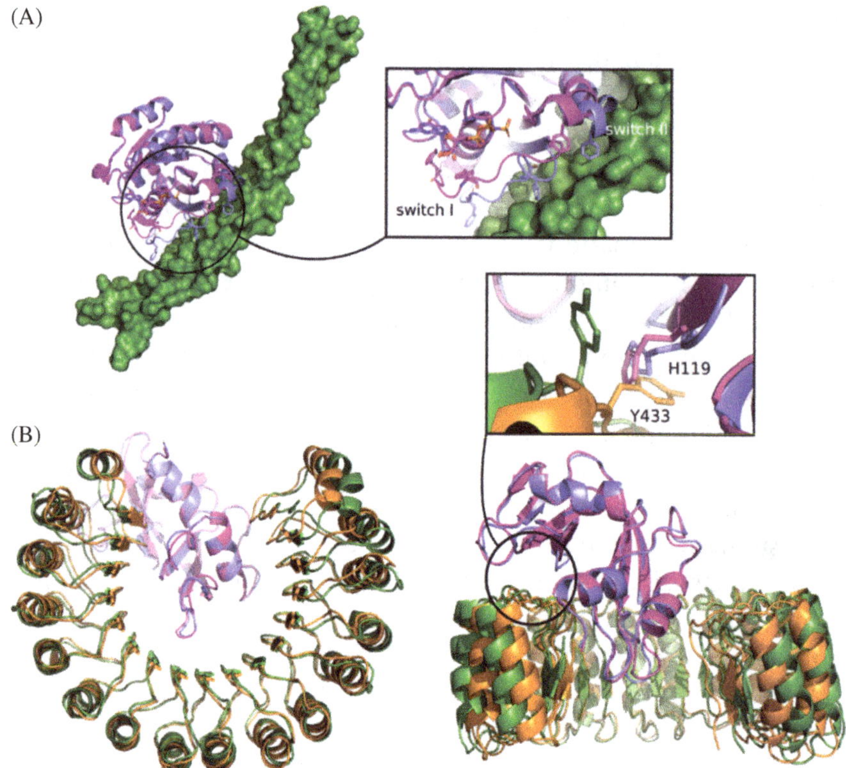

Figure 3.2.   Conformational changes observed upon protein–protein complex formation can be of local (e.g. side chain or loop motions) or global type. (A) Functional loop regions known as switch I and switch II in the small G-protein Rab8a adopt a closed arrangement in the unbound state (magenta, PDB: 4lhw), but adopt a more disordered conformation once bound to the Guanine nucleotide exchange factor, Rabin8 (blue and green, PDB: 4lhy). (B) Global backbone changes (opening/closing motion) are observed by comparing the cartoon representation of the Ribonuclease A inhibitor in the unbound (orange, PDB: 2bnh) vs. the bound forms (green, PDB: 1dfj). The unbound (magenta) vs. bound (blue) structures of RNAseA are also shown. In addition to the global changes also side changes are observed (stick models in the inset of panel B).

3.4. Inclusion of experimental and bioinformatics data during protein–protein docking

Like any other method, protein–protein docking approaches have limitations, such as false positives or the diversity of final predictions, especially when there are large proteins involved. Docking programs have improved in terms of accuracy by integrating low- or high-resolution experimental data to steer the docking engine toward the correct results [57] or to filter out false predictions at post-processing stages [58], although it might add to complexity of the workflow. The term "integrative modeling" is typically used for modeling studies that combine experimental and bioinformatics data from various sources to generate structural models of molecular assemblies [5, 59, 60]. A variety of experimental data can be used to guide docking of proteins and we will discuss only the most common types of data.

Cross-linking has been used increasingly in the recent years to supervise and validate protein docking data. Typically, mass spectrometry is used to identify crosslink sites on protein partners. These crosslinks can be included as upper-bound distance restraints during docking or for screening solutions. Vreven *et al.* improved the rank of the near-native prediction and also managed to refine the structure of symmetric complexes using hard cross-link distances as cutoffs [61]. In another study, distance restraints between specific residues obtained from cross-linking experiments served as spatial restraints to guide and evaluate the HADDOCK docking algorithm toward the correct predictions [62]. Similar approaches are available in the ATTRACT docking program [63]. Small-angle X-ray scattering in solution (SAXS) is another technique for characterizing structural and dynamics of biomolecules at low resolutions [5]. SAXS data, calculated as a form of convolution, has served in several docking programs as a scoring function [64]. Most of the current methods use SAXS data as a filter to refine and rank the final predictions [65–67], whereas other methods directly use them during the sampling stage [64, 68].

Although Cryo-EM experiments can nowadays provide atomic detail information of protein–protein complexes, often only low resolution (> 10 Å) is reached. This type of low-resolution electron-density envelope can nevertheless be very useful for evaluating the shape of the macromolecular complex. Besides of programs that are specific for fitting substructures into low-resolution CryoEM density, it has also been integrated into programs used for protein–protein docking. For example, the HADDOCK program can include CryoEM data in addition to other sources of experimental and bioinformatics restraints to generate putative models of complexes [69]. The generated solutions can then be scored based on HADDOCK scoring function. The ATTRACT approach has also the option to perform docking of an arbitrary number of protein molecules guided by low-resolution CryoEM density envelopes [70]. Based on this implementation, it was demonstrated that inclusion of low-resolution (15–20 Å) CryoEM data is highly efficient to guide docking to near-native geometries for the majority of complexes in a large benchmark set [71].

3.5. Flexible refinement and final scoring of docked complexes

Pons *et al.* have investigated the limitations of rigid docking strategies in combination with a rescoring step using rigid FFT-correlation-based docking and scoring with the pyDock approach [72]. The PyDock approach employs an electrostatic Coulomb contribution with a surface-area-based solvation term and a van der Waals term (optional). The protocol showed very good performance for most proteins that undergo minor conformational changes upon complex formation (<1 Å RMSD between unbound and bound structures) but unsatisfactory results for cases with significant binding-induced conformational changes or applications that involved homology-modeled proteins. Hence, improvement of the prediction accuracy of proposed binding modes is necessary for specific scoring indicating a coupling between realistic scoring and accurate prediction of the complex structure.

Basically, all docking approaches and also those methods that include some degree of conformational flexibility during the initial search employ a refinement and re-scoring step applied to a subset of docking solutions. Typically, the number of structures that undergo a refinement step is in the order of hundreds or thousands of the initial solutions and therefore computational efficiency of the refinement step is critical. The FireDock approach [73] uses a combination of rigid body moves and side chain optimization to refine docking solution obtained by PatchDock [32]. The refinement steps of the program HADDOCK consist of a series of energy minimizations and dynamics in dihedral variables followed by MD simulation in Cartesian coordinates and with the option to include explicit solvent [74]. Most FFT-based approaches such as CLUS-PRO or ZDOCK employ also energy minimization and sometimes also short MD simulations to remove sterical overlap and improve the surface complementarity [75, 76]. The Rosetta molecular modeling suite is frequently used for optimizing the geometry of protein–protein complexes [38, 77]. The Rosetta program typically uses internal dihedral angles as conformational variables. It can be used for flexible refinement of just the side chains at the interface but also in combination with "backrub" motions to modify the backbone geometry [38, 78]. The approach is used for refinement of both solutions from FFT correlation docking and solutions obtained directly from a coarse-grained Rosetta protein–protein docking approach.

Often the local refinement of docked complexes using energy minimization or MD approaches does not move the partner structures significantly from the starting geometry (it just optimizes the interface of the initial geometry). In many cases especially when the unbound protein structures differ from the conformation in the bound complex, none of the solutions come very close to the native binding geometry. Often not even "acceptable" solutions as defined by the Capri criteria (Table 3.1) are reached (the best solutions differ from the native by an RMSD of the interface >4 Å). In such cases it is desired that the refinement step does not only improve the interface structure but also moves the partners globally closer together. The iATTRACT approach [79] is a docking refinement strategy that tries to combine minimization in the global translational and rotational variables with a full atomic flexibility of only the predicted interface region. In this way small side chain movements can trigger larger scale whole body movements of partner proteins (outlined and illustrated in Figure 3.3).

With the improving performance of computer hardware, it is possible to perform longer MD simulations on putative docked complexes for further refinement under realistic conditions including full

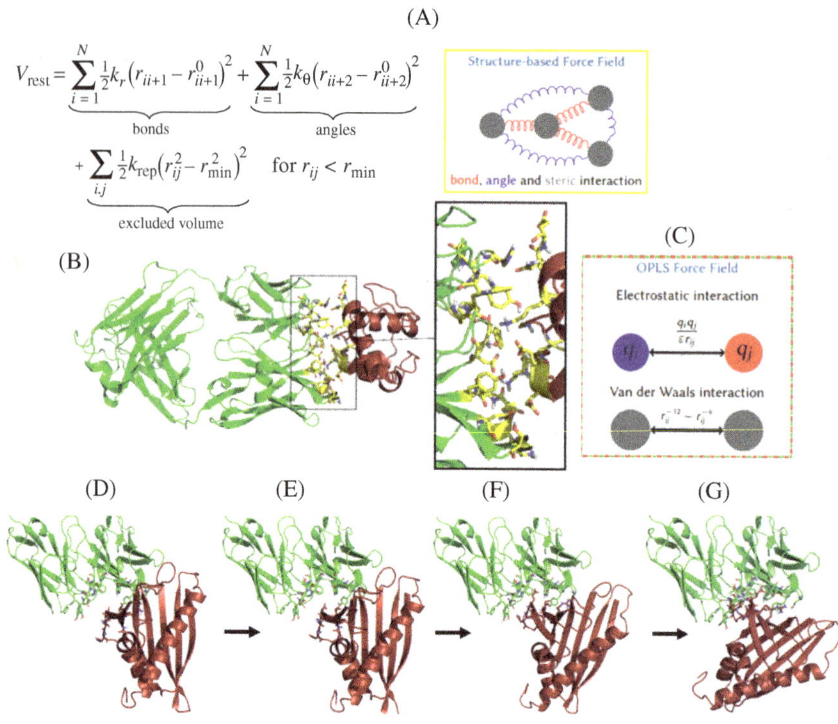

Figure 3.3. Refinement of protein–protein docking geometries by iATTRACT [79]. In iATTRACT full mobility of the interface residues (within 5–10 Å) is combined with the global mobility of the complete partners in terms of overall rotation and translation. The intra-molecular interactions of the interface atoms are described by an elastic network model (indicated in A and illustrated in B, receptor protein in green cartoon, ligand protein in red cartoon, interface as yellow stick model). The intermolecular interactions at the interface (yellow stick model) are described by a standard force field (OPLS [85], indicated in C). Small changes at the interface can trigger large-scale rearrangements in rotation and translation of a predicted start complex (D, one protein in green, second protein as red cartoon) as illustrated in a series of energy-minimization snapshots (E–G).

flexibility of partners and explicit solvent molecules [80–82]. However, even in long MD simulations protein–protein complexes can be trapped for a long time in non-specific "sticky" complexes and not move forward to reach a near-native state [82]. Among several advanced sampling options replica-exchange simulations (REMD) with an added biasing potential to the partners in the replicas

(H-REMD or BP-REMD approach) are possibilities to avoid such trapping in non-specific geometries and siege a broader range of conformations. The exchange of replicas allows also sampling of relevant states in the reference replica without biasing [83, 84]. This technique has shown promising results for rapidly refining docked solutions including full flexibility of binding partners.

Many docking methodologies distinguish between scoring of docking solutions after an initial search (or even after the refinement) and a final re-scoring using a more sophisticated evaluation [54]. The initial scoring based, for example, on surface complementarity is sufficient to eliminate unlikely docking solutions and can be used to restrict the relevant complexes to a few hundred or thousand complexes for more accurate evaluation. The final scoring can be based on a physics-based force field function that includes van der Waals interactions, electrostatic Coulomb interactions and solvation contributions. For example, the HADDOCK [40] and Rosetta [37] protein–protein docking programs, pyDock approach [58] but also the ATTRACT [86] approach (after atomistic refinement) use physics-based force fields as basis for final scoring of the docking results. Typically, the various force field terms are assigned weights in the final score and these weights are optimized on a benchmark set of known complexes and a large set of incorrect decoy complexes.

As an alternative to a force field scoring it is also possible to use knowledge-based or statistical potentials for scoring docking solutions [6, 17, 87–90]. A large variety of statistical potentials has been developed and only the principles of the methodology can be considered. The idea is to look at the observed probability for finding residues/atoms (or pairs of atoms/residues) at a protein–protein interface vs. the expected probability based on a random distribution of the residues on a protein surface (taking into account the observed frequency of occurrence of an amino acid residue). This type of probability or probability ratio is typically evaluated for different distances between residues or atoms at the interface and can be related to an effective free energy function (by taking the logarithm of the probability ratio). Most of the designed statistical potentials are based on the pairwise contacts or distance between residues [88, 90]. However, the concept can in principle also be extended to effective multibody potentials. Recently, new methods for optimizing such potentials have been used to further improve statistical potentials [91] based in part on machine learning applications [92, 93].

In most cases, scoring of complexes is performed on single docked structures (e.g. cluster representatives) by just calculating the interaction between partner molecules (based on a force field or knowledge-based potential). However, complexes in reality are not static structures. To account for the ensemble character of the complexes, the MMPBSA/MMGBSA (Molecular Mechanics Poisson Boltzmann/Generalized Born Surface Area) approaches can be employed [94]. In this case an ensemble of docked conformations in the vicinity of the starting complex is generated using MD simulations. This ensemble is then analyzed using a molecular mechanics force field (similar to the single structure scoring) combined with the Poisson-Boltzmann or Generalized Born approach to implicitly account for solvation effects. This technique has been used quite successfully for scoring protein–protein complexes [95] and was also used to score protein–peptide complexes [96]. It is, however, more demanding than scoring based on single structures.

It is important to note that basically all scoring methods are based on the interaction between partners. However, protein binding is not only driven by the interaction between partners but by the binding-free energy. The binding-free energy includes many effects besides of the effective interaction between partners. For example, it is influenced by the energy of deforming the unbound structure into the bound structure, by the entropic cost of reducing the rotational and translational freedom of one partner relative to the other and by the change in conformational entropy of the partners (usually a restriction of conformational mobility). In principle, all these effects can be calculated in free energy simulations of protein–protein binding using restraint MD simulations [97–99]. In such calculations, one typically restraints the conformations of the partners to stay close to the starting structure in the bound (or predicted) complex and includes restraints to keep the relative orientation of the partners. Subsequently, dissociation of the partners is achieved along a separation distance coordinate by adding an appropriate biasing potential, and the associated free energy (work) along the dissociation path is obtained. Finally, simulations are performed at the bound and dissociated states to calculate the free energy of releasing the restraints. Such methods are computationally very demanding but have been recently used to evaluate single complexes [100] or a set of complexes using a coarse-grained model [101]. Very recently, the methodology was tested systematically on 20 test systems and including 50 decoy complexes for each test case in combination with an implicit solvent description [102]. It was concluded that the performance is indeed slightly better than scoring based on interaction energies, but further developments are necessary to improve accuracy and convergence of the methodology.

3.6. Conclusion and future developments

Protein–protein docking approaches have seen steady progress in recent years that has been monitored regularly by the community-wide docking challenge CAPRI. Docking methods and docking servers are frequently used also by non-expert users working in the field of protein–protein interactions. In addition to prediction by protein–protein docking, complex structures for many of the natural protein–protein interactions can be generated based on template-based modeling methods followed by structural refinement. Even in cases with no appropriate template it is possible in many cases to include experimental information or data from bioinformatics to restrict or guide the search for protein–protein docking searches. Hence, the prediction of dimeric protein–protein interactions is a highly evolved field with many available approaches and many successful applications. However, the prediction of weak interactions that may form the basis for an assembly of several proteins to form multiprotein complexes is still extremely challenging. These assemblies mediate many cellular functions and may also undergo rapid changes in the cell due to the association and dissociation of sub-elements. Experimental data such as low-resolution density from CryoEM, Cryo-tomography or *in vivo* crosslinking data could be combined with docking methods to obtain structural models of larger molecular assemblies in the cell. These approaches may allow in the not too distant future the structural modeling of the assembly and disassembly of many multiprotein complexes even in a crowed cell-type environment.

References

1. Nooren, I.M.A. NEW EMBO MEMBER'S REVIEW: Diversity of protein–protein interactions. *EMBO J.* 2003 Jul 15;22(14):3486–3492.
2. Marsh, J.A. and Teichmann, S.A. Structure, dynamics, assembly, and evolution of protein complexes. *Annu Rev Biochem.* 2015;84(1):551–575.
3. Johnson, G.T., *et al.* cellPACK: A virtual mesoscope to model and visualize structural systems biology. *Nat Methods.* 2015 Jan;12(1): 85–91.
4. Zacharias, M. Accounting for conformational changes during protein–protein docking. *Curr Opin Struct Biol.* 2010 Apr 1;20(2):180–186.
5. Soni, N. and Madhusudhan, M.S. Computational modeling of protein assemblies. *Curr Opin Struct Biol.* 2017 Jun;44:179–189.
6. Shoemaker, B.A. and Panchenko, A.R. Deciphering protein–protein interactions. Part II. Computational methods to predict protein and domain interaction partners. *PLOS Comput Biol.* 2007 Apr 27;3(4):e43.
7. Mosca, R., Céol, A., and Aloy. P. Interactome3D: Adding structural details to protein networks. *Nat Methods.* 2013 Jan;10(1):47–53.
8. Bai, X., McMullan, G., and Scheres, S.H.W. How cryo-EM is revolutionizing structural biology. *Trends Biochem Sci.* 2015 Jan 1;40(1):49–57.
9. Elmlund, D., Le, S.N., and Elmlund, H. High-resolution cryo-EM: The nuts and bolts. *Curr Opin Struct Biol.* 2017 Oct 1;46:1–6.
10. Güntert, P. Automated structure determination from NMR spectra. *Eur Biophys J.* 2008 Sep 20;38(2):129.
11. Clore, G.M. and Gronenborn, A.M. Determining the structures of large proteins and protein complexes by NMR. *Trends Biotechnol.* 1998 Jan 1;16(1):22–34.
12. Im, W., *et al.* Challenges in structural approaches to cell modeling. *J Mol Biol.* 2016 Jul 31;428(15):2943–2964.
13. Bonvin, A.M. Flexible protein–protein docking. *Curr Opin Struct Biol.* 2006 Apr 1;16(2):194–200.

14. Kundrotas, P.J., *et al.* Dockground: A comprehensive data resource for modeling of protein complexes. *Protein Sci.* 2018;27(1):172–181.

15. Kundrotas, P.J., Zhu, Z., Janin, J., *et al.* Templates are available to model nearly all complexes of structurally characterized proteins. *Proc Natl Acad Sci.* 2012 Jun 12;109(24):9438–9441.

16. Porter, K.A., Desta, I., Kozakov, D., *et al.* What method to use for protein–protein docking? *Curr Opin Struct Biol.* 2019 Jan 31;55:1–7.

17. Gromiha, M.M., Yugandhar, K., and Jemimah, S. Protein–protein interactions: Scoring schemes and binding affinity. *Curr Opin Struct Biol.* 2017 Jun;44:31–38.

18. Koshland, D.E. Das Schüssel-Schloß-Prinzip und die Induced-fit-Theorie. *Angew Chem.* 1994;106(23–24):2468–2472.

19. Csermely, P., Palotai, R., and Nussinov, R. Induced fit, conformational selection and independent dynamic segments: An extended view of binding events. *Trends Biochem Sci.* 2010 Oct 1;35(10):539–546.

20. Nussinov, R., Ma, B., and Tsai, C.-J. Multiple conformational selection and induced fit events take place in allosteric propagation. *Biophys Chem.* 2014 Feb 1;186:22–30.

21. Geng, C., Xue, L.C., Roel-Touris, J., *et al.* Finding the $\Delta\Delta G$ spot: Are predictors of binding affinity changes upon mutations in protein–protein interactions ready for it? *Wiley Interdiscip Rev Comput Mol Sci.* 2019 Jun 14;0(0):e1410.

22. Moal, I.H. and Fernández-Recio, J. SKEMPI: A structural kinetic and energetic database of mutant protein interactions and its use in empirical models. *Bioinformatics.* 2012 Oct 15;28(20):2600–2607.

23. Moreira, I.S., Martins, J.M., Coimbra, J.T.S., *et al.* A new scoring function for protein–protein docking that identifies native structures with unprecedented accuracy. *Phys Chem Chem Phys.* 2015;17(4):2378–2387.

24. Katchalski-Katzir, E., *et al.* Molecular surface recognition: Determination of geometric fit between proteins and their ligands by correlation techniques. *Proc Natl Acad Sci.* 1992 Mar 15;89(6):2195–2199.

25. Mashiach, E., *et al.* An integrated suite of fast docking algorithms. *Proteins.* 2010 Nov 15;78(15):3197–3204.

26. Andrusier, N., Mashiach, E., Nussinov, R., *et al.* Principles of flexible protein–protein docking. *Proteins Struct Funct Bioinforma.* 2008;73(2):271–289.

27. Chen, R., Li, L., and Weng, Z. ZDOCK: An initial-stage protein-docking algorithm. *Proteins Struct Funct Bioinforma.* 2003;52(1):80–87.

28. Kozakov, D., *et al.* The ClusPro web server for protein–protein docking. *Nat Protoc.* 2017 Feb;12(2):255–278.

29. Ritchie, D.W. and Kemp, G.J.L. Protein docking using spherical polar Fourier correlations. *Proteins Struct Funct Bioinforma.* 2000;39(2): 178–194.

30. Padhorny, D., *et al.* Protein–protein docking by fast generalized Fourier transforms on 5D rotational manifolds. *Proc Natl Acad Sci.* 2016 Jul 26;113(30):E4286–E4293.

31. Venkatraman, V., Sael, L., and Kihara, D. Potential for protein surface shape analysis using spherical harmonics and 3D Zernike descriptors. *Cell Biochem Biophys.* 2009 Jul 1;54(1):23–32.

32. Schneidman-Duhovny, D., Inbar, Y., Nussinov, R., *et al.* PatchDock and SymmDock: Servers for rigid and symmetric docking. *Nucleic Acids Res.* 2005 Jul 1;33(suppl_2):W363–W367.

33. Zacharias, M. ATTRACT: Protein–protein docking in CAPRI using a reduced protein model. *Proteins Struct Funct Bioinforma.* 2005; 60(2):252–256.

34. Ruiz, M.E.E, de Beauchêne, I.C., and Ritchie, D.W. EROS-DOCK: Protein–protein docking using exhaustive branch-and-bound rotational search. *Bioinformatics* [Internet]. 2019 May 29 [cited 2019 May 29]. Available from https://academic.oup.com/bioinformatics/ advance-article/doi/10.1093/bioinformatics/btz434/5498285.

35. Zacharias, M. Protein–protein docking with a reduced protein model accounting for side-chain flexibility. *Protein Science*, Wiley Online Library [Internet]. 2003 [cited 2019 Jun 26]. Available from https:// onlinelibrary.wiley.com/doi/full/10.1110/ps.0239303.

36. Schneider, S., Saladin, A., Fiorucci, S., *et al.* ATTRACT and PTOOLS: Open source programs for protein–protein docking. In: Baron R, ed. *Computational drug discovery and design* [Internet]. New York, NY: Springer; 2012 [cited 2019 Jun 26]:221–232 (Methods in molecular biology). Available from https://doi.org/10.1007/978-1-61779- 465-0_15.

37. Chaudhury, S., *et al.* Benchmarking and analysis of protein docking performance in Rosetta v3.2. *PLOS ONE.* 2011 Feb 8;6(8):e22477.

38. Kuroda, D. and Gray, J.J. Pushing the backbone in protein–protein docking. *Structure.* 2016 Oct 4;24(10):1821–1829.

39. Moal, I.H. and Bates, P.A. SwarmDock and the use of normal modes in protein–protein docking. *Int J Mol Sci.* 2010 Oct;11(10): 3623–3648.

40. Dominguez, C., Boelens, R., and Bonvin, A.M.J.J. HADDOCK: A protein–protein docking approach based on biochemical or biophysical information. *J Am Chem Soc.* 2003 Feb 19;125(7):1731–1737.

41. Deplazes, E., Davies, J., Bonvin, A.M.J.J., *et al.* Combination of ambiguous and unambiguous data in the restraint-driven docking of flexible peptides with HADDOCK: The binding of the Spider Toxin PcTx1 to the acid sensing ion channel (ASIC) 1a. *J Chem Inf Model.* 2016 Jan 25;56(1):127–138.

42. Carter, P., Lesk, V.I., Islam, S.A., *et al.* Protein–protein docking using 3D-Dock in rounds 3, 4, and 5 of CAPRI. *Proteins Struct Funct Bioinforma.* 2005;60(2):281–288.

43. Mandell, J.G., *et al.* Protein docking using continuum electrostatics and geometric fit. *Protein Eng Des Sel.* 2001 Feb 1;14(2):105–113.

44. Tovchigrechko, A. and Vakser, I.A. GRAMM-X public web server for protein–protein docking. *Nucleic Acids Res.* 2006 Jul 1; 34(suppl_2):W310–W314.

45. Macindoe, G., Mavridis, L., Venkatraman, V., *et al.* HexServer: An FFT-based protein docking server powered by graphics processors. *Nucleic Acids Res.* 2010 Jul 1;38(suppl_2):W445–W449.

46. Fernández-Recio, J., Totrov, M., and Abagyan, R. ICM-DISCO docking by global energy optimization with fully flexible side-chains. *Proteins Struct Funct Bioinforma.* 2003;52(1):113–117.

47. Heifetz, A., Katchalski-Katzir, E., and Eisenstein, M. Electrostatics in protein–protein docking. *Protein Sci.* 2002;11(3):571–587.

48. Ohue, M., *et al.* MEGADOCK 4.0: An ultra–high-performance protein–protein docking software for heterogeneous supercomputers. *Bioinformatics.* 2014 Nov 15;30(22):3281–3283.

49. Chowdhury, R., *et al.* Protein–protein docking with F2Dock 2.0 and GB-Rerank. *PLOS ONE.* 2013 Jun 3;8(3):e51307.

50. Ramírez-Aportela, E., López-Blanco, J.R., and Chacón, P. FRODOCK 2.0: Fast protein–protein docking server. *Bioinformatics.* 2016 Aug 1;32(15):2386–2388.

51. Jiménez-García, B., Pons, C., and Fernández-Recio, J. pyDockWEB: A web server for rigid-body protein–protein docking using electrostatics and desolvation scoring. *Bioinformatics.* 2013 Jul 1;29(13): 1698–1699.

52. Mitra, P. and Pal, D. PRUNE and PROBE — two modular web services for protein–protein docking. *Nucleic Acids Res.* 2011 Jul;39(Web Server issue):W229–W234.

53. Janin, J. Assessing predictions of protein–protein interaction: The CAPRI experiment. *Protein Sci.* 2005;14(2):278–283.

54. Lensink, M.F., Méndez, R., and Wodak, S.J. Docking and scoring protein complexes: CAPRI 3rd Edition. *Proteins Struct Funct Bioinforma.* 2007;69(4):704–718.

55. Lensink, M.F. and Wodak, S.J. Docking and scoring protein interactions: CAPRI 2009. *Proteins Struct Funct Bioinforma.* 2010;78(15):3073–3084.

56. Lensink, M.F., Velankar, S., and Wodak, S.J. Modeling protein–protein and protein–peptide complexes: CAPRI 6th edition. *Proteins Struct Funct Bioinforma.* 2017;85(3):359–377.

57. Xia, B., Vajda, S., and Kozakov, D. Accounting for pairwise distance restraints in FFT-based protein–protein docking. *Bioinformatics.* 2016 Nov 1;32(21):3342–3344.

58. Pallara, C., Jiménez-García, B., Romero, M., *et al.* pyDock scoring for the new modeling challenges in docking: Protein–peptide, homo-multimers, and domain–domain interactions [Internet]. Proteins: Structure, function, and bioinformatics. 2017 [cited 2019 Jun 24]. Available from https://onlinelibrary.wiley.com/doi/abs/10.1002/prot.25184.

59. Lasker, K., *et al.* Integrative structure modeling of macromolecular assemblies from proteomics data. *Mol Cell Proteomics.* 2010 Aug 1;9(8):1689–1702.

60. Schmidt, C., *et al.* Surface accessibility and dynamics of macromolecular assemblies probed by covalent labeling mass spectrometry and integrative modeling. *Anal Chem.* 2017 Feb 7;89(3):1459–1468.

61. Vreven, T., *et al.* Integrating cross-linking experiments with Ab initio protein–protein docking. *J Mol Biol.* 2018 Jun 8;430(12):1814–1828.

62. Orbán-Németh, Z., *et al.* Structural prediction of protein models using distance restraints derived from cross-linking mass spectrometry data. *Nat Protoc.* 2018;13(3):478–494.

63. de Vries, S.J., Schindler, C.E.M., Chauvot de Beauchêne, I., *et al.* Web interface for easy flexible protein–protein docking with ATTRACT. *Biophys J.* 2015 Feb 3;108(3):462–465.

64. Schneidman-Duhovny, D., Hammel, M., Tainer, J.,A., *et al.* FoXS, FoXSDock and MultiFoXS: Single-state and multi-state structural modeling of proteins and their complexes based on SAXS profiles. *Nucleic Acids Res.* 2016 Jul 8;44(W1):W424–W429.

65. Jiménez-García, B., Pons, C., Svergun, D.I., *et al.* pyDockSAXS: Protein–protein complex structure by SAXS and computational docking. *Nucleic Acids Res.* 2015 Jul 1;43(W1):W356–W361.

66. Xia, B., *et al.* Accounting for observed small angle X-ray scattering profile in the protein–protein docking server cluspro. *J Comput Chem.* 2015;36(20):1568–1572.

67. Sønderby, P., *et al.* Small-angle X-ray scattering data in combination with RosettaDock improves the docking energy landscape. *J Chem Inf Model.* 2017 Oct 23;57(10):2463–2475.

68. Schindler, C.E.M., de Vries, S.J., Sasse, A., *et al.* SAXS data alone can generate high-quality models of protein–protein complexes. *Structure.* 2016 Aug 2;24(8):1387–1397.

69. van Zundert, G.C.P., Melquiond, A.S.J., and Bonvin, A.M.J.J. Integrative modeling of biomolecular complexes: HADDOCKing with cryo-electron microscopy data. *Structure.* 2015 May 5;23(5):949–960.

70. de Vries, S.J. and Zacharias, M. ATTRACT-EM: A new method for the computational assembly of large molecular machines using Cryo-EM maps. *PLOS ONE.* 2012 Dec 14;7(12):e49733.

71. de Vries, S.J., Chauvot de Beauchêne, I., Schindler, C.E.M., *et al.* Cryo-EM data are superior to contact and interface information in integrative modeling. *Biophys J.* 2016 Feb 23;110(4):785–797.

72. Pons, C., Grosdidier, S., Solernou, A., *et al.* Present and future challenges and limitations in protein-protein docking. *Proteins.* 2010 Jan;78(1):95–108.

73. Mashiach, E., Schneidman-Duhovny, D., Andrusier, N., *et al.* FireDock: A web server for fast interaction refinement in molecular docking. *Nucleic Acids Res.* 2008 Jul 1;36(suppl_2):W229–W232.

74. de Vries, S.J., *et al.* HADDOCK versus HADDOCK: New features and performance of HADDOCK 2.0 on the CAPRI targets. *Proteins Struct Funct Bioinforma.* 2007;69(4):726–733.

75. Li, L., Chen, R., and Weng, Z. RDOCK: Refinement of rigid-body protein docking predictions. *Proteins Struct Funct Bioinforma.* 2003;53(3):693–707.

76. Comeau, S.R., Gatchell, D.W., Vajda, S., *et al.* ClusPro: An automated docking and discrimination method for the prediction of protein complexes. *Bioinformatics.* 2004 Jan 1;20(1):45–50.

77. Wang, C., Bradley, P., and Baker, D. Protein–protein docking with backbone flexibility. *J Mol Biol.* 2007 Oct 19;373(2):503–519.

78. Chaudhury, S. and Gray, J.J. Conformer selection and induced fit in flexible backbone protein–protein docking using computational and NMR ensembles. *J Mol Biol.* 2008 Sep 12;381(4):1068–1087.

79. Schindler, C.E.M, de Vries, S.J., and Zacharias, M. iATTRACT: Simultaneous global and local interface optimization for protein–protein

docking refinement. *Proteins Struct Funct Bioinforma.* 2015;83(2): 248–258.

80. Wang, T. and Wade, R.C. Implicit solvent models for flexible protein–protein docking by molecular dynamics simulation. *Proteins Struct Funct Bioinforma.* 2003;50(1):158–169.

81. Perilla, J.R., *et al.* Molecular dynamics simulations of large macromolecular complexes. *Curr Opin Struct Biol.* 2015 Apr 1;31:64–74.

82. Pan, A.C., *et al.* Atomic-level characterization of protein–protein association. *Proc Natl Acad Sci.* 2019 Mar 5;116(10):4244–4249.

83. Luitz, M.P. and Zacharias, M. Protein–ligand docking using Hamiltonian replica exchange simulations with soft core potentials. *J Chem Inf Model.* 2014 Jun 23;54(6):1669–1675.

84. Ostermeir, K. and Zacharias, M. Accelerated flexible protein–ligand docking using Hamiltonian replica exchange with a repulsive biasing potential. *PLOS ONE.* 2017 Feb 16;12(2):e0172072.

85. Martin, M.G. and Siepmann, J.I. Transferable potentials for phase equilibria. 1. United-atom description of n-alkanes. *J Phys Chem B.* 1998 Apr 1;102(14):2569–2577.

86. de Vries, S. and Zacharias, M. Flexible docking and refinement with a coarse-grained protein model using ATTRACT. *Proteins Struct Funct Bioinforma.* 2013;81(12):2167–2174.

87. Zhang, C., Liu, S., Zhu, Q., *et al.* A knowledge-based energy function for protein–ligand, protein–protein, and protein–DNA complexes. *J Med Chem.* 2005 Apr 1;48(7):2325–2335.

88. Chuang, G.-Y., Kozakov, D., Brenke, R., *et al.* DARS (Decoys As the Reference State) potentials for protein–protein docking. *Biophys J.* 2008 Nov 1;95(9):4217–4227.

89. Moreira, I.S., Martins, J.M., Coimbra, J.T.S., *et al.* A new scoring function for protein–protein docking that identifies native structures with unprecedented accuracy. *Phys Chem Chem Phys.* 2015;17(4):2378–2387.

90. Liu, S., Zhang, C., Zhou, H., *et al.* A physical reference state unifies the structure-derived potential of mean force for protein folding and binding. *Proteins Struct Funct Bioinforma.* 2004;56(1):93–101.

91. Sasse, A., de Vries, S.J., Schindler, C.E.M., *et al.* Rapid design of knowledge-based scoring potentials for enrichment of near-native geometries in protein–protein docking. *PLOS ONE.* 2017 Jan 24;12(1):e0170625.

92. Ballester, P.J. and Mitchell, J.B.O. A machine learning approach to predicting protein–ligand binding affinity with applications to molecular docking. *Bioinformatics.* 2010 May 1;26(9):1169–1175.

93. Feliu, E., Aloy, P., and Oliva, B. On the analysis of protein–protein interactions via knowledge-based potentials for the prediction of protein–protein docking. *Protein Sci.* 2011;20(3):529–541.

94. Wang, C., Greene, D., Xiao, L., *et al.* Recent developments and applications of the MMPBSA method. *Front Mol Biosci* [Internet]. 2018 [cited 2019 May 15];4. Available from https://www.frontiersin.org/articles/10.3389/fmolb.2017.00087/full.

95. Chen, F., *et al.* Assessing the performance of the MM/PBSA and MM/GBSA methods. 6. Capability to predict protein–protein binding free energies and re-rank binding poses generated by protein–protein docking. *Phys Chem Chem Phys.* 2016;18(32):22129–22139.

96. Spiliotopoulos, D., *et al.* dMM-PBSA: A new HADDOCK scoring function for protein–peptide docking. *Front Mol Biosci* [Internet]. 2016 [cited 2019 Jun 27];3. Available from https://www.frontiersin.org/articles/10.3389/fmolb.2016.00046/full.

97. Gilson, M.K., Given, J.A., Bush, B.L., *et al.* The statistical-thermodynamic basis for computation of binding affinities: A critical review. *Biophys J.* 1997 Mar;72(3):1047–1069.

98. Mobley, D.L. and Gilson, M.K. Predicting binding free energies: Frontiers and benchmarks. *Annu Rev Biophys.* 2017;46(1):531–558.

99. Woo, H.-J. and Roux, B. Calculation of absolute protein–ligand binding free energy from computer simulations. *Proc Natl Acad Sci.* 2005 May 10;102(19):6825–6830.

100. Gumbart, J.C., Roux, B., and Chipot, C. Efficient determination of protein–protein standard binding free energies from first principles. *J Chem Theory Comput.* 2013;9(8):3789–3798.

101. May, A., *et al.* Coarse-grained versus atomistic simulations: Realistic interaction free energies for real proteins. *Bioinformatics.* 2014 Feb 1;30(3):326–334.

102. Siebenmorgen, T. and Zacharias, M. Evaluation of predicted protein–protein complexes by binding free energy simulations. *J Chem Theory Comput.* 2019 Mar 12;15(3):2071–2086.

Chapter 4

Binding affinity of protein–protein complexes: experimental techniques, databases and computational methods

Sherlyn Jemimah[†], Yugandhar Kumar[†] and M. Michael Gromiha*

Department of Biotechnology, Bhupat and Jyoti Mehta School of Biosciences, Indian Institute of Technology Madras, Chennai 600036, Tamil Nadu, India

Protein–protein interactions (PPIs) are important for several cellular processes. The principles governing the recognition of protein–protein complexes are investigated using the information on binding-site residues at the interface, specific non-covalent interactions as well as binding affinity. Experimentally, binding affinity of PPIs is studied with isothermal titration calorimetry, surface plasmon resonance and fluorescence-based techniques. On the other hand, computational methods have been developed to understand/predict the binding affinity and the effects of mutations. In this chapter, with a brief introduction on the concept of binding affinity, we outline the experimental techniques to explore the binding affinity of protein–protein complexes along with computational analysis and prediction of binding affinity. Further, we illustrate the utilities and applications of PROXiMATE database for binding affinity of protein–protein complexes and their mutants as well as effects of mutations on binding affinity.

---

*gromiha@iitm.ac.in
[†]Contributed equally to this work.

4.1. Introduction

Protein–protein interactions (PPIs) play major roles in several cellular processes such as signal transduction, immunity, cellular transport, metabolic pathways. The functions of protein–protein complexes are explored with three-dimensional structure determination using X-ray crystallography, NMR spectroscopy and electron microscopy, binding affinity measurements such as isothermal titration calorimetry (ITC) and surface plasmon resonance (SPR), and interacting partners using yeast 2-hybrid systems, phage display, Förster resonance energy transfer (FRET) and co-immunoprecipitation. These techniques have been reviewed in detail in the literature (Phizicky and Fields, 1995; Stites, 1997; Xing *et al.*, 2016). Utilizing the information, several computational investigations have been carried out to identify the binding-site residues, preferred residues at the interface, influence of non-covalent interactions and so on (Gromiha *et al.*, 2009, Chakrabarti and Janin, 2002; Levy, 2010). Further, gaining knowledge about affinity-based regulation of protein functions deepens our understanding on mechanisms of biomolecular interactions. Protein–protein binding affinity is the strength of interaction between two or more interacting protein molecules and it is one of the major factors that drive PPIs and dictates the associated functional impacts. The principles underlying protein–protein binding affinity, databases and reliable prediction methods have been extensively reviewed (Kastritis and Bonvin, 2013; Gromiha and Yugandhar, 2017; Yugandhar and Gromiha, 2017; Gromiha *et al.*, 2017).

Altered binding affinity due to mutations may lead to disease phenotypes. Rapid progress has been made in the recent years to estimate and predict the binding affinity between interacting proteins in native state and upon mutations (Moal *et al.*, 2011; Moal *et al.*, 2014; Yugandhar and Gromiha, 2014a; Geng *et al.*, 2019). These studies greatly improve our knowledge about the regulation of biomolecular interactions.

In this chapter, we will review the key concepts in protein–protein binding affinity along with bioinformatics resources available for the study of binding affinity of proteins. Further, we examine the effect of mutations on binding affinity and demonstrate PROXiMATE

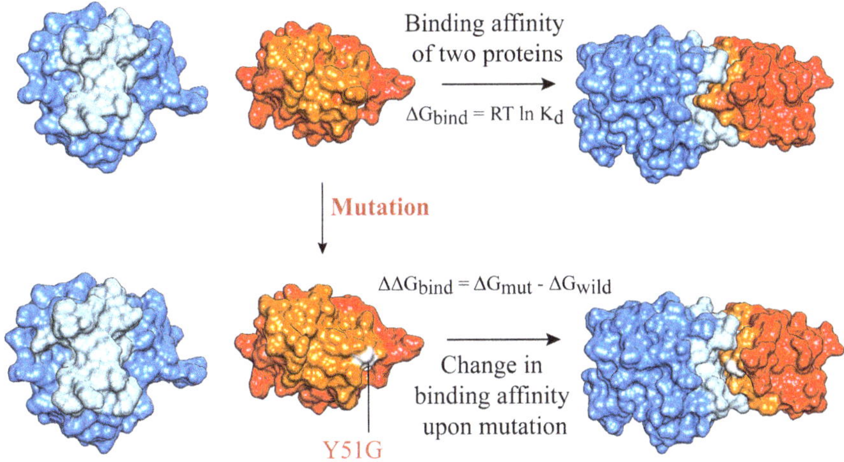

Figure 4.1. Binding affinity of a protein–protein complex and change in affinity upon mutation.

database with features and applications. In addition, the additivity of multiple mutations in terms of change in binding affinity will be discussed. Figure 4.1 illustrates the contents of the chapter in terms of binding affinity and mutational effects.

4.2. Protein–protein binding affinity

In the cell, proteins collide with each other multiple times in the course of diffusion. If they are in a conformation close to the native complex structure, they may then proceed to form the protein–protein complex (Schreiber *et al.*, 2009). Non-covalent interactions help in stabilizing the bound form of the complex. The complex may associate and dissociate dynamically, as represented below:

$$A + B \rightleftharpoons AB$$

where A and B represent the two interacting proteins. The equilibrium constant K is defined using the concentrations of the two proteins and their complex, [A], [B] and [AB], respectively, and it is given by

$$K = K_a = \frac{[AB]}{[A][B]}$$

The equilibrium constant is equated to the association constant, which is used to calculate the binding free energy as:

$$\Delta G = -RT \ln K_a$$

The dissociation constant K_d is the inverse of the association constant and usually denotes the binding affinity of the protein–protein complex:

$$K_d = \frac{1}{K_a} = \frac{[A][B]}{[AB]}$$

Thus, binding free energy can be written using the dissociation constant as follows (Kuriyan *et al.*, 2012):

$$\Delta G = -RT \ln K_d$$

4.3. Experimental methods to estimate protein–protein binding affinity

In general, all experimental methods to determine binding affinity are time-consuming and require specialized, expensive equipment as well as the purified mutated/wild-type proteins of interest. Each method has its own advantages and limitations, which will be explained in detail.

4.3.1. *Surface plasmon resonance*

Surface plasmons are the surface electromagnetic waves that propagate in a direction parallel to the metal–external medium interface. These are very sensitive to any change of this boundary, such as the adsorption of molecules to the metal surface. SPR works on this principle and it is one of the most widely used techniques to characterize the kinetics of PPIs (Willander and Al-Hilli, 2009). In order to determine the affinity parameters for a binary protein–protein complex, one of the interacting proteins is immobilized onto the sensor surface. The other protein is injected into the aqueous solution through the flow channel. As the injected protein binds to the immobilized partner, the accumulation of proteins on the surface results in an increase in the refractive index. This change is measured using an optical detection unit and the result is plotted as response signal versus time. After a defined time, a solution without the protein sample is injected that dissociates the bound complex between the immobilized protein and the partner. During this dissociation process, a decrease in SPR signal is observed. From these data, kinetic constants are derived. Absolutely no requirement for fluorescent or radio-labeled tags is one of the main advantages of this technique that facilitates its compatibility with both colored and opaque solutions. However, the complex and costly experimental setup remains to be a potential drawback of this technique.

4.3.2. *Isothermal titration calorimetry*

The word "calorimetry" comes from the Greek words, *calor*: heat, and *metry*: measure. It is one of the most commonly used calorimetric approaches to study PPIs, which measures the heat uptake or release in the process of a biomolecular interaction (Ladbury and Chowdhry, 1996). The experimental procedure consists of successive additions of protein A to a solution of protein B, the latter contained in a reaction cell. Each such addition results in a specific number of protein–protein complexes, as dictated by the binding affinity that could be observed by monitoring the heat uptake or release. One of the additional benefits from ITC compared to the other methods is that besides measuring binding affinity, it also provides the measures for enthalpy, entropy and heat capacity of the interaction. However, it might not be very effective when employed for either very-high- or very-low-affinity complexes (Kastritis and Bonvin, 2013). The necessity of binding partners to be soluble in the same buffer system and requirement of large quantities of proteins are other potential drawbacks of this technique.

4.3.3. *Fluorescence-based techniques*

In most of the fluorescence-based methods such as FRET and fluorescence (de)polarization (FP), competitive binding assays are used in which a labeled ligand molecule is bound and subsequently displaced by any of the competitive inhibitors (Masi *et al.*, 2010). As per the experimental procedure, a small amount of the labeled ligand is first bound to protein A and is subsequently displaced by titrating the unlabeled protein B. In this way, the inhibition constant K_i of the unlabeled protein can be measured, from which the dissociation constant (K_d) can be derived. Advantages of these methods include the option for deriving additional information such as binding distance between fluorophore and the protein, structural data, etc. Usage of labels might alter the binding behavior of the complex and hamper the actual calculation of binding affinity. This issue, along with other potential sources of errors such as the presence of non-binding contaminants or contaminants that might enhance binding, must be treated carefully during measurements (Klotz, 1985).

4.4. Important features influencing the binding affinity

Investigations on sequences and structures of protein–protein complexes along with their binding affinity revealed the importance of specific features that influence the affinity.

4.4.1. *Structure-based features*

4.4.1.1. *Buried surface area*

Buried surface area (BSA) is the surface area of the interacting proteins that lose solvent accessibility upon complex formation. Solvent accessible surface area (SASA) or simply known as accessible surface area (ASA), which is measured in $Å^2$, is theoretically estimated by rolling a probe (typically of water, with radius 1.4 Å) along the surface of the protein.

According to the model proposed by Chothia and Janin, BSA is the principal feature to be related to the binding affinity, and even more explicitly to the intrinsic bond energy (Chothia and Janin, 1975). It is a macroscopic parameter for hydrophobic interactions of proteins, and its magnitude has been estimated to be 0.025 kcal mol per 1 Å^2 of the hydrophobic surface removed from contact with water (Kastritis and Bonvin, 2013). Apart from being a favorable attraction between hydrophobic surfaces of proteins, this hydrophobic interaction also represents the entropy gain of the water molecules released upon the complex formation (Kastritis and Bonvin, 2013).

Kastritis *et al.* (2011) demonstrated that for a set of 70 rigid complexes (interface RMSD < 1 Å), BSA alone showed a correlation of 0.54. However, the correlation vanished for complexes with interface RMSD > 1 Å. This suggests the dominant role played by the interface area in determining the binding affinity and especially in proteins undergoing less or negligible conformational changes. Swapna *et al.* (2012) reported that the rigid residues at the interfaces make a significant contribution toward the stabilization of the interface structure in the unbound state. Janin (2014) proposed a minimal model and showed that interface information along with conformational changes could explain the binding affinity on a diverse data set. In addition, the non-interacting surface is shown to be important in modulating the protein–protein binding affinity (Kastritis *et al.*, 2014). Specifically, polar residues on the surface and charged residues are major determinants in estimating the affinity, and their combinations with other conventional interface properties showed a correlation of 0.50.

4.4.1.2. *Hot spots and anchor residues*

A relatively small number of interface residues, known as warm spots and hot spots, account for majority of the binding energy (Cunningham and Wells, 1989; Bogan and Thorn, 1998). Hot and warm spots are defined as the amino acid residues, whose mutation to alanine causes destabilization of the bound state by ≥ 4 and 1–2 kcal/mol, respectively (Kastritis and Bonvin, 2013). On the contrary, null spots do not result in such differences in the free energy. The contribution of an amino acid residue toward the binding free energy is experimentally estimated using alanine scanning mutagenesis (Cunningham and Wells, 1993; Clackson and Wells, 1995). Mutation of a typical amino acid residue to alanine essentially minimizes the side chain, leaving only the β-carbon atom. Further, the kinetic analyses can provide information about the role played by individual amino acid residues in protein binding. Nevertheless, it is noteworthy that mutating the reference residue to glycine might theoretically be a better option, as the whole side chain would be removed. However, mutation to glycine is not ideal, as it might introduce global or local conformational changes to the protein (Kastritis and Bonvin, 2013).

Typical characteristic features of hot spot residues include occlusion from solvent (Bogan and Thorn, 1998; DeLano, 2002), higher conservation when compared to the non-hotspots (Hu *et al.*, 2000), differential amino acid composition than that of the non-hotspot residues (Ofran and Rost, 2007) and preferential localization in the central region of the interface region (Bogan and Thorn, 1998). Typically, the hot spots that are buried at the interface are surrounded by polar regions with high packing density (Halperin *et al.*, 2004). Even though, the hydrophobic interactions are not the absolute determinant for binding, the bulkier residues tend to be found frequently in hot spots and they possess large surface areas (Janin, 2009).

4.4.1.3. *Non-covalent interactions and binding affinity*

Gromiha *et al.* (2009) developed an energy-based approach to identify the binding sites and interacting pairs in heterodimeric protein complexes. They derived the binding propensity of 20 amino acid residues and showed that charged and aromatic residues are important for binding in protein–protein complexes. These residues influence to form cation–π, electrostatic and aromatic interactions at the interface. The comparison of hotspot residues and experimental binding affinity of typical protein–protein complexes upon mutations emphasizes the importance of these interactions and showed a good agreement with computational observations.

4.4.2. *Sequence-based features*

Apart from the features derived from three-dimensional structures, physicochemical properties extracted from the primary sequences of interacting proteins have also been demonstrated to be useful in protein–protein binding affinity predictions (Cunningham and Wells, 1989, 1993; Bogan and Thorn, 1998). These key features include sequence composition, predicted binding-site residues and propensity of secondary structural elements. The sequence-based descriptors are derived from pre-computed property values for the 20 amino acid residues, available in AAIndex (Kawashima and Kanehisa, 2000) as well as in the literature (Gromiha, 2005), and predicted binding-site residues using the tools available in the literature such as PredictProtein (Yachdav *et al.*, 2014) and BSPred (Mukherjee and Zhang, 2011).

4.5. Computational resources for binding affinity

Few databases have been developed for the binding affinity of protein–protein complexes. PDBBind is one among them, which has binding affinity data for about 2000 protein–protein complexes. Kastritis *et al.* (2011) published a binding affinity benchmark database for 144 protein–protein complexes containing wild-type binding affinity and links to high-resolution structures of the complexes, and it has been updated with additional data by Vreven *et al.* (2015).

On the other hand, a number of computational methods have been developed to predict the binding affinity of protein–protein complexes. Moal *et al.* (2011) used a consensus approach with more than 10 structural parameters on the benchmark data set of binding affinity of protein–protein complexes. Moal *et al.* (2014) developed the CCharPPI method which uses scoring functions to predict the affinity. Yugandhar and Gromiha (2014a) developed a sequence-based method to predict the binding affinity using functional information. The Web server has been described in detail in the following section.

4.5.1. *PPA-Pred Web server*

PPA-Pred (https://www.iitm.ac.in/bioinfo/PPA_Pred/) predicts the absolute binding affinity between any two interacting proteins from their amino acid sequences. It accepts the primary sequences of the interacting proteins in FASTA format (Figure 4.2A) followed by selecting one of the functional classes from the available drop-down menu. In the output, the predicted binding affinity is displayed as both ΔG (binding free energy) and K_d (dissociation constant) (Figure 4.2B).

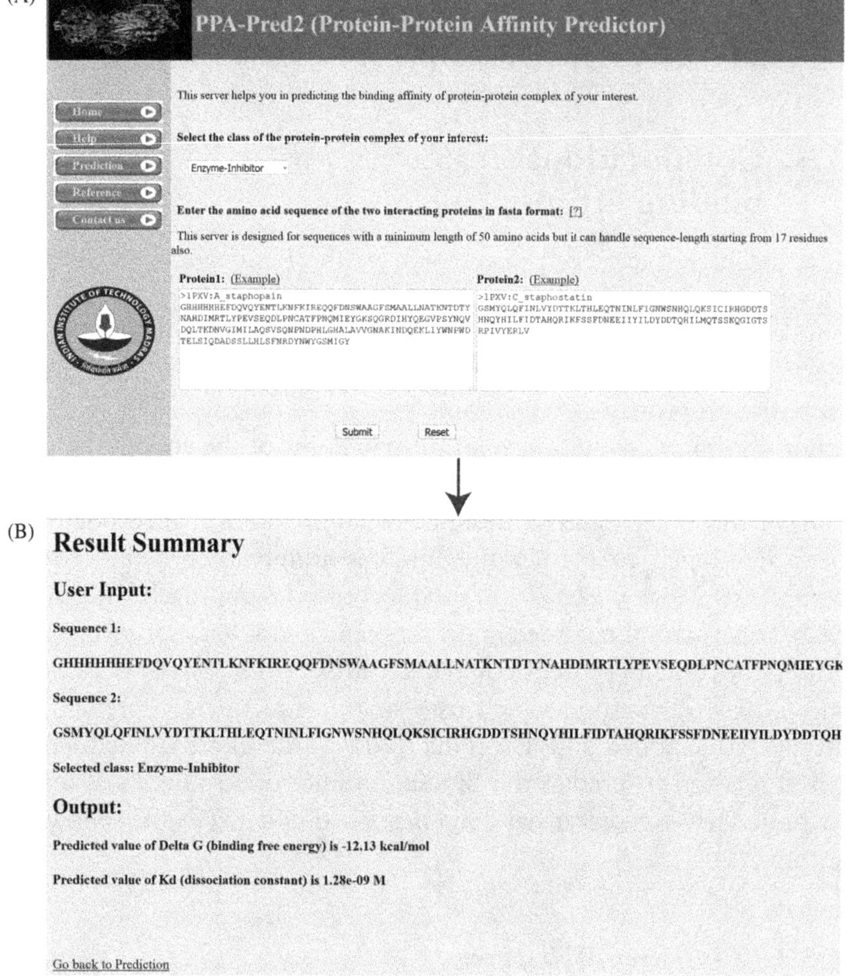

Result Summary

User Input:

Sequence 1:

GHHHHHHEFDQVQYENTLKNFKIREQQFDNSWAAGFSMAALLNATKNTDTYNAHDIMRTLYPEVSEQDLPNCATFPNQMIEYGK

Sequence 2:

GSMYQLQFINLVYDTTKLTHLEQTNINLFIGNWSNHQLQKSICIRHGDDTSHNQYHILFIDTAHQRIKFSSFDNEEIIYILDYDDTQH

Selected class: Enzyme-Inhibitor

Output:

Predicted value of Delta G (binding free energy) is -12.13 kcal/mol

Predicted value of Kd (dissociation constant) is 1.28e-09 M

Go back to Prediction

Figure 4.2.   PPA-Pred Web server, showing the (A) example query and (B) prediction results.

4.6. Analysis of PPI networks based on binding affinity

Studying PPIs using networks provides deep insights into the interaction patterns that could play crucial roles in various biological pathways. Incorporating additional aspects such as binding affinity further improves to effectively capturing the role of specificity in the interaction networks. The currently available expensive and time-intensive experimental techniques preclude analysis of proteome-wide interaction networks based on binding affinities. However, robust and reliable sequence-based computational prediction methods can be utilized for such analysis. In this regard, Yugandhar and Gromiha (2014b) developed a machine learning algorithm for the classification of protein–protein complexes into "high-" or "low-"affinity complexes based on their predicted interaction strength, and a regression model to predict the absolute affinity (Yugandhar and Gromiha, 2014a). Utilizing both of those methods, Yugandhar and Gromiha (2016) performed proteome-wide affinity predictions for the interactions for five organisms (Eukaryotic: Human, *Drosophila* and *Sacromyces cerevisiae*; Prokaryotic: *E. coli* and *H. pylori*). Their analysis revealed some key insights such as the relationship between binding affinity and hub status of interacting proteins in a network, enrichment of specific network motifs with high and low affinity interactions and the association between various physicochemical properties of interacting proteins and their binding affinities.

4.7. Effect of mutations on binding affinity

Mutations in protein–protein complexes can affect the binding affinity by disrupting bonds and causing conformational changes. The change in binding free energy caused by mutation is denoted as

$$\Delta\Delta G = \Delta G_{\text{mutant}} - \Delta G_{\text{wild-type}}$$

where, $\Delta G_{\text{wild-type}}$ and ΔG_{mutant} are the binding free energy of wild-type and mutant proteins, respectively.

4.7.1. *Databases for change in binding affinity upon mutation*

The compilation of experimental data on binding affinities is essential to investigate the factors influencing the binding affinity and develop computational tools. Currently, several databases on protein–protein binding affinity upon mutation are available in the literature, which include ASEdb, Alanine Scanning Energetics database (Thorn and Bogan, 2001), PINT (Kumar and Gromiha, 2006), SKEMPI (Moal and Fernández-Recio, 2012; Jankauskaitė *et al.*, 2018) and PROXiMATE (Jemimah *et al.*, 2017), and are described in the literature (Gromiha *et al.*, 2017). The utilities and applications of PROXiMATE database (Jemimah *et al.*, 2017) are described below:

(a)   Navigate to https://www.iitm.ac.in/bioinfo/PROXiMATE/ index.html.

(b)   Go to the search page, where users have several options such as interacting protein names, PDB ID, type of mutation, experimental technique, functional class, range of binding affinity values, author names, etc. (Figure 4.3A).

(c)   Output shows the search results and an example for barnase–barstar complex is displayed in Figure 4.3B.

(d)   Users can download the results and navigate to each entry by clicking on the entry number. The information for each entry is organized into four sections (Figure 4.3C) viz. summary (which contains details of the interaction, experiment, relative ASA, secondary structure, and thermodynamic and kinetic data), complex structure (which displays the complex structure if PDB ID is available), interaction (which displays the STRING

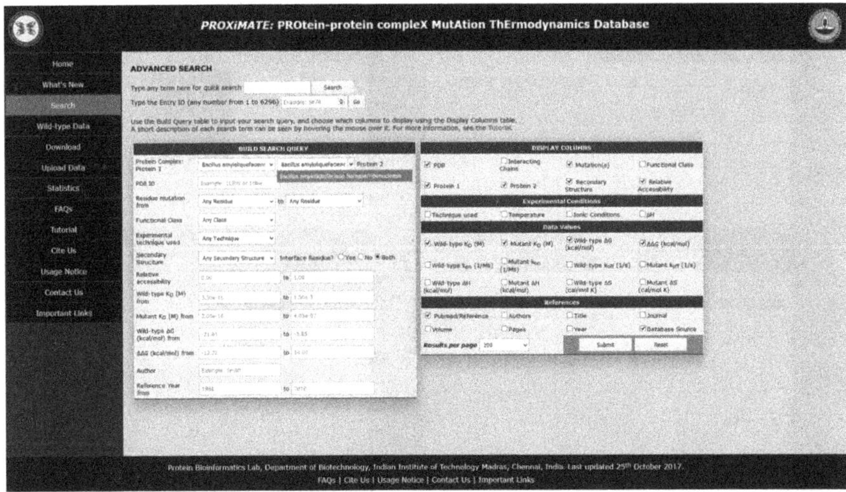

Figure 4.3A. Search page to build a search query with an example for barnase–barstar complex.

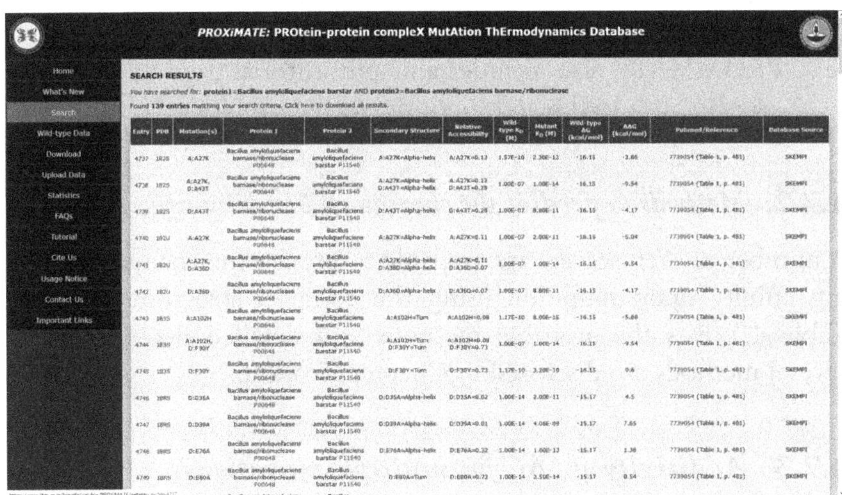

Figure 4.3B. Search results for the example query (barnase–barstar complex).

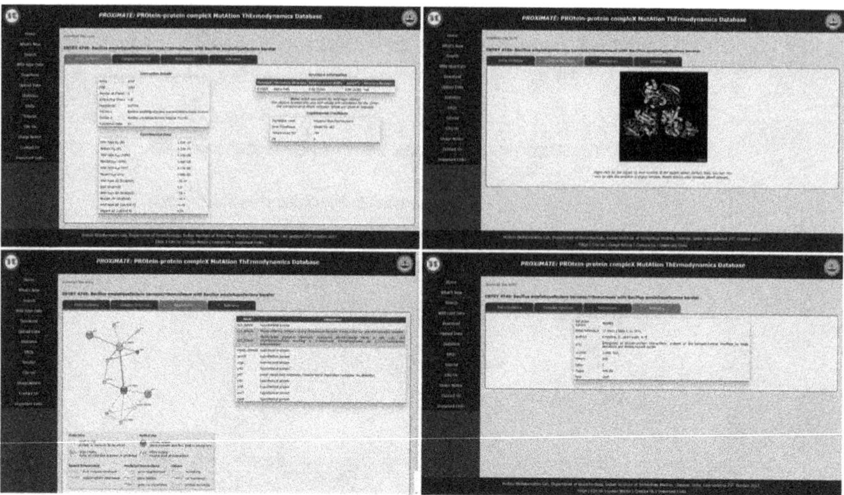

Figure 4.3C. Information for each entry, organized into four sections viz. summary, complex visualization, interaction and reference.

network) and the reference (including the exact location of the data).

(e) PROXiMATE also includes a helpful tutorial page to guide the users in advanced search options, with examples.

4.7.2. *Methods to predict the change in binding affinity*

A number of Web servers are available to predict the change in binding affinity upon mutation using the complex structure as input. Table 4.1 lists the available methods and detailed descriptions of several methods are discussed in Chapters 5 and 6.

4.7.3. *Additivity of* $\Delta\Delta G$ *for multiple mutations*

Additivity in binding affinity of protein–protein complexes refers to the change in free energy of binding ($\Delta\Delta G_{bind}$) for double (or multiple) mutations is approximately equal to the sum of their corresponding single mutation $\Delta\Delta G_{bind}$ values.

Table 4.1. Web servers for predicting the change in binding free energy of protein–protein complexes upon mutation.

Method	Web server URL/Standalone	Features	Reference
BeAtMuSiC	http://babylone.ulb.ac.be/beatmusic/index.php	Statistical potentials derived from sequence- and structure-based features	Dehouck *et al.* (2013)
ELASPIC	http://www.kimlab.org/software/elaspic	Sequence conservation, intra- and intermolecular contacts, SASA and FoldX-generated energy descriptors	Berliner *et al.* (2014)
BindProf	http://zhanglab.ccmb.med.umich.edu/BindProf/	Interface profile score, shape complementarity and sequence-based features	Brender and Zhang (2015)
BindProfX	https://zhanglab.ccmb.med.umich.edu/BindProfX/	Updated version of BindProf, uses pseudocounts to improve prediction	Xiong *et al.* (2017)
SAAMBE	http://compbio.clemson.edu/saambe_webserver/	van der Waals, solvation and Coulomb energy, entropy, hydrophobicity, SASA, hydrogen bonds and interface area	Petukh *et al.* (2015, 2016)
MutaBind	http://www.ncbi.nlm.nih.gov/projects/mutabind/	van der Waals energy, solvation energy, free energy change due to unfolding, SASA, conservation score and structural constraints due to proline	Li *et al.* (2016)
mCSM	http://biosig.unimelb.edu.au/mcsm/protein_protein	Graph-based signatures	Pires *et al.* (2014)
mCSM-AB	http://biosig.unimelb.edu.au/mcsm_ab/	Graph-based signatures, specifically for antigen–antibody complexes	Pires and Ascher (2016)
ZEMu	No Web server available. MacroMoleculeBuilder can be downloaded from https://simtk.org/projects/rnatoolbox	Uses MacroMoleculeBuilder to equilibrate a 5 Å zone around the mutation, subject to the forces of the physics zone, which is 12 Å around the mutation	Dourado and Flores (2014)
Flex ddG	Rosetta can be found at https://www.rosettacommons.org/software/license-and-download	Uses Rosetta's backrub protocol to sample conformational changes caused by the mutation and calculates the $\Delta\Delta G$ value using an energy function	Barlow *et al.* (2018)
iSEE	Source code can be found at https://github.com/haddocking/iSee	Uses HADDOCK server to generate the mutant complex and energy-based features for a random forest model which predicts the $\Delta\Delta G$ value	Geng *et al.* (2019)

Consider two mutations X and Y. The changes in binding affinity caused by the individual mutations are $\Delta\Delta G_X$ and $\Delta\Delta G_Y$, respectively. The change in double mutant binding free energy ($\Delta\Delta G_{X,Y}$) to that of the single mutants is given by (Wells, 1990):

$$\Delta\Delta G_{X,Y} = \Delta\Delta G_X + \Delta\Delta G_Y + \Delta G_I$$

where ΔG_I is the coupling energy and indicates the extent to which the interaction between the two mutations affects the measured binding free energy. It represents both intermolecular and intramolecular interactions (Horovitz, 1987). When there are no significant interactions between the two mutations, ΔG_I becomes negligible. Thus, the equation simplifies to

$$\Delta\Delta G_{X,Y} \approx \Delta\Delta G_X + \Delta\Delta G_Y$$

The comparison of 379 double mutants with the sum of their respective single mutants showed a correlation of 0.90 with mean absolute deviation of 0.86 kcal/mol between them (Jemimah and Gromiha, 2018). However, about 25% of double mutants deviated more than 1 kcal/mol and are termed as nonadditive.

Further, Jemimah and Gromiha (2018) explored the role of different structure-based features for nonadditivity of double mutants and reported that atom contacts and distance between the two mutations are important for nonadditivity. The distribution of additive and nonadditive mutants based on minimum distance between heavy atoms in the two mutations (Figure 4.4A) showed that the nonadditive mutations tend to be closer to each other compared to additive mutants. In addition, the distribution of atom contacts using a distance cutoff of 6 Å between heavy atoms of the two mutations (Figure 4.4B) revealed that nonadditive mutants have more number of contacts between the mutants than additive mutants. Hence, the coupling energy term becomes significant due to the interactions between the two mutations, which leads to cooperative effects and significant deviations from the additive $\Delta\Delta G$ value (Jemimah and Gromiha, 2018).

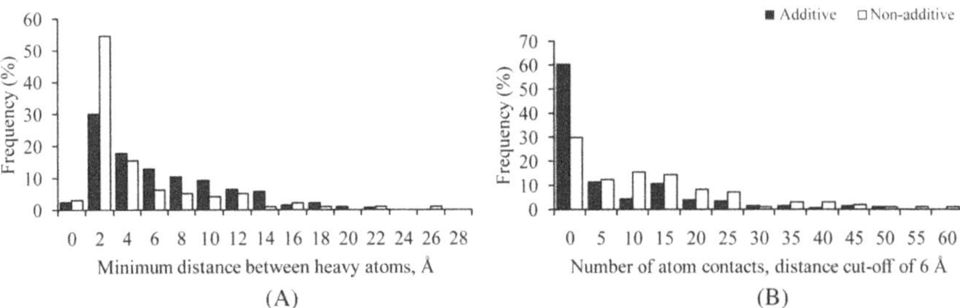

Figure 4.4. Distribution of additive and nonadditive mutants for (A) minimum distance between heavy atoms in the two mutations and (B) number of atom contacts between the two mutations, within a distance cutoff of 6 Å; *X*-axis values denote the starting value of the bin. Figure was taken from Jemimah and Gromiha (2018).

4.8. Conclusion

PPIs are important for several cellular functions. The binding affinity of PPIs deepens our understanding of protein interactions and their functional aspects. In this chapter, we reviewed the experimental methods for determining the binding affinity of protein–protein complexes and various structure- and sequence-based features, which influence the binding affinity. Further, we have discussed the functionalities and applications of PPA-Pred for predicting the binding affinity from protein sequences and functional class. In the second part, we focused on mutants and discussed the availability of different databases and methods with an emphasis on PROXiMATE database. The concept of additivity of mutants on binding affinity has been outlined and the analysis showed that the change in binding affinity upon the double mutation is usually additive except if the mutations are close to each other and/or able to make many contacts with each other.

Acknowledgments

We thank the Bioinformatics Facility, Department of Biotechnology and Indian Institute of Technology Madras for computational facilities. This work is supported by Department of Science and Technology,

India, to MMG [EMR/2016/001476] and Ministry of Human Resources Development, India, to SJ.

References

Barlow, K.A., *et al.* Flex ddG: Rosetta ensemble-based estimation of changes in protein–protein binding affinity upon mutation. *J Phys Chem B.* 2018;122(21):5389–5399.

Berliner, N., Teyra, J., Çolak, R., *et al.* Combining structural modeling with ensemble machine learning to accurately predict protein fold stability and binding affinity effects upon mutation. *PLOS ONE.* 2014;9(9): e107353.

Bogan, A.A., and Thorn, K.S. Anatomy of hot spots in protein interfaces. *J Mol Biol.* 1998;280(1):1–9.

Brender, J.R., and Zhang, Y. Predicting the effect of mutations on protein–protein binding interactions through structure-based interface profiles. *PLoS Comput. Biol.* 2015;11(10):e1004494.

Chakrabarti, P., and Janin, J. Dissecting protein–protein recognition sites. *Proteins.* 2002;47(3):334–343.

Chothia, C., and Janin, J. Principles of protein–protein recognition. *Nature.* 1975;256(5520):705.

Clackson, T., and Wells, J.A. A hot spot of binding energy in a hormone-receptor interface. *Science.* 1995;267(5196):383–386.

Cunningham, B.C., and Wells, J.A. High-resolution epitope mapping of hGH-receptor interactions by alanine-scanning mutagenesis. *Science.* 1989;244(4908):1081–1085.

Cunningham, B.C., and Wells, J.A. Comparison of a structural and a functional epitope. *J Mol Biol.* 1993;234(3):554–563.

Dehouck, Y., Kwasigroch, J.M., Rooman, M., *et al.* BeAtMuSiC: Prediction of changes in protein–protein binding affinity on mutations. *Nucleic Acids Res.* 2013;41(W1):W333–W339.

DeLano, W.L. Unraveling hot spots in binding interfaces: Progress and challenges. *Curr Opin Struct Bio.* 2002;12(1):14–20.

Dourado, D.F., and Flores, S.C. A multiscale approach to predicting affinity changes in protein–protein interfaces. *Proteins.* 2014;82(10): 2681–2690.

Geng, C., Vangone, A., Folkers, G.E., *et al.* iSEE: Interface structure, evolution, and energy based machine learning predictor of binding affinity changes upon mutations. *Proteins.* 2019;87(2):110–119.

Gromiha, M.M. A statistical model for predicting protein folding rates from amino acid sequence with structural class information. *J Chem Inf Model.* 2005;45(2):494–501.

Gromiha, M.M., Yokota, K., and Fukui, K. Energy based approach for understanding the recognition mechanism in protein–protein complexes. *Mol BioSyst.* 2009;5(12):1779–1786.

Gromiha, M.M., and Yugandhar, K. Integrating computational methods and experimental data for understanding the recognition mechanism and binding affinity of protein–protein complexes. *Prog Biophys Mol Biol.* 2017;128:33–38.

Gromiha, M.M., Yugandhar, K., and Jemimah, S. Protein–protein interactions: Scoring schemes and binding affinity. *Curr Opin Struct Biol.* 2017;44:31–38.

Halperin, I., Wolfson, H., and Nussinov, R. Protein–protein interactions: Coupling of structurally conserved residues and of hot spots across interfaces. Implications for docking. *Structure.* 2004;12(6):1027–1038.

Horovitz, A. Non-additivity in protein–protein interactions. *J Mol Biol.* 1987;96(3):733–735.

Hu, Z., Ma, B., Wolfson, H., *et al.* Conservation of polar residues as hot spots at protein interfaces. *Proteins.* 2000;39(4):331–342.

Janin, J. Basic principles of protein–protein interaction. In: *Computational Protein–Protein Interactions.* Florida, FL: CRC Press; 2009:1–20.

Janin, J. A minimal model of protein–protein binding affinities. *Protein Sci.* 2014;23(12):1813–1817.

Jankauskait, J., Jiménez-García, B., Dapk nas, J., *et al.* SKEMPI 2.0: An updated benchmark of changes in protein–protein binding energy, kinetics and thermodynamics upon mutation. *Bioinformatics.* 2018;35(3):462–469.

Jemimah, S., and Gromiha, M.M. Exploring additivity effects of double mutations on the binding affinity of protein protein complexes. *Proteins.* 2018;86(5):536–547.

Jemimah, S., Yugandhar, K., and Michael Gromiha, M. PROXiMATE: A database of mutant protein–protein complex thermodynamics and kinetics. *Bioinformatics.* 2017;33(17):2787–2788.

Kastritis, P.L., and Bonvin, A.M. On the binding affinity of macromolecular interactions: Daring to ask why proteins interact. *J R Soc Interface.* 2013;10(79):20120835.

Kastritis, P.L., *et al.* A structure based benchmark for protein–protein binding affinity. *Protein Sci.* 2011;20(3):482–491.

Kastritis, P.L., *et al.* Proteins feel more than they see: Fine-tuning of binding affinity by properties of the non-interacting surface. *J Mol Biol.* 2014;426(14);2632–2652.

Kawashima, S., and Kanehisa, M. AAindex: Amino acid index database. *Nucleic Acids Res.* 2000;28(1):374.

Klotz, I.M. Ligand — receptor interactions: Facts and fantasies. *Q Rev Biophys.* 1985;18(3):227–259.

Kumar, M.S., and Gromiha, M.M. PINT: Protein–protein interactions thermodynamic database. *Nucleic Acids Res.* 2006;34(suppl_1):D195–D198.

Kuriyan, J., Konforti, B., and Wemmer, D. Molecular recognition: The thermodynamics of binding. In *The Molecules of Life: Physical and Chemical Principles.* 2012;1:531–548.

Ladbury, J.E., and Chowdhry, B.Z. Sensing the heat: The application of isothermal titration calorimetry to thermodynamic studies of biomolecular interactions. *Chem Biol.* 1996;3(10):791–801.

Levy, E.D. A simple definition of structural regions in proteins and its use in analyzing interface evolution. *J Mol Biol.* 2010;403(4):660–670.

Li, M., Simonetti, F.L., Goncearenco, A., *et al.* MutaBind estimates and interprets the effects of sequence variants on protein–protein interactions. *Nucleic Acids Res.* 2016;44(W1):W494–W501.

Masi, A., Cicchi, R., Carloni, A., *et al.* (2010). Optical methods in the study of protein–protein interactions. In: *Integrins and Ion Channels.* New York, NY: Springer; 2010: 33–42.

Moal, I.H., Agius, R., and Bates, P.A. Protein–protein binding affinity prediction on a diverse set of structures. *Bioinformatics.* 2011;27(21): 3002–3009.

Moal, I.H., and Fernández-Recio, J. SKEMPI: A structural kinetic and energetic database of mutant protein interactions and its use in empirical models. *Bioinformatics.* 2012;28(20):2600–2607.

Moal, I.H., Jiménez-García, B., and Fernández-Recio, J. CCharPPI web server: Computational characterization of protein–protein interactions from structure. *Bioinformatics.* 2014;31(1):123–125.

Mukherjee, S., and Zhang, Y. Protein–protein complex structure predictions by multimeric threading and template recombination. *Structure.* 2011;19(7):955–966.

Nikam, R., Yugandhar, K., and Gromiha, M.M. Discrimination and prediction of protein–protein binding affinity using deep learning approach. *Lect. Notes Comp. Sci.* 2018 Aug;10955:809–815.

Ofran, Y., and Rost, B. Protein–protein interaction hotspots carved into sequences. *PLoS Comput Biol.* 2007;3(7):e119.

Petukh, M., Dai, L., and Alexov, E. SAAMBE: Webserver to predict the charge of binding free energy caused by amino acids mutations. *Int J Mol Sci.* 2016;17(4):547.

Petukh, M., Li, M., and Alexov, E. Predicting binding free energy change caused by point mutations with knowledge-modified MM/PBSA method. *PLoS Comput Biol.* 2015;11(7):e1004276.

Phizicky, E.M., and Fields, S. Protein–protein interactions: Methods for detection and analysis. *Microbiol Mol Biol Rev.* 1995;59(1):94–123.

Pires, D.E., and Ascher, D.B. mCSM-AB: A web server for predicting antibody–antigen affinity changes upon mutation with graph-based signatures. *Nucleic Acids Res.* 2016;44(W1):W469–W473.

Pires, D.E., Ascher, D.B., and Blundell, T.L. mCSM: Predicting the effects of mutations in proteins using graph-based signatures. *Bioinformatics.* 2014;30(3):335–342.

Schreiber, G., Haran, G., and Zhou, H.X. Fundamental aspects of protein–protein association kinetics. *Chem Rev.* 2009;109(3):839–860.

Stites, W.E. Protein–protein interactions: Interface structure, binding thermodynamics, and mutational analysis. *Chemical Rev.* 1997;97(5):1233–1250.

Swapna, L.S., Bhaskara, R. M., Sharma, J., *et al.* Roles of residues in the interface of transient protein–protein complexes before complexation. *Sci Rep.* 2012;2:334.

Thorn, K.S., and Bogan, A.A. ASEdb: A database of alanine mutations and their effects on the free energy of binding in protein interactions. *Bioinformatics.* 2001;17(3):284–285.

Vreven, T., *et al.* Updates to the integrated protein–protein interaction benchmarks: Docking benchmark version 5 and affinity benchmark version 2. *J Mol Biol.* 2015;427(19):3031–3041.

Wells, J.A. Additivity of mutational effects in proteins. *Biochemistry.* 1990;29(37):8509–8517.

Willander, M., and Al-Hilli, S. Analysis of biomolecules using surface plasmons. In: *Micro and Nano Technologies in Bioanalysis.* Totowa, NJ: Humana Press; 2009: 201–229.

Xing, S., Wallmeroth, N., Berendzen, K.W., *et al.* Techniques for the analysis of protein–protein interactions in vivo. *Plant Physiol.* 2016;171(2):727–758.

Xiong, P., Zhang, C., Zheng, W., *et al.* BindProfX: Assessing mutation-induced binding affinity change by protein interface profiles with pseudo-counts. *J Mol Biol.* 2017;429(3):426–434.

Yachdav, G., *et al.* PredictProtein — An open resource for online prediction of protein structural and functional features. *Nucleic Acids Res.* 2014;42(W1):W337–W343.

Yugandhar, K., and Gromiha, M.M. Protein–protein binding affinity prediction from amino acid sequence. *Bioinformatics.* 2014a;30(24):3583–3589.

Yugandhar, K., and Gromiha, M.M. Feature selection and classification of protein–protein complexes based on their binding affinities using machine learning approaches. *Proteins.* 2014b;82(9):2088–2096.

Yugandhar, K., and Gromiha, M.M. Analysis of protein–protein interaction networks based on binding affinity. *Curr Protein Pept Sci.* 2016;17(1): 72–81.

Yugandhar, K., and Gromiha, M.M. Computational approaches for predicting binding partners, interface residues, and binding affinity of protein–protein complexes. In: *Prediction of Protein Secondary Structure.* New York, NY: Humana Press; 2017: 237–253.

Chapter 5

Mutational effects on protein–protein interactions

Jackson Weako*, Attila Gursoy[†,§] and Ozlem Keskin[‡,¶]

*Computational Science and Engineering,
†Computer Science and Engineering,
‡Chemical and Biological Engineering,
Koç University, Sariyer/Istanbul 34450, Turkey

Interatomically, protein's ability to enact careful and tight interactions with its biomolecular partners is pivotal necessitating proper biological function. Protein–protein interactions (PPIs) are paramount in biological processes as manifested by their tenacious cellular duties. Protein interfaces are the contact regions between proteins. The interface residues are under constant evolutionary restriction than surface residues. Variations in interface residues which alter binding affinity of proteins may lead to pronounce perturbation or absolute annulment of their functions, potentially resulting in diseases. Hence, wealth of investigations on mutational consequences of PPIs has evolved. The availability of both experimental and computational techniques to assess the effects of mutations on protein–protein binding affinities is essential for varieties of biomedical applications, spurring establishment of various mutational databases. Here, handy experimental and computational methods for detecting/predicting consequences of mutations on PPIs have

§agursoy@ku.edu.tr
¶okeskin@ku.edu.tr

been explored. Also, protocols and features of proteins utilized by these techniques have been elaborated and updates on mutational databases have been provided. Finally, case studies on mutations emerging at PPI interfaces and their involvements in human-related diseases such as cancer have been provided.

5.1. Introduction

Most biological processes are modulated through assembly of proteins. Protein–protein interactions (PPIs) or protein associations play vital roles in biomolecular processes including cell signaling, cell proliferation, catalysis, apoptosis and metabolism [1]. Physically, PPIs are mediated by their interface residues, which are well conserved than their counterparts, non-interface [2, 3]. Due to the conservation of interface residues and/or hotspots, unsurprisingly, mutations at PPI interfaces are associated abnormal biological processes [3]. These mutations can be deleterious or benign depending on the chemistry of mutated residue and its location within the proteins. Notwithstanding, assessment of PPI affinity and impacts of mutation on protein's stability and PPI affinity are burdensome due to the transitory and obligatory nature of their interactions [1]. Orthodoxically, experimental methods are gold standard to determine the effects of mutation on PPIs. However, experimental techniques are costly and labor-intensive; thus, a gamut of computational methodologies have been developed for predicting phenotypic and thermodynamic effects of mutations on PPIs. These computational techniques are classified into sequence based, structure based or combination of both [4].

In this chapter, we provide a review with classification, description and case study examples of single amino acid mutation effects on PPIs. At the heart of the chapter, we discuss how mutation of interface residues may perturb protein interactions, influence electrostatic features of proteins' surface and introduce baleful impacts resulting in alterations in 3-D structures, functions, stability, folding, alteration of their specificity and affinity to their binding partners, and developing drug resistance [5]. The chapter also incorporates mutations in proteins' interface residues that promote oncogenesis. Further, experimental and computational methodologies for foreseeing impact of mutations on PPIs are discussed with emphasis on protocols and features of proteins used in predictions. At this juncture, we want to emphasize that our coverage of effects of mutation on PPIs is non-exhaustive. Here, we embark on the effect of mutations (especially single amino acid mutation or variation) which is the nitty-gritty for understanding the significance of PPIs, genetic diseases and cancer developments. Finally, of note, we interchangeably use mutation and variation throughout the chapter.

5.2. Protein–protein interaction and mutation

PPIs are driven by hydrophobic effects, hydrogen bonds, electrostatic and π–π bonds. The flexible architecture of proteins enables them to exist in ensembles of conformations, especially those that are free in solutions. The backbone and side chain atoms of proteins do move; hence, they undergo additional conformational changes upon binding to their allies. Also, a huge number of proteins *in vivo* are presumed to be natively disordered in solution [6, 7]. In the disordered state, proteins do exist as numerous conformers. The roles of proteins considering their features, for example, interacting with other biomolecules, are greatly controlled by their 3-D structures. A variation in a protein's sequence of amino acids can alter its functions, folding, conformation ensemble and binding potential to other molecules [1].

The intrinsic nature of PPIs, interacting proteins assembling to form functional protein complexes that shape the cell and linking to pathways and larger protein networks make PPIs a key area of interest. A mutation in protein sequence can have significant impacts at different levels, ranging from function and structure of a single protein, ripple effects on proteins in contact to eventually phenotypic consequences at the cellular and beyond [8]. Mutations may alter proteins' primary sequences in several ways including insertion, deletion, substitution or truncation. These various forms of alterations can, in turn, have a far-reaching effect on the protein's structural conformation, interaction and stability at the cellular level.

The mutations at genome level can be non-synonymous or synonymous. Alteration in protein stability and gain or loss of interacting partners are significant consequences of non-synonymous mutations in proteins, which can lead to changes in cellular behavior of the organism. For instance, certain mutations on the tumor repressor protein p53 can stimulate its oncogenic properties in the cell [9]. Essentially, mutations that "constitutively" activate or stabilize oncoproteins are not unlikely to induce complex "downstream activities" which maybe illustrated by basic invigoration of the associated signaling pathway [10]. For example, assessments of the consequences of cancer-causing Ras mutations at the network level have disclosed complicated negative and positive feedback mechanisms [11]. As stated, amino acid substitutions can have important consequences on proteins structure, functions, interactions and stabilities. Specific alterations in physico-chemical properties of mutated residues are essential determinant of the resulting consequences. For instance, a change in the charge of amino acids usually destabilizes proteins [12]. Mutations that occur at the interaction site can cause tremendous changes, including hydrophobic destabilization, loss of electrostatic features, salt bridges, alterations in main-chain protein conformations, loss of hydrogen bonds and steric hindrances [13]. Studies have revealed that protein interactions impart not only selectivity but also "sensitivity" to several biological events, which occurs either by cooperative gathering of distinct multi-protein or cooperative protein folding and binding [3]. The elucidation that the interface of a protein contains key residues for driving the formation of a protein complex has "spurred" medicinal chemists to design drug molecules to target these interactions. Unsurprisingly, single mutations at PPIs can have important effects on protein–protein binding affinity and stability and may also lead to drug resistance [3, 14].

5.2.1. *Mutations of interface and non-interface residues*

Genetic variations play crucial roles in evolution through introduction of diversity into the genomes. This induced diversity can be specifically advantageous or may lead to alteration in protein's binding affinity and stability causing destabilization/overstabilization of PPIs. Protein's binding affinity is evolutionarily restrained. We can deduce that the likelihood of a residue which favorably impacts binding to be at an interface site is more than finding an unfavorable residue. The interface residues and non-interface residues contribute unequally to the binding free energy; therefore, mutations in these two regions may not have the same consequences on PPI and binding affinity [15]. Non-interface residues might have allosteric effect, whereas the interface residues will have direct effect. Figure 5.1 shows the interface region between two proteins. The ubiquity of PPIs and their essential role in cellular processes have implicated PPIs as a prime prospect in modulating disease processes [16, 17]. For example, a thorough examination of the structural nature of variations in cancer showed that mutations at

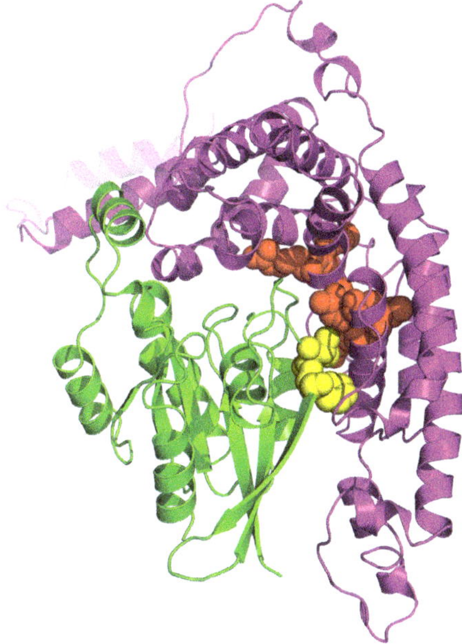

Figure 5.1. A representative interface between two proteins. Interface residues are shown as cpk model: VAL 36 and Phe 37 (yellow) in chain C (green color) and Tyr 1106, Arg 1107, Asn 1110, Val 1114, Ala 1142, Lys 1143 and Gln 1146 (red) in chain E (purple color) of Cdc42·GTP bound to the GTPase-activating protein (GAP)-related domain (GRD) of IQGAP2 (PDB ID is 5CJP).

protein–protein interface regions drive cancer development [18]. Protein interfaces are not uniform areas and are usually categorized into two distinct divisions: "core" and "rim" [19]. These two regions differ from each other in terms of physiochemical properties and evolutionary conservations. The interfaces of proteins have more point mutations compared to non-interfaces, which leads to diseases [13]. Teng *et al.* [20] conducted a study on multitude of protein complexes and uncovered that variations occurring at proteins' interfaces may change the binding free energy in protein complexes compared to benign single amino acid variations (SAVs). *In silico* and *in vitro* studies have revealed that energetic "hotspots" are residues that lead to a ≥ 2.0 kcal/mole reduction in binding energy after mutating to Ala. The hotspot residues are enriched in arginine, tyrosine and tryptophan [21, 22]. Change in binding free energy upon mutation is usually computed by the difference in binding energy between the mutant and the wild type: $\Delta\Delta G = \Delta G_m - \Delta G_w$, where the subscripts w and m, respectively, are the wild type and the mutant. Studies have disclosed that disease-causing mutations arise significantly in the interface's core as compared to the rim because interface core is well noted to establish PPIs affinity and stability. Among homologous proteins, the interface core residues are well conserved because of the peculiar functional and structural role like residues in hydrophobic core of proteins. Hence, amino acid substitutions in these regions are poorly tolerated compared with residue substitutions that occur in the rim and surface [13, 23]. So disease-causing amino acid substitutions are mostly found in buried and interface core than rim and non-interacting surface regions. On the other hand, although interface rim and non-interacting surface regions are not enriched in harmful amino acid substitutions, still their contribution to disease cannot be ignored [22].

Ozdemir *et al.* [24] performed a molecular dynamics (MD) simulation to investigate how a mutation in energetic hotspot residues of GRD2 (Ex-domain of GRD of IQGAP2) may affect its binding to Cdc42 and Rac1. Tyrosine 1106 is an interface residue (hotspot) in GRD2, and upon *in silico* mutation to Alanine (Y1106A), they observed a complete abolishment of interactions of GRD2 with both Cdc42 and Rac1. The binding free energy (ΔG) of the wild type (Cdc42–GRD2) was −43.1 ± 14 kcal/mole and that of the mutant complex was −10.9 ± 11.5 kcal/mole, hence weakening the interaction. Similarly, the ΔG for the wild type (Rac1-GRD2) was −8.1 ± 10.2 kcal/mole and for the mutant, it was +5.3 ± 11.8 kcal/mol, hence thermodynamically unfavorable. Figure 5.2 shows the wild-type and mutated complex structures. The mutated structure is obtained as a result of MD simulations. These results were consistent with their experimental findings, thus hypothesizing that the Tyr1106 residue is important, and its mutation provides a possible reason for the abolishment of GRD2 interactions with RAC1 and Cdc42.

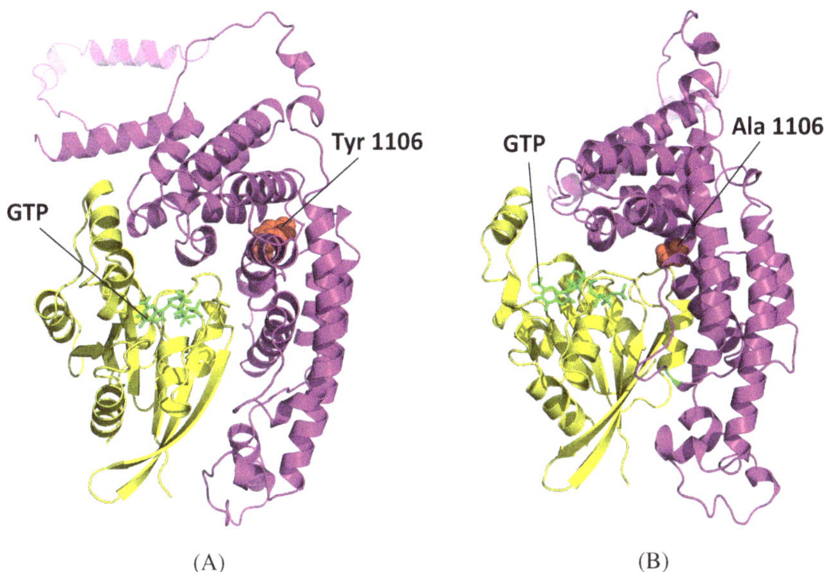

Figure 5.2. (A) The wild-type residue (Tyr1106) as cpk model colored red in chain E (purple), chain C (yellow) and GTP (stick model colored green) of Cdc42·GTP bound to the GRD of IQGAP2. (B) The mutant alanine residue: Tyr1106Ala is shown as cpk model, Colored red in the interface region between chains E (purple) and C (yellow), GTP (stick model colored green) of Cdc42·GTP bound to the GRD of IQGAP2 (PDB code 5CJP).

From a biological networks viewpoint, SAVs at protein interfaces can totally distort not only a specific protein but also entire biological pathway [13]. A large-scale structure analysis study by Alessia and Micheal [22] revealed that mutations in codons that code for Arginine result in the substitution of Arginine with amino acid residues with dissimilar chemical properties. Prtukh *et al.* [23], investigating the disease-causing mutations that occur frequently, found that a large proportion of mutations involved in the alteration from Arginine to Cysteine, Proline and Tryptophan. Among the most frequent delete-rious amino acid substitutions in proteins are alterations of Gly to Arg, Asp, Gln and Val [23]. Biochemically, Gly is the simplest amino acid; obviously, substitution of Gly to other amino acids can affect proteins' 3-D structures. The chemistry of Tryptophan is unique and has a functional role in PPIs contributing stacking interactions, and its substitution to other amino acids is poorly accepted. Due to its bulky side chain, replacements of Trp are not unlikely to create cavi-ties in protein's structure, which may affect structural stability and binding of protein [22].

Regardless of protein's "structural environment," mutations of interface residues in several PPIs have appeared to contribute signifi-cantly to diseases [22, 25]. Notwithstanding, for some protein inter-faces that lack characterized hotspots, for example, viral capsids, mutations of essential residues strikingly affect protein complex gath-ering [26]. For instance, Rincon *et al.* reported that mutation of a residue at "inter-pentamer interface" of "mouth-and-foot" disease may ablate entire virus assembly [27]. Additionally, Pappalardo *et al.* proclaimed that only a couple of mutations in Reston, the main Ebola viruses non-pathogenic in people deranges viral–human PPIs, and maybe re-established through mutation prompting contamination of human cells [28].

The consequences of PPI SAVs have been reviewed by Yates and Sternberg [29] from a structural perspective; they provide several examples that mutations at PPIs have consequences not only by dis-rupting interactions directly but also by changes in post-translational alteration locales, Parkinson's disease, for example [30], and inher-ently disordered areas, notwithstanding the outcomes caused by interface "specificity switching." Furthermore, *in vitro* and *in vivo* investigations inclusive of basic analysis have revealed that mutations at PPI interfaces prompt "shifting in interface specificity" [31]. The interface specificity switching further illustrates remarkability of variations at PPI interfaces with regard to impacting cellular processes. For example, "antibody-escape mutations" observed in HIV1 wherein Nabs (neutralizing antibodies) inhibit viral envelop development place constraint on HIV1 so that resistance to Nabs eventually occurs. The antibody-escape phenomenon for most part happens by specific mutations specifically ablating Nab binding directly or indirectly while maintaining the power to gather viral envelope [32–34]. Hence, such plasticity occurring at PPI interfaces further convolutes or prompts huge therapeutic challenges with respect to handling "drug resistance" at what are as of now difficult target proteins for medicinal drugs.

Ozdemir *et al.* [35] applied a statistical and structural analysis of SAVs in "singlet hotspots" in protein–protein interfaces. Using SKEMPI [36] and PIFACE [37] data sets, they observed 1643 out of 4441 SAVs that were associated with contacting residues of 172 proteins (234 unique chains). They identified 151 hot regions consisting of 885 amino acid residues of all the interfaces studied, corresponding to 5.66 residues per hot region (clusters of hotspots). Moreover, they observed that destabilizing SAVs occur significantly in "singlet hotspots," and "hot regions" as compared to "non-hotspot regions." Their result demonstrated that destabilizing SAV occurs in hotspots, particularly in singlet hotspots and neutral SAV maybe found in non-hotspots.

5.3. Mutation on PPIs and genetic diseases

Mutations in human genome lead to numerous genetic and Mendelian diseases. Biologists have undertaken several investigations to understand the nature of hereditary diseases [38]. These investigations have led to the establishment of databases, for example OMIM [39] and UniProt [40], altogether containing over 30,000 experimentally verified mutations. A multitude of known disease-causing variations is attributed to non-synonymous single nucleotide polymorphism (nsSNPs) in coding regions of genes. However, roles play by stop and nonsense mutations in hereditary diseases cannot be underestimated.

Bockler and Bateman [38] conducted a study to predict interaction hotspots and interface mutations that lead to diseases. In 264 protein complexes, they identified 1428 mutations linked with protein interactions using a set of criteria with 25,322 mutations derived from OMIM [39] and UniProt [40]. Approximately, 4% of the total mutations [25, 32] was associated with PPIs. Further, they were able to identify three diseases, Griscelli syndrome type 2, Adrenocorticotropin hormone deficiency (AHD) and Baller–Gerold syndrome (BGS), which are linked to changes in PPIs.

Griscelli syndrome type 2, a fatal disease, is often associated with abnormal hair and skin pigmentation in human. Occasionally, the disease has been considered as immunodeficiency due to the absence of gamma-globulin and inadequate lymphocyte stimulation [41]. Bocker and Bateman [38] reported that Trp73Gly substitution in Rab-27A resulted in the replacement of Trp which is well conservedand in the interface of the protein. Strong evidence has suggested that indeed Rab-27A associates with Myophillin, as a result, the substitution of Trp to Gly seems to vitally affect the transport of vesicle by reducing the binding affinity of Rab-27A to Myophillin [42].

AHD has been categorized by diminishment of the pituitary hormone, adrenocorticotropin, and steroids. The symptoms often associated with AHD include low blood pressure, anorexia, weight loss, among others. Lamolet *et al.* [43] have disclosed that Ser128Phe mutations in the "T-box" transcription regulator TBX19 leads to dominant loss of functional phenotype. A homolog of T-Box domain with PDB from Xenopus laevis Brachyury transcription regulator crystal structure has been resolved (PDB ID: 1XBR) [44]. It shares about 81% sequence identity with the human TBX19 protein. Analysis of the crystal structure revealed that the mutated residue falls right in the dimerization interface core. The above mutation substitutes serine, a polar amino acid, with an aromatic amino acid, phenylalanine. A research conducted by Pulichino *et al.* [45] showed that Ser128Phe

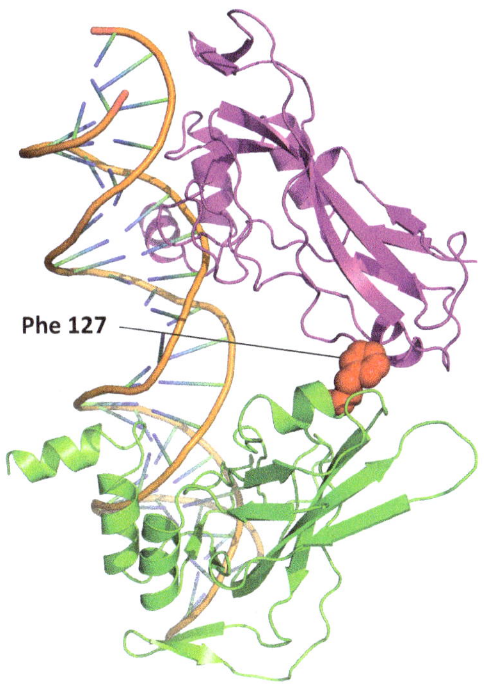

Figure 5.3. Mutation of a polar amino acid, serine, to an aromatic amino acid, phenylalanine (Ser127Phe). It falls right in the dimerization interface core between the chains A (green) and B (purple) of the T domain–DNA complex of the Brachyury transcription factor (PDB ID: 1XBR).

(Ser127Phe in PDB) substitution does not enhance DNA binding. Figure 5.3 displays the position of the Phe128 (Phe127 in PDB) at the dimer interface.

BGS is an uncommon congenital condition identified by distinct abnormality of skull, and bones of the hands and forearms. Phenotypically, BGS often imbricates with other diseases such as Rothmund–Thompson syndrome and Saethre–Chotzen syndrome (SCS). Seto *et al.* [46] reported a case of BGS that embodied features of SCS. In that study, they identified and reported that Ile156Val of H-Twist protein was responsible for SCS. Experimental studies with Y2H have proclaimed that losing H-Twist or dimerizing ability of E12 caused SCS [47].

5.4. Mutation in protein interfaces and cancer

Cancer is categorized by the accumulation of multitude of genetic variations and the alteration of a large number of biological phenomena [48]. Cancer has been considered as the most complicated disease in the history of mankind, and more than 200 different forms have been identified. Each type has unique descriptions and molecular profiles requiring different therapeutic strategies [49]. Mutations in interface residues of proteins involved in key biological pathways have been linked to cancer developments. In network analysis, node-and-edge description is routinely used; wherein nodes and edges are representations of proteins and their interactions, respectively. However, this simplified version of pathways is unable to provide structural interaction details; hence, construction of a structural pathway based on structural models of PPIs is preferred [50, 51]. To understand interleukin-10's (IL-10)'s structural network and its association with cancer and inflammation, Ozbabacan *et al.* [52] used several mutational databases, Web servers and a well-tested tool, Protein Interaction by Structural Matching (PRISM) [53] to construct structural pathway of IL-10. Approximately, 4% of all the mutations in the databases were predicted computationally to be associated with protein interactions. Further, they identified 879 missense and coding-silent mutations within COSMIC database [54] and were mapped to 29 target proteins whose structures are resolved and deposited into the PDB [55]. They noted four key protein interactions that were associated with oncogenesis in their reconstructed structural IL-10-centered network. Their findings were supported by experiments and computationally mutating each wild-type residue to its respective oncogenic mutant and further analyzed the consequences on authentic interaction. Of note, except for IL-10 an IL-10RA (IL-10 receptor A), they modeled all protein structures with PRISM, and their corresponding interface residues agreed with experimental findings. Tanikawa *et al.* showed the crucial role of IL-10 in cancer development. They propounded that the lack of IL-10, a cytokine, leads to an increase in the production of IL-1 promoting cancer in rat [56]. Also, mutations in IL-10 or its receptors can disturb possible interactions; thereby, hindering signaling process of IL-10. For example, nonsense mutation (E41) within IL-10RB terminates the interaction of IL-10 with IL-10RB, thus arresting IL-10 signaling process. The mutation above results in loss of most of IL-10RB and large region of the interface between IL-10 and IL-10RB. Moreover, analysis of The Cancer Genome Atlas (TCGA) [57] data has revealed that the above mutation occurred in lung adenocarcinoma with about 2% frequency [58]. Additionally, the substitution of R198W in IL-10RB from the COSMIC database abolishes IL-10RB and IL-10 interaction. The mutation has been noticed in "endometrioid carcinoma," and as reported by their study, the mutation falls directly alongside the interface of IL-10RB. Consequently, the blocking of IL-10 signaling can boost inflammation and increase regulatory T cells (T_{regs}) and myeloid-derived suppressor cells (MDSCs), which further inhibit cancer immunity, thus promoting tumor growth [59]. Computationally, they induced a nonsense mutation (Q56*) in IL-10, which is considered to be associated with lung adenocarcinoma [58, 60] that possibly disrupts the interaction of IL-10 with α-2 macroglobulin (A2M), facilitating IL-10 recruitment to the site of inflammation [60] and further protects it from proteolysis [61]. They observed that nonsense variation causes loss of interface between IL-10 and A2M in totality.

In a similar line, they executed computational analyses to observe the impact of mutation on A2M, and Beta-amyloid precursor protein (APP) interaction. Their results showed possible link between the dissolution of the interaction and the cancer. They realized that R945Q, associated with colorectal cancer [62], was within the interface of the predicted complex (A2M–APP). Further still, they noticed that R945 in human A2M is an energetic hotspot residue and its mutation to Gln resulted in alteration in binding free energy. Hence, weakening the interaction which further proposes to weaken A2M–APP interaction may promote carcinogenesis. The observation is backed with recent findings that APP is overexpressed in various cancers due to its growth-promoting ability [63, 64].

In addition, they scrutinized A2M and Kallikren-13 (KLK13) interaction structurally and their possible implication in the development of cancer. Withal, analysis of the COSMIC database disclosed that R263L mutation in KLK13 has been observed in carcinoma. Also, a study has shown KLK13 and A2M (serine-protease inhibitor) can form a complex [65]. However, PRISM was unable to predict favorable synergy between the mutant and the KLK13–A2M complex suggesting that this mutation disrupts the interaction. This result was consistent with another predictor, HotPoint server [66]. This observation led them to postulate that the disruption of the KLK13–A2M interaction promotes KLK13 interaction with other proteins in lieu of A2M, causing cleavage in considerable components of extracellular matrix and further promoting tumorigenesis, metastasis and invasion [52].

Engin *et al.* [67] undertook a similar computational task to investigate the involvement of PPIs in cancer metastasis. They generated phenotype-specific sub-networks of PPIs and identified appropriate edges to the seed genes (metastasis-causing genes) accountable for brain and lung metastasis in primary tumors (breast cancer). Their work was influenced by Massague and his colleagues [68] who earlier on identified these seed genes. The study was intended to discern molecular details of brain and lung metastasis in breast-cancer patients. Their findings after thorough analyses of protein interfaces in their networks and mutations occurring on relevant proteins within breast-cancer metastasis sub-network proposed that key protein–protein interfaces may modulate metastasis, wherein certain variation may selectively alter interactions. In their PPI networks, two interface-associated mutations were reported. After predicting modeled complexes of EGFR, ERBB4, HBEGF and EREG using PRISM, they realized that EGFR and ERBB4 interact with HBEGF as well as EREG. Indeed, a study has revealed that HBEGF gene plays a crucial role in brain metastasis in breast cancer [69], whereas EREG gene mediates lung metastasis in breast cancer [70]. The interface residues of their modeled complexes agreed and overlapped reasonably well when they were compared with the original complexes in PDB [55]. At 102th position EREG amino acid sequence requires an SNP (p.R102L) in few cancer patients derived from COSMIC and was found to be in the interface region of the predicted complex, EREG–ERBB4. Additionally, several other mutations were observed such as the five mutations in the interface of ERBB4 that are observed in cancer patients obtained from the

COSMIC database. These analyses and observations led them to suggest that these mutations may weaken or strengthen their interactions but most probably altering their functions. Taking altogether, they concluded that a link between metastasis development and observed mutations exists in the above proteins.

Along the same vein, they also observed that ELANE interacts with a seed gene, CSF3, in brain metastasis sub-networks (BMSN) [69], and interacts with VCAM1 in lung metastasis sub-networks [70], hence switching its interaction partners. ELANE was observed to have mutants that coincide within interfaces: pV98L, p.V101L, p.V10M and p.S126L — all derived from the UniProt. The various polymorphisms occurring at the 101th position correspond to energetic hotspots within interface region betwixt ELANE and CSF3 complex, whereas the polymorphisms at 98th and 126th positions were found within the "interface region" of VCAM1 and ELANE. Consequently, these observed variations in amino acid led them to insinuate that these mutations may affect the interactions of ELANE with both CSF3 and VCAM1, which may further be linked with growth of metastasis in breast cancer victims.

Along similar lines, Acuner-Ozabacan and co-workers [71] carried out a computational study on "structural pathway" of IL-1 signaling initiation to reveal oncogenic mutation mechanisms and SNPs in cancer and inflammation. Their aim was to construct a structural pathway of IL-1 and relate oncogenic mutations and SNPs to protein complexes within the pathways. They reported that the allotment of tumorigenic mutations and SNPs in the experimental and predicted structures of protein complexes in IL-1 constructed pathways was significantly found in interface regions and adjacent residues in that order. For example, they observed three oncogenic mutations and an SNP that were found directly on predicted interfaces of MKK4–JNK2 complex and JNK3 or at an adjacent residue. Further analysis uncovered MKK4 Ser251 and Gly268 in JNK2 were interface residues. The interface residues are essential as p.Ser251Asn links to "metastatic melanoma" and p.Gly268Ala, an SNP. Similarly, in MKK4–JNK3, Arg154, energetic-hotspot residue of MKK4, substituting to Trp (p.Arg154Trp) leads to adenocarcinoma. Furthermore, Gln142, a nearby residue of MKK4, substituting to Leu (p.Gln142Leu) links lung-squamous-cell carcinoma. Also, computationally induced mutations in predicted models and their relation to cancer developments were reported. Their findings resulted in hypothesizing: computational modeling of PPIs at multi-scale can elucidate accurate, structural and atomic-level details of signaling pathways at cellular level and may depict mechanism of "disease-causing" mutations [71].

5.5. Experimental measurement of mutational effects on PPIs strength

An efficient way to measure the effects of mutations on PPIs is to do an experiment. Intrinsically, the aim of understanding the consequences of variations on PPIs has led to the development of experimental methods. Here, we present descriptions of some of the experimental methods currently used for measuring the impact of amino acid variations on PPIs.

The isothermal titration calorimetry (ITC) is a proficient and direct method of measuring the effects of variants on the thermodynamics of binding [72]. The ITC has the potential to decompose the binding energy into both the entropic and enthalpic terms, which is not readily made available by other methods. A disadvantage of the ITC method is that it involves a lot of technicalities susceptible to experimental errors when the method is not attended meticulously.

The surface plasmon resonance (SPR) is an experimental method that can determine the variation in the "refractive index" that occurs when a protein binds to the sensor surface [73, 74]. In synopsis, SPR and other surface-based techniques are advantageous when determining K_{off} and K_{on} (kinetics rates) by closely monitoring the disappearance of signals as the protein disengages from the surface during wash cycle succeeding injection terminations. Thorough understanding of the association rate is important while studying interaction under kinetic controls in lieu of thermodynamic controls. For instance, several proteins may battle for the same receptor active site, in which case the association rate becomes essential.

Spectroscopy methods (SPMs) verify PPIs in solution. The requirement for K_d measurement of a protein complex is the method's potential to offer significant signal to noise at concentration near K_d value. Routinely, one component is held at a steady concentration while titrating the other component in stepwise manner for approximating the coupling parameters to fit the obtained data. The superiority of SPMs is labeling various interacting partners with NMR-active isotopes or fluorescent probes. Also, the same interaction can be designed *in vitro* using purified components, with background of other particles in the same solution mirroring *in vivo* environment. Typically, the expression of one or two collaborating partners tagged with different analogs of green fluorescent proteins verifies interaction *in vivo* [75].

5.6. Experimental databases

The wealth of data obtained from experimental methods has prompted the establishment of several manually curated databases to accumulate the changes in binding energy, binding kinetics, and thermodynamics upon mutations. The development, validation and evaluation of most computational methods have always relied on experimental data as a benchmark. Here, we present some reliable experimental databases.

SKEMPI [36, 76] serves as manually curated database encompassing information alterations in "thermodynamic parameters and kinetic rate constants" upon mutation at PPIs for proteins whose structures are resolved. The SKEMPI 2.0, current version, contains binding information on 7085 mutations, including alterations in kinetics for 1844, 443 and 440 mutations, respectively, for enthalpy and entropy changes which abrogate detectable binding. Also, the database consists of protein complexes; the number of mutations, the number of pairs of kinetic rate constants; the number of entropies and enthalpies; and the relevant PDB structure of the complex. Included also, is the proportion of mutations that are in the interior, surface, core, support and rim of the protein [77]. Finally, in the final columns of the database, the various experimental methods such as SPR, fluorescence (FL) and spectroscopic inhibition assay (IASP) are indicated. Of note, most of the data present in SKEMPI 2.0 were obtained from literature search, and 4%, 3% and 6%, respectively, from ABbind [78], PROXIMATE [79] and dbMPIKT [80]. SKEMP1.1 (first release) has been utilized in numerous computational studies, especially in developing scoring functions [81].

Like SKEMPI, PROXIMATE [79] is an experimental thermodynamic database containing missense mutations derived from literature and other sources. PROXIMATE accommodates 6296 mutations in 174 heterodimeric protein complexes including structures, functional information, linked with sequence, and structure databases and proteins' interaction networks. Although a vast majority of the entries in the database is single or alanine mutation, there are also multiple and "non-alanine" mutations. A peculiar feature of "PROXIMATE" is the provision of structural-and-network information, visualization of protein complexes and options to search, display, upload and download complexes. Importantly, "PROXIMATE" provides unbiased data set for training and validation. It serves as the basis for rapid development of algorithms that predict binding energy changes upon missense mutation. Moreover, the database can be used as a benchmark for studying disease-causing mutation in progression, diagnosis and treatment of numerous diseases, and provides drug targets for novel therapeutic strategies. Lastly, the database can serve as a source for identifying mutants that may exhibit an increase in binding energy to their interaction partners.

The dbMPIKT [80], kinetic and thermodynamic database of mutants' protein interactions, is derived from previous databases and curated literature. The database incorporates data from BID, SKEMPI, AB-bind and literature mining from PubMed, NCBI totaling 5291 mutations from 245 protein complexes. The protein complexes include heterodimeric complexes, enzyme inhibitors and antigen – antibodies. The dbMPIKT database contains information on mutation, which includes the original amino acid, chain identifier, specific location of mutant amino acid in sequence and name of mutant amino acid; kinetic data, which comprises association (K_{on}) and dissociation (K_{off}) rate constants; thermodynamic data, alterations ΔG and differences in $\Delta\Delta G$; the experimental methods SPR and ITC for determining the affinity of PPIs and the temperature — all of which servers as its unique attributes. An intriguing feature of the database is its capacity to assist researchers on hotspot prediction of PPIs, development of efficient scoring functions, drug discovery to target mutant PPIs and promote the study on PPIs because mutations in PPIs are often linked to human diseases, for example, Alzheimer's disease and cancer.

Antibody-Bind (AB-Bind), a reliable database, holds 1101 variants for experimentally determined $\Delta\Delta G$ of 32 protein complexes [78]. The database includes mutational information from antibody – antigen, antibody effector and antibody-like protein assemblies having resolved structures. Unlike other databases that include only alanine mutations, AB-Bind compiles both alanine and non-alanine mutations. Utilizing the data set in AB-Bind, performance of protein scoring potentials can be evaluated for predicting $\Delta\Delta G$ upon mutations. Additionally, the database allows computational benchmarking studies about existing approaches and may accelerate scoring function and algorithm developments. Table 5.1 depicts a summary of the mutational databases discussed above.

Table 5.1.   Summary of mutational databases.

Database	Number of protein complexes	Number of mutations	URL/Reference
SKEMPI 2.0	345	7085	https://life.bsc.es/pid/skempi2/
PROXIMATE	174	6296	http://www.iitm.ac.in/bioinfo/PROXiMATE/
dbMPIKT	245	5291	http://DeepLearner.ahu.edu.cn/web/dbMPIKT/
AB-bind	32	1101	https://github.com/sarahsirin/AB-Bind-Database

5.7. Computational methods

Studies of the impact of sequence variants on PPIs can provide a wealth of information for further understanding the nature of PPIs. From the early days in the study of effects of mutations on PPIs, the experimental techniques have been phenomenal in predicting the effects of variants on PPIs. Although some of the experimental techniques such as site-directed mutagenesis [82] are fast and inexpensive, other methods such as ITC and SPR are costly and labor-intensive. Consequently, a gamut of computational methods has been developed to rapidly predict mutational impacts on PPIs. Most methods rely on machine learning techniques, for instance "neural networks," "support vector machines" or "random forest" to combine different features that characterized amino acids and their vicinities [83–85]. The proteins' features include information on sequence conservation; structural, for example, solvent accessibility, number of contacts; secondary structures; physiochemical parameters such as electrostatic charge and hydrophobicity of a residue; or energetic parameters. The computational techniques can be grouped into sequence based, structure based or both.

Sequence-based method predictions heavily rely on conservation score that describes the recurrence of an amino acid being found at a specific location in protein or at the "genome-level" alignments [86–88]. The sequence-based methods remain the "de facto standard" to annotate recently discovered mutants due to their speed and scalability [89]. However, they have some liabilities such as their accuracy and kind of data being provided. For example, sequence-based method can only predict whether a certain variation maybe deleterious and withhold further information on why the variation may be harmful. Consequently, acting upon their predictions is tedious, for instance, by designing small molecules to inhibit the effect of harmful mutations or mutations found in cancer cells [4].

Structure-based methods usually predict consequences of variations on structure and function of protein. They use key attributes that describe the 3-D structures of protein. Structure-based methods seem to be accurate; however, they are not computationally inexpensive due to free energy calculations. The free energy calculations require modeling "structural transition" from the wild-type protein to the variant protein utilizing various integration methods for calculating the transitional state energy [90, 91]. Structure-based methods should have leverage on predicting the impact of missense mutations than sequence-based methods, because the effects are directly linked to alterations in the structures and functions of proteins. However, structure-based methods rely on the crystal structures of the proteins to be mutated manually [4]. Due to limitations and discrepancies in sequence-based and structure-based techniques, the scope of methods has been developed combining both methods to accurately predict deleterious and structural impacts of mutations on PPIs [92, 93]. These methods have been termed "meta-predictor," because they combine results of various sequence-based and structure-based methods using "machine learning" algorithms trained on relevant data set [36, 94]. One example is ELASPIC [95]. Below are some of these computational methods described above to assess consequences of mutations on PPIs, some of which are sequence based, structure based or both.

MutaBind is an authentic computational method and Web server for predicting effect of missense mutation at PPIs [96]. MutaBind uses a "molecular mechanics" force field, "statistical potentials" as well as fast "side-chain optimization" algorithms to predict effects of point mutations on PPIs. The predictor assesses the consequences of sequence variations and "disease-causing mutations" at interface and non-interface of PPIs, and quantifies the quantitative alterations in binding affinity upon missense mutation. A mutation has the potential to alter the components of PPIs, affect solvation of complex, exchange folding free energy of interacting proteins and demoralize binding hotspot regions. MutaBind utilizes key features of the protein such as the Vander Waals interaction energies for wild-type and mutant protein complexes, and solvent accessible areas that significantly contribute to $\Delta\Delta G$. The predictor has the propensity to produce model-mutant proteins and approximate confidence of its predictions. Furthermore, the predictor considers a mutation as deleterious if $\Delta\Delta G$ is greater than or equal to 1.5 kcal/mole. Also, if $\Delta\Delta G$ is (+), the associated mutation is said to be destabilizing thus decreasing binding affinity. On the other hand, if $\Delta\Delta G$ is negative which leads to an increase in binding affinity, then the mutation is stabilizing and neutral otherwise. MutaBind has been well tested and validated with numerous types of "cross-validation" and autonomous test sets from "26th Critical Assessment of Predicted Interactions," CAPRI [97] compared to other techniques. MutaBind has wide applications in protein biology such as determining "potential-driver mutations" in cancer, assessing impact of mutations on the survivability of protein in evolution, designing a potential drug molecule for a target enzyme and protein design.

FoldX [98] predicts effect of amino acid change on stability, folding, binding affinity and, "dynamics" of proteins and nucleic acids using empirical, "physics-based" force field considering negligible backbone shifting. FoldX force field consists of linear combination of empirical expressions to calculate ΔG. The empirical expressions include the desolvation term contributions from hydrophobic and polar groups, VdW and electrostatic terms. A distinctive feature of FoldX is its crude entropy calculation to obtain measure of ΔG. FoldX can be used in target-oriented mutagenesis to identify beneficial mutation sites that enhance protein stability [98, 99].

ELASPIC is a genome era ensemble machine learning approach for predicting stability effects upon mutations in "domain cores and domain-domain interfaces." ELASPIC integrates semi-empirical energy terms, sequence conservation and several molecular features with a "Stochastic Gradient Boosting of Decision Trees," SGDT [100] algorithms for accurate prediction. Unlike other methods that are dependent only on sequence and structure information for predictions, ELASPIC also incorporates a homology modeling technique for proteome-wide usage. The predictor has been evaluated and validated to differentiate benign mutations from disease-associated mutations. ELASPIC not only shows supremacy in predictions but also opens a state-of-art perspective in proteome-wide identification of disease mutations, prediction of stability alterations in proteins cores and interfaces, and unravels the molecular details behind genetic variations. The predictor has "user-friendly" Web server online for predicting consequences of mutations on protein's stability and binding affinity of protomer whose homology model maybe generated [95, 101].

SIFT algorithm is one of the earlier algorithms developed to estimate potential effect of amino acid substitutions on function and stability of protein. The SIFT algorithm is based on sequence homology to compute a probability that amino acid replacement will have adverse consequence on protein's function and binding affinity. The primary assumption is evolutionary-conserved regions in protein-like interface tend to be less tolerant of mutations. As a result, any amino acid insertion, substitution or deletion in these regions has a high tendency to affect protein function and stability [87]. Two prominent data sets, the human variation (HumVar) and human divergence (HumDiv) from UniProtKB [40] were used to evaluate the performance of SIFT. SIFT has broad applications in genetic research, Mendelian disease, cancer, infectious diseases and beyond. For example, Alessia and Michael [22] used SIFT to evaluate the residue conservation and physiochemical properties of "disease-causing" SAVs and "polymorphisms" occurring in different locations in proteins. They reported that SAVs considered as inimical were remarkably higher for disease-related SAVS situated in buried regions in comparison to disease-related SAVs in different regions, whereas number of detrimental SAVs considered tolerant was remarkably higher for surface regions in comparison to disease-related SAVs in different regions.

BeAtMuSic is one of the top-performer methods that predict alterations in protein's binding affinity on mutations. As coarse-grained predictor, BeAtMuSic heavily depends on "statistical potentials" obtained from resolved protein's structures. It provides impact of mutation depending on strength of the interaction at the interface as well as stability of entire protein complex. The predictor uses "coarse-grained" representations of the protein structures, which enhances fast evaluation of all possible mutations that could occur in the protein complex [102].

Polyphen2 [86] is a Web server to predict effects of disease-associated amino acid replacements on protein's stability and function. Considering structural and comparative evolutionary nature of proteins, Polyphen2 can estimate the probability of missense mutation to be deleterious based on the combination of properties. Polyphen2 can analyze huge data such as data produced by Next-generation Sequencing (NGS) projects.

BindProfX [103] is a predictor developed for understanding effects of amino acid variations on binding affinity of protein–protein interfaces at gene level. The inclusion of statistical energy function based on Boltzmann distribution is the distinctive feature of BindProfX compared to the previous version, BindProf. The approach here is based on structural-based profiles of protein–protein interfaces and $\Delta\Delta G$ is often computed as logarithm of relative probability of variant's residues over wild-type residues in "interface alignment matrix."

SAAMBE is a combination of sequence-based and structural-based methods for predicting $\Delta\Delta G$ and alteration of respective energy components. The Web server of SAAMBE [104] provides downloadable "energy-minimized" 3-D model of wild-type and mutant-type proteins. SAAMBE performance was evaluated against more than 1300 mutations in 43 proteins from SKEMPI with experimentally determined changes in ΔG. The "Pearson correlation coefficient" of the predictor was 0.62. Elsewhere, other computational techniques have been developed and the number of predictors continues to increase. In Table 5.2, we provide key features and URL address of computational methods discussed above.

5.7.1. *The drawbacks to computational methods*

The limitations of these bona-fide-computational methods cannot be disregarded, despite being primed in predicting the effects of mutations on PPIs. A major bottleneck with existing methods is their ineptitude to predict with speed and accuracy, the effect of $\Delta\Delta G$ upon mutation [105]. For example, BeAtMuSiC

Table 5.2. Key features and URL for computational methods.

Method	Features	URL
MutaBind	Conservation score, solvation energy, VdW energy, ΔG alteration due to unfolding, SASA and proline's structural constraints	http://www.ncbi.nlm.nih.gov/projects/mutabind/.
FoldX	Empirical, physics-based force field, desolvation, Van der Waals energy, electrostatic, hydrophobicity and polarity	http://foldx.embl.de/
ELASPIC	Intermolecular and intramolecular contacts, sequence conservation, SASA and FoldX-generated energy descriptors, homology modeling, and, Stochastic Gradient of decision Trees	http://elaspic.kimlab.org
SIFT	Sequence homology and evolutionary conservation	http://sift-dna.org
BeAtMuSiC	Statistical potentials derived from resolved protein structures, and coarse-grained proteins structure representations	http://babylone.ulb.ac.be/beatmusic
Polyphen 2	Structural and comparative evolutionary nature of proteins	http://genetics.bwh.harvard.edu/pph2/
BindProfX	Interface profiles scores, interface alignment matrix, log-odd likelihood scores and statistical energy derived from Boltzmann distribution	http://zhanglab.ccmb.med.umich.edu/BindProfX/
SAAMBE	Entropy, hydrophobicity, hydrogen bonds, SASA, VdW and interface area	http://compbio.clemson.edu/saambe_webserver/

[102] does predict $\Delta\Delta G$ within a few seconds, but is prone to immense error as reviewed by Gromiha *et. al.* [105]. On the other hand, methods such as ELASPIC and MutaBind can provide better accurate prediction but require a longer time, thus computationally expensive. In the same vein, BindProf [15] is biased in predicting the effect of mutations in complex structure because of its inability to automatically detect the interacting chains. As a result, BindProf can only predict $\Delta\Delta G$ on one side

of the protein interface. Another issue hindering computational techniques is their constant maintenance. For example, the binding interface database (BID) is no longer maintained [106]. Finally, some of these methods do not provide a very user-friendly environment. The needs for algorithm developers to combat these bottlenecks (speed, accuracy, maintenance and user-friendly interface) of available methods are essential for not only improving their performances but also developing new methods to predict $\Delta\Delta G$ in protein complexes.

Numerous mutational databases have been established as repositories for various forms of mutation occurring in the human population. These databases are comprehensive and provide in-depth understanding of the causes and effects of mutations. They have benchmarked several computational techniques and studies that strive to mirror experimental methods. In this section, we provide brief descriptions of the key mutational databases that are essential to the research community (Table 5.3).

Table 5.3.   General mutational databases.

Database	Compositions	Number of mutation	URL
OMIM	Comprehensive repository and curated information on genes and phenotypes	More than 15,500 genes, 26,200 allelic mutations and 7800 hereditary phenotypes	(https://omim.org
HGMD	Comprehensive archive of germline mutations in nuclear genes	Over 141,000 different lesions discovered in more than 5700 different genes	http://www.hgmd.org
COSMIC	All-inclusive depository on somatic mutations in cancer from over 26,000 curated scientific papers	6 million coding mutations from 1.4 million tumor samples	https://cancer.sanger.ac.uk
TCGA	Catalog and discover major "cancer-causing" genome alterations through large-scale genome sequence and integrated multi-dimensional analyses	Over 11,000 tumors from 33 of most prevalent forms of cancer	https://www.genome.gov/17516564/the-cancer-genome-atlas/

5.8. Conclusion

Understanding the molecular impacts of mutations at PPI interfaces is significant because of the central role PPIs play in biological processes. The widespread "disease-causing" mutations that change PPIs make mutations occurring at PPIs as key targets for therapeutic intervention. Resultingly, gamut of computational techniques has been developed to assess repercussion of mutations on PPIs by predicting $\Delta\Delta G$ but, is limited in the computing power needed for conformational sampling and scoring and is yet to mirror the experimental methods. For example, several computational techniques are noted for predicting effect of binding affinity change upon mutation as reported in scientific literature, but their performances showed that they have not reached the black box level; hence, there is a need for improvement in their performances. Also, the prevalence of mutations especially the ones that alter PPI interfaces has led to establishing mutational databases such as ABbind, SKEMPI, OMIM and COSMIC, benchmarking several computational methods. Mutations at protein–protein interfaces have contributed eminently to several diseases including cancer, genetic disorder, among others. Structural knowledge and quantification of the consequences of PPIs will enhance the design of better PPI repressors and stabilizers for targeting definite disease states and addressing situations where further variations in interface promote drug resistance.

References

1. Keskin, O., Gursoy, A., Ma, B., *et al.* Principles of protein–protein interactions: What are the preferred ways for proteins to interact? *Chem Rev.* 2008;108(4):1225–1244.
2. Chelliah, V., Chen, L., Blundell, T.L., *et al.* Distinguishing structural and functional restraints in evolution in order to identify interaction sites. *J Mol Biol.* 2004;342(5):1487–1504.
3. Jubb, H.C., Pandurangan, A.P., Turner, M.A., *et al.* Mutations at protein–protein interfaces: Small changes over big surfaces have large impacts on human health. *Prog Biophys Mol Biol.* 2017;128:3–13.

4. Strokach, A., Corbi-Verge, C., Teyra, J., *et al.* Predicting the effect of mutations on protein folding and protein–protein interactions. *Methods Mol Biol.* 2019;1851:1–17.

5. Nagasundaram, N., Zhu, H., Liu, J., *et al.* Analysing the effect of mutation on protein function and discovering potential inhibitors of CDK4: Molecular modelling and dynamics studies. *PLOS ONE.* 2015;10(8):e0133969.

6. Fong, J.H., Shoemaker, B.A., and Panchenko, A.R. Intrinsic protein disorder in human pathways. *Mol Biosyst.* 2012;8(1):320–326.

7. van der Lee, R., Buljan, M., Lang, B., *et al.* Classification of intrinsically disordered regions and proteins. *Chem Rev.* 2014;114(13): 6589–6631.

8. Bowler, E.H., Wang, Z., and Ewing, R.M. How do oncoprotein mutations rewire protein–protein interaction networks? *Expert Rev Proteomics.* 2015;12(5):449–455.

9. Muller, P.A.J., and Vousden, K.H. p53 mutations in cancer. *Nat Cell Biol.* 2013;15(1):2–8.

10. Kreeger, P.K., and Lauffenburger, D.A. Cancer systems biology: A network modeling perspective. *Carcinogenesis.* 2010;31(1):2–8.

11. Kreeger, P.K., Mandhana, R., Alford, S.K., *et al.* RAS mutations affect tumor necrosis factor-induced apoptosis in colon carcinoma cells via ERK-modulatory negative and positive feedback circuits along with non-ERK pathway effects. *Cancer Res.* 2009;69(20):8191–8199.

12. Nishi, H., Tyagi, M., Teng, S., *et al.* Cancer missense mutations alter binding properties of proteins and their interaction networks. *PLOS ONE.* 2013;8(6):e66273.

13. David, A., Razali, R., Wass, M.N., *et al.* Protein–protein interaction sites are hot spots for disease-associated nonsynonymous SNPs. *Hum Mutat.* 2012;33(2):359–363.

14. Winter, A., Higueruelo, A.P., Marsh, M., *et al.* Biophysical and computational fragment-based approaches to targeting protein–protein interactions: Applications in structure-guided drug discovery. *Q Rev Biophys.* 2012;45(4):383–426.

15. Brender, J.R., and Zhang, Y. Predicting the effect of mutations on protein–protein binding interactions through structure-based interface profiles. *PLoS Comput Biol.* 2015;11(10):e1004494.

16. Strong, M., and Eisenberg, D. The protein network as a tool for finding novel drug targets. *Prog Drug Res.* 2007;64:191, 3–215.

17. Blundell, T.L., Burke, D.F., Chirgadze, D., *et al.* Protein–protein interactions in receptor activation and intracellular signalling. *Biol Chem.* 2000;381(9–10):955–959.

18. Porta-Pardo, E., Garcia-Alonso, L., Hrabe, T., *et al.* A pan-cancer catalogue of cancer driver protein interaction interfaces. *PLoS Comput Biol.* 2015;11(10):e1004518.

19. Bahadur, R.P., Chakrabarti, P., Rodier, F., *et al.* Dissecting subunit interfaces in homodimeric proteins. *Proteins.* 2003;53(3):708–719.

20. Teng, S., Madej, T., Panchenko, A., *et al.* Modeling effects of human single nucleotide polymorphisms on protein–protein interactions. *Biophys J.* 2009;96(6):2178–2188.

21. Thorn, K.S., and Bogan, A.A. ASEdb: A database of alanine mutations and their effects on the free energy of binding in protein interactions. *Bioinformatics.* 2001;17(3):284–285.

22. David, A., and Sternberg, M.J. The contribution of missense mutations in core and rim residues of protein–protein interfaces to human disease. *J Mol Biol.* 2015;427(17):2886–2898.

23. Petukh, M., Kucukkal, T.G., and Alexov, E. On human disease — causing amino acid variants: Statistical study of sequence and structural patterns. *Hum Mutat.* 2015;36(5):524–534.

24. Ozdemir, E.S., Jang, H., Gursoy, A., *et al.* Unraveling the molecular mechanism of interactions of the Rho GTPases Cdc42 and Rac1 with the scaffolding protein IQGAP2. *J Biol Chem.* 2018;293(10):3685–3699.

25. Nemethova, M., Radvanszky, J., Kadasi, L., *et al.* Twelve novel HGD gene variants identified in 99 alkaptonuria patients: Focus on 'black bone disease' in Italy. *Eur J Hum Genet.* 2016;24(1):66–72.

26. White, R.R., Ponsford, A.H., Weekes, M.P., *et al.* Ubiquitin-dependent modification of skeletal muscle by the parasitic nematode, Trichinella spiralis. *PLoS Pathog.* 2016;12(11):e1005977.

27. Rincon, V., Rodriguez-Huete, A., and Mateu, M.G. Different functional sensitivity to mutation at intersubunit interfaces involved in consecutive stages of foot-and-mouth disease virus assembly. *J Gen Virol.* 2015;96(9):2595–2606.

28. Pappalardo, M., Julia, M., Howard, M.J., *et al.* Conserved differences in protein sequence determine the human pathogenicity of Ebolaviruses. *Sci Rep.* 2016;6:23743.

29. Yates, C.M., and Sternberg, M.J. The effects of non-synonymous single nucleotide polymorphisms (nsSNPs) on protein–protein interactions. *J Mol Biol.* 2013;425(21):3949–3963.

30. Muda, K., Bertinetti, D., Gesellchen, F., *et al.* Parkinson-related LRRK2 mutation R1441C/G/H impairs PKA phosphorylation of LRRK2 and disrupts its interaction with 14-3-3. *Proc Natl Acad Sci.* 2014;111(1):E34–E43.

31. Kortemme, T., Joachimiak, L.A., Bullock, A.N., *et al.* Computational redesign of protein–protein interaction specificity. *Nat Struct Mol Biol.* 2004;11(4):371–379.

32. Wei, X., Decker, J.M., Wang, S., *et al.* Antibody neutralization and escape by HIV-1. *Nature.* 2003;422(6929):307–312.

33. Mascola, J.R. The cat and mouse of HIV-1 antibody escape. *PLoS Pathog.* 2009;5(9):e1000592.

34. Pires, D.E., and Ascher, D.B. mCSM-AB: A web server for predicting antibody-antigen affinity changes upon mutation with graph-based signatures. *Nucleic Acids Res.* 2016;44(W1):W469–W473.

35. Ozdemir, E.S., Gursoy, A., and Keskin, O. Analysis of single amino acid variations in singlet hot spots of protein–protein interfaces. *Bioinformatics.* 2018;34(17):i795–i801.

36. Moal, I.H., and Fernandez-Recio, J. SKEMPI: A Structural Kinetic and Energetic database of Mutant Protein Interactions and its use in empirical models. *Bioinformatics.* 2012;28(20):2600–2607.

37. Cukuroglu, E., Gursoy, A., Nussinov, R., *et al.* Non-redundant unique interface structures as templates for modeling protein interactions. *PLOS ONE.* 2014;9(1):e86738.

38. Schuster-Bockler, B., and Bateman, A. Protein interactions in human genetic diseases. *Genome Biol.* 2008;9(1):R9.

39. Amberger, J., Bocchini, C.A., Scott, A.F., *et al.* Online Mendelian Inheritance in Man (OMIM). *Nucleic Acids Res.* 2009;37(Database issue):D793–D796.

40. UniProt, C. The universal protein resource (UniProt). *Nucleic Acids Res.* 2007;35(Database issue):D193–D197.

41. Klein, C., Philippe, N., Le Deist, F., *et al.* Partial albinism with immunodeficiency (Griscelli syndrome). *J Pediatr.* 1994;125(6 Pt 1):886–895.

42. Strom, M., Hume, A.N., Tarafder, A.K., *et al.* A family of Rab27-binding proteins. Melanophilin links Rab27a and myosin Va function in melanosome transport. *J Biol Chem.* 2002;277(28):25423–25430.

43. Lamolet, B., Pulichino, A.M., Lamonerie, T., *et al.* A pituitary cell-restricted T box factor, Tpit, activates POMC transcription in cooperation with Pitx homeoproteins. *Cell.* 2001;104(6):849–859.

44. Muller, C.W., and Herrmann, B.G. Crystallographic structure of the T domain-DNA complex of the Brachyury transcription factor. *Nature.* 1997;389(6653):884–888.

45. Pulichino, A.M., Vallette-Kasic, S., Couture, C., *et al.* Human and mouse TPIT gene mutations cause early onset pituitary ACTH deficiency. *Genes Dev.* 2003;17(6):711–716.

46. Seto, M.L., Lee, S.J., Sze, R.W., *et al.* Another TWIST on Baller-Gerold syndrome. *Am J Med Genet.* 2001;104(4):323–330.

47. El Ghouzzi, V., Legeai-Mallet, L., Aresta, S., *et al.* Saethre-Chotzen mutations cause TWIST protein degradation or impaired nuclear location. *Hum Mol Genet.* 2000;9(5):813–819.

48. Hanahan, D., and Weinberg, R.A. Hallmarks of cancer: The next generation. *Cell.* 2011;144(5):646–674.

49. Hanahan, D., and Weinberg, R.A. The hallmarks of cancer. *Cell.* 2000;100(1):57–70.

50. Mosca, R., Ceol, A., and Aloy, P. Interactome3D: Adding structural details to protein networks. *Nat Methods.* 2013;10(1):47–53.

51. Tuncbag, N., Gursoy, A., and Keskin, O. Prediction of protein–protein interactions: Unifying evolution and structure at protein interfaces. *Phys Biol.* 2011;8(3):035006.

52. Acuner-Ozbabacan, E.S., Engin, B.H., Guven-Maiorov, E., *et al.* The structural network of Interleukin — 10 and its implications in inflammation and cancer. *BMC Genomics.* 2014;15(suppl 4):S2.

53. Ogmen, U., Keskin, O., Aytuna, A.S., *et al.* PRISM: Protein interactions by structural matching. *Nucleic Acids Res.* 2005;33(Web Server issue):W331–W336.

54. Tate, J.G., Bamford, S., Jubb, H.C., *et al.* COSMIC: The catalogue of somatic mutations in cancer. *Nucleic Acids Res.* 2019;47(D1):D941-D947.

55. Berman, H.M., Westbrook, J., Feng, Z., *et al.* The protein data bank. *Nucleic Acids Res.* 2000;28(1):235–242.

56. Tanikawa, T., Wilke, C.M., Kryczek, I., *et al.* Interleukin — 10 ablation promotes tumor development, growth, and metastasis. *Cancer Res.* 2012;72(2):420–429.

57. Tomczak, K., Czerwinska, P., and Wiznerowicz, M. The Cancer Genome Atlas (TCGA): An immeasurable source of knowledge. *Contemp Oncol (Pozn).* 2015;19(1A):A68–A77.

58. Cerami, E., Gao, J., Dogrusoz, U., *et al.* The cBio cancer genomics portal: An open platform for exploring multidimensional cancer genomics data. *Cancer Discov.* 2012;2(5):401–404.

59. Merlo, P., and Cecconi, F. XIAP: Inhibitor of two worlds. *EMBO J.* 2013;32(16):2187–2188.

60. Borth, W. Alpha 2-macroglobulin. A multifunctional binding and targeting protein with possible roles in immunity and autoimmunity. *Ann N Y Acad Sci.* 1994;737:267–272.

61. Zhevago, N.A., and Samoilova, K.A. Pro- and anti-inflammatory cytokine content in human peripheral blood after its transcutaneous (in vivo) and direct (in vitro) irradiation with polychromatic visible and infrared light. *Photomed Laser Surg.* 2006;24(2):129–139.

62. Forbes, S.A., Bindal, N., Bamford, S., *et al.* COSMIC: Mining complete cancer genomes in the catalogue of somatic mutations in cancer. *Nucleic Acids Res.* 2010;39(Database issue):D945–D950.

63. Meng, J.Y., Kataoka, H., Itoh, H., *et al.* Amyloid beta protein precursor is involved in the growth of human colon carcinoma cell in vitro and in vivo. *Int J Cancer.* 2001;92(1):31–39.

64. Cavga, A., Karahan, N., Keskin, O., *et al.* Taming oncogenic signaling at protein interfaces: Challenges and opportunities. *Curr Top Med Chem.* 2015;15(20):2005–2018.

65. Kapadia, C., Yousef, G.M., Mellati, A.A., *et al.* Complex formation between human kallikrein 13 and serum protease inhibitors. *Clin Chim Acta.* 2004;339(1–2):157–167.

66. Tuncbag, N., Keskin, O., and Gursoy, A. HotPoint: Hot spot prediction server for protein interfaces. *Nucleic Acids Res.* 2010;38(Web Server issue):W402–W406.

67. Engin, H.B., Guney, E., Keskin, O., *et al.* Integrating structure to protein–protein interaction networks that drive metastasis to brain and lung in breast cancer. *PLOS ONE.* 2013;8(11):e81035.

68. Kim, M.Y., Oskarsson, T., Acharyya, S., *et al.* Tumor self-seeding by circulating cancer cells. *Cell.* 2009;139(7):1315–1326.

69. Bos, P.D., Zhang, X.H., Nadal, C., *et al.* Genes that mediate breast cancer metastasis to the brain. *Nature.* 2009;459(7249):1005–1009.

70. Minn, A.J., Gupta, G.P., Siegel, P.M., *et al.* Genes that mediate breast cancer metastasis to lung. *Nature.* 2005;436(7050):518–524.

71. Acuner Ozbabacan, S.E., Gursoy, A., Nussinov, R., *et al.* The structural pathway of interleukin 1 (IL-1) initiated signaling reveals mechanisms of oncogenic mutations and SNPs in inflammation and cancer. *PLoS Comput Biol.* 2014;10(2):e1003470.

72. Velazquez-Campoy, A., Leavitt, S.A., and Freire, E. Characterization of protein–protein interactions by isothermal titration calorimetry. *Methods Mol Biol.* 2004;261:35–54.

73. Douzi, B. Protein–protein interactions: Surface plasmon resonance. *Methods Mol Biol.* 2017;1615:257–275.

74. Zhao, H., and Beckett. D. Kinetic partitioning between alternative protein–protein interactions controls a transcriptional switch. *J Mol Biol.* 2008;380(1):223–236.

75. Berggard, T., Linse, S., and James, P. Methods for the detection and analysis of protein–protein interactions. *Proteomics.* 2007;7(16):2833–2842.

76. Jankauskaite, J., Jimenez-Garcia, B., Dapkunas, J., *et al.* SKEMPI 2.0: An updated benchmark of changes in protein–protein binding energy, kinetics and thermodynamics upon mutation. *Bioinformatics.* 2019;35(3):462–469.

77. Levy, E.D. A simple definition of structural regions in proteins and its use in analyzing interface evolution. *J Mol Biol.* 2010;403(4): 660–670.

78. Sirin, S., Apgar, J.R., Bennett, E.M., *et al.* AB-Bind: Antibody binding mutational database for computational affinity predictions. *Protein Sci.* 2016;25(2):393–409.

79. Jemimah, S., Yugandhar, K., and Michael Gromiha, M. PROXiMATE: A database of mutant protein–protein complex thermodynamics and kinetics. *Bioinformatics.* 2017;33(17):2787–2788.

80. Liu, Q., Chen, P., Wang, B., *et al.* dbMPIKT: A database of kinetic and thermodynamic mutant protein interactions. *BMC Bioinformatics.* 2018;19(1):455.

81. Moal, I.H., and Fernandez-Recio, J. Intermolecular contact potentials for protein–protein interactions extracted from binding free energy changes upon mutation. *J Chem Theory Comput.* 2013;9(8):3715–3727.

82. Xu, Z., Colosimo, A., and Gruenert, D.C. Site-directed mutagenesis using the megaprimer method. *Methods Mol Biol.* 2003;235:203–207.

83. Capriotti, E., Fariselli, P., and Casadio, R. A neural-network-based method for predicting protein stability changes upon single point mutations. *Bioinformatics.* 2004;20(suppl 1):i63–i68.

84. Cheng, J., Randall, A., and Baldi, P. Prediction of protein stability changes for single-site mutations using support vector machines. *Proteins.* 2006;62(4):1125–1132.

85. Capriotti, E., Fariselli, P., Calabrese, R., *et al.* Predicting protein stability changes from sequences using support vector machines. *Bioinformatics.* 2005;21(suppl 2):ii54–ii58.

86. Adzhubei, I., Jordan, D.M., and Sunyaev, S.R. Predicting functional effect of human missense mutations using PolyPhen-2. *Curr Protoc Hum Genet.* 2013;Chapter 7:Unit7.20.

87. Sim, N.L., Kumar, P., Hu, J.,*et al.* SIFT web server: Predicting effects of amino acid substitutions on proteins. *Nucleic Acids Res.* 2012;40(Web Server issue):W452–W457.

88. Li, B., Krishnan, V.G., Mort, M.E., *et al.* Automated inference of molecular mechanisms of disease from amino acid substitutions. *Bioinformatics.* 2009;25(21):2744–2750.

89. Dorfman, R., Nalpathamkalam, T., Taylor, C., *et al.* Do common in silico tools predict the clinical consequences of amino-acid substitutions in the CFTR gene? *Clin Genet.* 2010;77(5):464–473.

90. Benedix, A., Becker, C.M., de Groot, B.L., *et al.* Predicting free energy changes using structural ensembles. *Nat Methods.* 2009;6(1):3–4.

91. Shirts, M.R., and Mobley, D.L. An introduction to best practices in free energy calculations. *Methods Mol Biol.* 2013;924:271–311.

92. Dehouck, Y., Grosfils, A., Folch, B., *et al.*. Fast and accurate predictions of protein stability changes upon mutations using statistical potentials and neural networks: PoPMuSiC-2.0. *Bioinformatics.* 2009;25(19):2537–2543.

93. Baugh, E.H., Simmons-Edler, R., Muller, C.L., *et al.* Robust classification of protein variation using structural modelling and large-scale data integration. *Nucleic Acids Res.* 2016;44(6):2501–2513.

94. Kumar, M.D., Bava, K.A., Gromiha, M.M., *et al.* ProTherm and ProNIT: Thermodynamic databases for proteins and protein-nucleic acid interactions. *Nucleic Acids Res.* 2006;34(Database issue): D204–D206.

95. Witvliet, D.K., Strokach, A., Giraldo-Forero, A.F., *et al.* ELASPIC web-server: Proteome-wide structure-based prediction of mutation effects on protein stability and binding affinity. *Bioinformatics.* 2016;32(10):1589–1591.

96. Li, M., Simonetti, F.L., Goncearenco, A., *et al.* MutaBind estimates and interprets the effects of sequence variants on protein–protein interactions. *Nucleic Acids Res.* 2016;44(W1):W494–W501.

97. Janin, J., Henrick, K., Moult, J., *et al.* CAPRI: A critical assessment of PRedicted interactions. *Proteins.* 2003;52(1):2–9.

98. Schymkowitz, J., Borg, J., Stricher, F., *et al.* The FoldX web server: An online force field. *Nucleic Acids Res.* 2005;33(Web Server issue): W382–W388.

99. Buss, O., Rudat, J., and Ochsenreither, K. FoldX as protein engineering tool: Better than random based approaches? *Comput Struct Biotechnol J.* 2018;16:25–33.

100. Guo, H., Wang, J., Ao, W., *et al.* SGB-ELM: An advanced stochastic gradient boosting-based ensemble scheme for extreme learning machine. *Comput Intell Neurosci.* 2018;2018:4058403.

101. Berliner, N., Teyra, J., Colak, R., *et al.* Combining structural modeling with ensemble machine learning to accurately predict protein fold stability and binding affinity effects upon mutation. *PLOS ONE.* 2014;9(9):e107353.

102. Dehouck, Y., Kwasigroch, J.M., Rooman, M., *et al.* BeAtMuSiC: Prediction of changes in protein–protein binding affinity on mutations. *Nucleic Acids Res.* 2013;41(Web Server issue):W333–W339.

103. Xiong, P., Zhang, C., Zheng, W., *et al.* BindProfX: Assessing mutation-induced binding affinity change by protein interface profiles with pseudo-counts. *J Mol Biol.* 2017;429(3):426–434.

104. Petukh, M., Dai, L., and Alexov, E. SAAMBE: Webserver to predict the charge of binding free energy caused by amino acids mutations. *Int J Mol Sci.* 2016;17(4):547.

105. Gromiha, M.M., Yugandhar, K., and Jemimah, S. Protein–protein interactions: Scoring schemes and binding affinity. *Curr Opin Struct Biol.* 2017;44:31–38.

106. Fischer, T.B., Arunachalam, K.V., Bailey, D., *et al.* The binding interface database (BID): A compilation of amino acid hot spots in protein interfaces. *Bioinformatics.* 2003;19(11):1453–1454.

Chapter 6
Predicting the consequences of mutations

Hemant Kumar and Julia M. Shifman*

*Department of Biological Chemistry,
The Alexander Silberman Institute of Life Sciences
Hebrew University of Jerusalem, Jerusalem, Israel*

Protein–protein interactions (PPIs) are involved in all critical cellular functions, and mutations in PPIs frequently lead to phenotypic changes that might cause a disease. In this chapter, the various effects that could result from mutations in PPIs such as destabilization/ stabilization of a complex, appearance/disappearance of posttranslational modifications, switching on/off of signaling pathways and coevolutionary effects have been summarized. Furthermore, the various computational tools that have been developed by the research community for the prediction of these mutational effects based on protein sequence, structure and PPI network analysis have been described. These computational tools help us obtaining better understanding of how PPIs evolve and how various mutations could lead to serious diseases such as cancer and neurodegenerative disease. Ultimately such predictions facilitate discovery of drugs targeting specific PPIs and PPI-regulated networks.

---

*jshifman@mail.huji.ac.il

6.1. Introduction

Interactions between two or more proteins control virtually all functions in the cell including signaling and metabolic pathways, transcription and translation, cell division and death, and macromolecular assembly [1–3]. Individual protein usually participates in a number of different interactions, being a part of a large protein–protein interaction (PPI) network that defines cellular function. Even a single mutation could profoundly affect binding affinity (K_d) of a PPI, thus disrupting or sometimes creating a new PPI. Single mutations in one PPI are translated to modification of the whole PPI network, altering cellular function and frequently leading to disease [4, 5]. Mutations in various PPIs have been documented as a major cause for diseases such as cancer, viral and bacterial infections, and neurodegenerative diseases [5, 6].

The key for understanding PPI-related mutations and their relationship to diseases lies in our ability to calculate changes in binding free energy due to mutations ($\Delta\Delta G_{bind}$) and to predict how these changes would affect the whole PPI network. Such predictions would help us to obtain a comprehensive picture of PPI evolution, to facilitate identification of disease-causing mutations and to suggest strategies for drug design [7]. In this chapter, we will review the existing methodology for predicting mutational effects in PPIs and give examples of applications of this methodology to most recent discoveries in the field.

6.2. Not all mutations are made equal

Protein–protein interfaces frequently extend over a large surface area and involve many residues [8]. Not all interface mutations, however, have the same effect on PPI binding affinity and on the ability to cause disease; the location and the nature of the mutation could play a significant role here. For example, mutations in the core of the binding interface are usually more deleterious and disease causing compared to mutations occurring at the rim of the binding interface [9]. The more deleterious effect was also documented for mutations in structurally ordered regions of PPIs compared to disordered protein regions [10].

These findings are directly related to the non-equal effect of variously placed mutations on the binding free energy ($\Delta\Delta G_{\text{bind}}$), which is directly related to changes in binding affinity of the mutated and wild-type PPI:

$$\Delta\Delta G_{\text{bind}} = kT \cdot \ln\left(\frac{K_{\text{d}}^{\text{MUT}}}{K_{\text{d}}^{\text{WT}}}\right).$$

Early studies on PPIs revealed that only a small subset of all the positions within the binding interface are responsible for most of the binding free energy [11]. Such positions, termed hotspots [12], are usually highly conserved [13], located in the center of the interface [8, 14] and are frequently occupied by tryptophan, tyrosine and arginine [8]. Hotspots are often clustered together, forming hot regions [15, 16], where mutations are coupled to each other [17]. As hotspots and hot regions contribute the most to the binding free energy of the PPI, directing inhibitors to these regions present an attractive strategy for drug design [18, 19].

As opposed to hotspots, cold spots are positions in PPIs occupied by non-optimal amino acids with high potential for affinity improvement [20]. Such positions might play an important role in PPI evolution [21] and are crucial for engineering of protein-based therapeutics. Furthermore, they may support low-affinity PPIs that need quick dissociation due to their cellular function [22]. Distribution and structural context of cold spots is still largely unknown and is the subject of future studies.

6.3. Predicting changes in binding free energy due to mutation

Change in K_{d} or $\Delta\Delta G_{\text{bind}}$ in an individual PPI is the major factor that determines whether the mutation would substantially change PPI function and lead to disease. Various experimental methods to quantify the effect of mutations on $\Delta\Delta G_{bind}$ are available such as isothermal calorimetry (ITC), surface plasmon resonance (SPR), fluorescence spectroscopy, stopped flow and enzyme assays [23–25]. However, measuring $\Delta\Delta G_{bind}$ is a time-consuming and expensive process that becomes impractical on a high-throughput scale. Thus, computational predictions of $\Delta\Delta G_{bind}$ in individual PPIs are an attractive strategy for substituting lengthy experiments. In this direction, a large number of computational tools for predicting the absolute value of binding free energy of binding (ΔG_{bind}) as well as $\Delta\Delta G_{bind}$ have been developed in the last decade [26–34]. Many of the methodologies have been converted into Web servers (summarized in Table 6.1).

The majority of the computational approaches rely on experimental data sets of $\Delta\Delta G_{bind}$ values to train their energy function. Here, one has to be careful to include only data measured by reliable biophysical methods and at similar conditions. Several databases compiling $\Delta\Delta G_{bind}$ in various PPIs have been reported [43–49]. Among them, the SKEMPI database is the largest available up to date with the collection of binding affinity data for 7085 mutations in PPIs with known 3D structures and reliable binding affinity measurement methods [43, 49].

The approaches for $\Delta\Delta G_{bind}$ prediction could be generally divided into two categories: sequence based and structure based. Some of the more recent works combine the two approaches to improve accuracy and/or speed of calculations.

6.3.1. *Sequence-based prediction of $\Delta\Delta G_{bind}$*

Sequence-based methods rely solely on the information about the sequence of the interaction partners and on the nature of mutations. Such methods use hundreds of features that describe each amino acid in the wild-type and in the mutated protein (such as hydrophobicity, charge, size, disorder propensity and others). Powerful machine learning techniques are used to train an energy function on the set of available experimental $\Delta\Delta G_{bind}$ values. Using such an approach, PPA-Pred was developed for the prediction of PPI binding affinity [50]. This approach was able to achieve high correlation with experimental data but only if different energy functions were utilized for different functional classes of PPIs. Sequence-based approaches are attractive when no structural information on the PPI is available and thus allow us to make predictions on the proteomic scale. Yet, these methods have been mostly applied to discriminate between binders and non-binders [51, 52] or to predict binding specificity of various peptide-binding domains [53] rather than for exact $\Delta\Delta G_{bind}$ predictions. However, some recent approaches combine sequence-based and structure-based features to improve method performance for $\Delta\Delta G_{bind}$ predictions (see the following section).

6.3.2. *Structure-based prediction of* $\Delta\Delta G_{bind}$

All structure-based predictions use 3D structure of the PPI complex as an input to compute a number of molecular interactions or features that are linearly combined into one energy function. Just as in sequence-based predictions, the weight in front of each term is optimized using a data set of experimental $\Delta\Delta G_{bind}$ values [54]. Possible features include van der Waals interactions, hydrogen bonds, buried surface area, polar and non-polar contacts, conformational changes, electrostatic interactions, inter-residue contacts and others [31, 55, 56]. Different methodologies explore and emphasize different features in attempt to improve accuracy and/or calculation speed. Although earlier approaches predict $\Delta\Delta G_{bind}$ solely based on a few physically based features important for PPI binding [28, 57–59], more recent approaches tend to rely on statistical potentials and complicated energy functions obtained through machine learning [31]. For example, SAAMBE combines a physically based component with statistical terms derived from sequence characteristics of PPIs [29]. BeAtMuSiC uses a set of statistical potentials adapted to a coarse-grained representation of proteins to predict $\Delta\Delta G_{bind}$ within short running times [31]. Several recent approaches explore the possibility of predicting $\Delta\Delta G_{bind}$ solely from the number of interfacial contacts [60] or a number of differently classified contacts [61]. The advantage of these approaches is their simplicity and short computational time, but they have to be yet tested on large data sets of mutants.

A separate approach for predicting $\Delta\Delta G_{bind}$ in antibody–antigen complexes, mCSM, was developed by optimizing the energy function on experimental data for antibody-antigen complexes only [26]. Zhang and colleagues developed methodology that combines structure-based, sequence-based and evolutionary information from interface alignments to train their function for $\Delta\Delta G_{bind}$ predictions (see BindProf and BindProfX Web servers) [27]. A similar combined approach has been used by Bonvin and colleagues to develop iSEE for $\Delta\Delta G_{bind}$ predictions [33].

Although most structure-based methods utilize a fixed-backbone approximation for their predictions, backbone flexibility could be also incorporated into the model either through allowing small backbone perturbations called backrub movement [62] or by a calculation on the ensemble of backbone structures [63]. Modeling backbone flexibility becomes necessary when mutation substitutes a small residue with a large residue and creates a clash. However, incorporating backbone flexibility for all $\Delta\Delta G_{bind}$ predictions is not always beneficial and is definitely more time consuming [64].

6.3.3. *Predicting global interface features*

Several tools have been developed to predict global characteristics of protein–protein interfaces such as hotspots and cold spots of binding. Predicting hotspots is actually easier than predicting exact $\Delta\Delta G_{bind}$ values, because hotspots are characterized by a number of pronounced features such as high conservation, high burial from solvent, high contact number, etc. (see Table 6.1).Although most methods for hotspot predictions are structure based [37, 39, 65–68], some approaches utilize sequence information only [69].

Our group has been developing methods for predicting protein cold spots of binding, positions with high potential to increase binding affinity upon mutations. Predictions of protein cold spots could be done by systematically scanning the whole binding interface with 20 amino acids and calculating $\Delta\Delta G_{bind}$ for all interfacial mutations (see PPIscan Web server) [35]. Additional methods for cold spot predictions could be based on identifying positions with highly

Table 6.1. Summary of Web servers for predicting mutational effects in PPIs.

Web server	Based on	URL	Note	References
Mcsm-AB	Structure	http://biosig.unimelb.edu.au/mcsm_ab/	For antibody/antigen interactions.	[26]
BindProfX	Sequence/ Structure/ Evolution	https://zhanglab.ccmb.med.umich.edu/ BindProfX/	Predicts $\Delta\Delta G$ using structure–profile-based approach.	[27]
MutaBind	Structure/ Evolution	https://www.ncbi.nlm.nih.gov/research/ mutabind/	Predicts $\Delta\Delta G$ using molecular mechanics force fields.	[28]
SAAMBE	Sequence/ Structure	http://compbio.clemson.edu/ saambe_webserver/	Predicts $\Delta\Delta G$.	[29]
ELASPIC	Structure	http://elaspic.kimlab.org/	Makes homology models for any protein.	[30]
BeAtMuSiC	Structure	http://babylone.ulb.ac.be/beatmusic/	Coarse-grained protein representation.	[31]
FoldX	Structure	http://foldxsuite.crg.eu/	Predicts $\Delta\Delta G$ using an empirical force field.	[34]
PPIscan	Structure	http://ppi-scan.cs.huji.ac.il	Predicts $\Delta\Delta G$ for all possible interface mutations and cold spots.	[35]
HotRegion	Structure	http://prism.ccbb.ku.edu.tr/hotregion/	Predicts hotspot clusters.	[36]
HotPoint	Structure	http://prism.ccbb.ku.edu.tr/hotpoint/	Predicts hotspots.	[37]
Robetta	Structure	http://robetta.bakerlab.org	Predicts hotspots using Ala mutagenesis.	[38]
KFC2	Structure	http://kfc.mitchell-lab.org/	Predicts hotspots based on solvation, atomic density and plasticity	[39]
DrugScore	Structure	http://cpclab.uni-duesseldorf.de/dsppi/	Predicts hotspots using Ala mutagenesis.	[32]
SNP-IN	Structure	http://andromeda.rnet.missouri.edu/ snpintool/	Classifies mutations into neutral, beneficial and deleterious.	[40]
WS-SNPs&GO	Sequence/ Structure	http://snps.biofold.org/snps-and-go/	Predicts impact of mutation using protein functional annotation.	[41]
SuSPect	Sequence/ Structure	http://www.sbg.bio.ic.ac.uk/~suspect/	Predicts phenotypic effects of missense mutations.	[42]

unfavorable interactions in the wild-type PPI or positions where a new interaction across the binding interface could be created [21].

6.4. Predicting the effects of post-translational modifications

The existence of some PPIs depends on post-translational modifications (PTMs). For example, phosphorylation on serine, threonine and tyrosine is commonly used to switch on/off PPIs in various signaling pathways [70]. For PTM-dependent interactions, mutations at residues that carry the PTM would result in a significant effect on the affinity of this particular interaction and are likely to affect the function of the whole PPI network. As the exact PTM patterns in the cell depend on various conditions and are difficult to measure, only recent studies have started to address mutational effects of the PTM-dependent interactions, revealing interesting patterns [71]. For example, phosphorylated proteins are identified frequently in more centrally located network spots, whereas glycosylated proteins are located at the network periphery [40, 72]. Reimand *et al.* predict that ~70% of all single mutations potentially change phosphorylation pattern and rewire biological pathways [73]. Mutations to negatively charged residues could serve as phosphomimetics and result in constitutive activation of certain signaling pathways leading to cancer [74]. Some recent works present approaches to visualize PTMs in the context of PPI networks and PPI structure [75, 76]. As several new methods for large-scale prediction of PTMs emerge [77], future studies should integrate PTM predictions with PPI network data to obtain a broader picture of how PTMs affect various protein signaling pathways in health and disease [78].

6.5. Predictions of coevolving mutations in PPIs

Coevolving or correlated mutations are two (or more) mutations that occur simultaneously during evolution to preserve PPI binding function. If a mutation in one binding partner happens to be deleterious for PPI complex formation, it could be compensated by another mutation either in the same protein or in the partner protein [79]. Correlated mutations could be computationally predicted through multiple sequence alignments of partner proteins from different species and computing a correlation function [80]. Analysis of correlated mutations has been repeatedly used to predict pairs of interacting proteins, to discriminate between different docking solutions and to predict binding interfaces of known interacting proteins [81]. Coevolution analysis could be also used to predict disease-associated mutations [82].

6.6. Applications of computational studies

6.6.1. *Designing mutations that alter binding affinity and specificity*

Rational approaches for changing binding affinity through mutations could be utilized for altering the PPI signaling network in the cell and for various biotechnological applications. Protein design studies by multiple groups demonstrated that various protocols could be utilized to enhance PPI binding affinity [83–88]. Additional studies focused on altering PPI binding specificity [35]. It has been shown that multi-specific proteins could be converted into proteins with single specificity through computational protein design [89–95], and orthogonal pairs of PPIs could be designed [96]. On the other hand, for some biotechnological applications, it is necessary to create proteins with multi-target recognition capabilities [97, 98]. Mutations could be introduced into PPIs to increase affinity to only one physiological state of a target protein [99]. Furthermore, computational means have been also applied to design focused combinatorial libraries that are further screened to achieve unprecedented enhancements in binding affinity and specificity [100–103]. In summary, there are many successful examples of how predictions in $\Delta\Delta G_{\mathrm{bind}}$ could be used to alter PPI properties and to engineer binders with specific characteristic that could be used for drug design or other biotechnological applications.

6.6.2. *Predicting disease-causing mutations*

Ultimately, we would like to predict whether a certain mutation in a particular PPI is disease-causing. This is largely determined by the location of a particular PPI within the whole PPI network. If K_d of the PPI is substantially changed and the PPI is highly connected in the network, the mutation is likely to be disease-causing [104]. However, if the mutated PPI is at the network periphery, the disruption of an individual PPI should not have a profound effect on the cellular function. PPI network analysis thus has become an important tool for predicting the effect of disease-causing mutations [105–108]. For such predictions, however, an extensive understanding of the PPI network is required. To this end, scientists have been working on acquiring experimental information on PPI networks related to different diseases. For example, Human Cancer Pathway Protein Interaction Network (HCPIN) has been created as a curated database of interacting proteins related to cancer progression [109]. Furthermore, mapping of cancer-associated missense mutations into PPI interfaces revealed that the large proportion of such mutations are found in already known oncogenes, although many of the remaining PPIs while not known could be also important for cancer [110]. The majority of the disease-casing mutations serve to disrupt a particular PPI. For example, introduction of two missense mutations in two breast cancer genes (BRCA2 and PALB2) results in PPI interruption and cancer progression [111]. Yet, not all disease-associated mutations are deleterious for PPIs. Some mutations on the contrary increase the interaction strength and activity through the PPI network [112]. Furthermore, mutations were shown to cause a disease through an appearance of a new PPI [113] or through stimulating aggregation and amyloid fiber formation. For example, the Leu75Pro mutation in transthyretin disturbs H-bonding in a beta-sheet, leading to the formation of amyloid fibrils [114]. Likewise, Glu6Val mutation in β-hemoglobin results in protein aggregation and is the cause of sickle cell anemia [115]. A number of Web tools have been developed for predicting the harmful or benign effects of mutations (PolyPhen-2 [116], nsSNPAnalyzer [117], mutpred2 [118], PhD-SNP [119], SNIP-IN [40], WS-SNPs&GO [41], SuSPect [42], PON-P2 [120].

All the tools mentioned use machine learning for distinguishing the benign and harmful effects of mutations.

6.7. Future directions

In recent years, large efforts have been invested to advance the methodology for $\Delta\Delta G_{bind}$ prediction in pairwise protein complexes as well as in PPI networks. The most recent methodology utilizes the newest machine learning approaches to improve prediction accuracy and speed. Yet, when using machine learning, one must have a deep understanding of the molecular interactions underlying the protein–protein binding process as well as the accuracy and the limitations of experimental measurements. Computational method development will benefit from the large databases recently developed to accumulate experimental data on binding affinity measurements [49]. In addition, new experimental methodology based on deep sequencing has been recently reported, allowing us to produce $\Delta\Delta G_{bind}$ values for thousands of mutants of the same PPI in one experiment and opening a new way for studies of mutation effects in PPIs [121, 122]. This high-throughput methodology is likely to generate new comprehensive data sets for various PPIs, which would facilitate training and testing of computational approaches. Future studies should also focus on predicting globular interface features such as hotspots and cold spots of binding and on correlating such globular features to PPI binding affinity and PPI function. Additional computational approaches will be likely developed for incorporating PTMs into the big picture of PPI-mediated mutational changes. Predicting the mutational effects in PPIs and better understanding the mechanism of the disease-causing mutations facilitate the design of PPI inhibitors both from small molecules and from protein scaffolds.

References

1. Alberts, B. The cell as a collection of protein machines: preparing the next generation of molecular biologists. *Cell*. 1998 Feb 6;92(3): 291–294.

2. Jones, S., and Thornton, J.M. Principles of protein–protein interactions. *Proc Natl Acad Sci U S A.* 1996 Jan 9;93(1):13–20.

3. Nooren, I.M.A., and Thornton, J.M. Structural characterisation and functional significance of transient protein–protein interactions. *J Mol Biol.* 2003 Jan 31;325(5):991–1018.

4. Gonzalez, M.W., and Kann, M.G. Chapter 4: Protein Interactions and Disease. Lewitter F, editor. *PLoS Comput Biol* 2012 Dec 27;8(12): e1002819.

5. Ryan, D.P., and Matthews, J.M. Protein–protein interactions in human disease. *Curr Opin Struct Biol.* 2005 Aug;15(4):441–446.

6. Bowler, E.H., Wang, Z, and Ewing, R.M. How do oncoprotein mutations rewire protein–protein interaction networks? *Expert Rev Proteomics.* 2015;12(5):449–455.

7. Kar, G., Kuzu, G., Keskin, O., *et al.* Protein–protein interfaces integrated into interaction networks: Implications on drug design. *Curr Pharm Des.* 2012 Aug 23;18(30):4697–705.

8. Bogan, A.A., and Thorn, K.S. Anatomy of hot spots in protein interfaces. *J Mol Biol.* 1998 Jul 3;280(1):1–9.

9. David, A., and Sternberg, M.J.E. The contribution of missense mutations in core and rim residues of protein–protein interfaces to human disease. *J Mol Biol.* 2015 Aug 28;427(17):2886–2898.

10. Lu, H-C., Chung, S.S., Fornili, A., *et al.* Anatomy of protein disorder, flexibility and disease-related mutations. *Front Mol Biosci.* 2015 Aug 12;2:47.

11. Clackson, T., and Wells, J.A. A hot spot of binding energy in a hormone-receptor interface. *Science.* 1995 Jan 20;267(5196):383–386.

12. DeLano, W.L. Unraveling hot spots in binding interfaces: Progress and challenges. *Curr Opin Struct Biol.* 2002 Feb 12(1):14–20.

13. Ma, B., Elkayam, T., Wolfson, H., *et al.* Protein–protein interactions: Structurally conserved residues distinguish between binding sites and exposed protein surfaces. *Proc Natl Acad Sci.* 2003 May 13; 100(10):5772–5777.

14. Stanfield, R., Fieser, T., Lerner, R., *et al.* Crystal structures of an antibody to a peptide and its complex with peptide antigen at 2.8 A. *Science.* 1990 May 11;248(4956):712–719.

15. Keskin, O., Ma, B., and Nussinov, R. Hot regions in protein–protein interactions: The organization and contribution of structurally conserved hot spot residues. *J Mol Biol.* 2005 Feb 4;345(5): 1281–1294.

16. David, A., Razali, R., Wass, M.N., *et al.* Protein–protein interaction sites are hot spots for disease-associated nonsynonymous SNPs. *Hum Mutat.* 2012 Feb 1;33(2):359–363.

17. Vakser, I.A., Matar, O.G., Lam, C.F., *et al.* A systematic study of low-resolution recognition in protein — protein complexes. *Proc Natl Acad Sci U S A.* 1999 Jul 20;96(15):8477–8482.

18. Cukuroglu, E., Engin, H.B., Gursoy, A., *et al.* Hot spots in protein–protein interfaces: Towards drug discovery. *Prog Biophys Mol Biol.* 2014;116(2–3):165–173.

19. Guo, W., Wisniewski, J.A., and Ji, H. Hot spot-based design of small-molecule inhibitors for protein–protein interactions. *Bioorg Med Chem Lett.* 2014 Jun 1;24(11):2546–2554.

20. Sharabi, O., Shirian, J., and Shifman, J.M. Predicting affinity- and specificity-enhancing mutations at protein–protein interfaces. *Biochem Soc Trans.* 2013;41(5):1166–1169.

21. Shirian, J., Sharabi, O., and Shifman, J.M. Cold spots in protein binding. *Trends Biochem Sci.* 2016;41(9):739–745.

22. Cohen, A., Rosenthal, E., and Shifman, J.M. Analysis of structural features contributing to weak affinities of ubiquitin/protein interactions. *J Mol Biol.* 2017;429(22):3353–3362.

23. Ladbury, J.E., and Chowdhry, B.Z. Sensing the heat: The application of isothermal titration calorimetry to thermodynamic studies of biomolecular interactions. *Chem Biol.* 1996 Oct;3(10):791–801.

24. Willander, M., and Al-Hilli, S. Analysis of biomolecules using surface plasmons. *Methods Mol Biol.* 2009;544:201–229.

25. Masi, A., Cicchi, R., Carloni, A., *et al.* Optical methods in the study of protein–protein interactions. *Adv Exp Med Biol.* 2010;674:33–42.

26. Pires, D.E., and Ascher. D.B. mCSM-AB: A web server for predicting antibody–antigen affinity changes upon mutation with graph-based signatures. *Nucleic Acids Res.* 2016 Jul 8;44(W1):W469–W473.

27. Xiong, P., Zhang, C., Zheng, W., *et al.* BindProfX: Assessing mutation-induced binding affinity change by protein interface profiles with pseudo-counts. *J Mol Biol.* 2016;429:426–434.

28. Li, M., Simonetti, F.L., Goncearenco, A., *et al.* MutaBind estimates and interprets the effects of sequence variants on protein–protein interactions. *Nucleic Acids Res.* 2016 Jul 8;44(W1):W494–W501.

29. Petukh, M., Dai, L., and Alexov, E. SAAMBE: Webserver to predict the charge of binding free energy caused by amino acids mutations. *Int J Mol Sci.* 2016 Apr 12;17(4):547.

30. Witvliet, D.K., Strokach, A., Giraldo-Forero, A.F., *et al.* ELASPIC web-server: Proteome-wide structure-based prediction of mutation effects on protein stability and binding affinity. *Bioinformatics.* 2016; 32(10):1589–1591.

31. Dehouck, Y., Kwasigroch, J.M., Rooman, M., *et al.* BeAtMuSiC: Prediction of changes in protein–protein binding affinity on mutations. *Nucleic Acids Res.* 2013 Jul 1;41(W1):W333–W339.

32. Krüger, D.M., and Gohlke, H. DrugScorePPI webserver: Fast and accurate in silico alanine scanning for scoring protein–protein interactions. *Nucleic Acids Res.* 2010;38(Suppl 2):480–486.

33. Geng, C., Vangone, A., Folkers, G.E., *et al.* iSEE: Interface structure, evolution and energy-based machine learning predictor of binding affinity changes upon mutations. *Proteins.* 2019;87(2):110–119.

34. Guerois, R., Nielsen, J.E., and Serrano, L. Predicting changes in the stability of proteins and protein complexes: A study of more than 1000 mutations. *J Mol Biol.* 2002;320(2):369–387.

35. Sharabi, O., Erijman, A., and Shifman, J.M. Computational methods for controlling binding specificity. *Methods Enzymol.* 2013 Jan 1; 523:41–59.

36. Cukuroglu, E., Gursoy, A., and Keskin, O. HotRegion: A database of predicted hot spot clusters. *Nucleic Acids Res.* 2012 Jan 1;40(D1): D829–D833.

37. Tuncbag, N., Keskin, O., and Gursoy, A. HotPoint: Hot spot prediction server for protein interfaces. *Nucleic Acids Res.* 2010 Jul 1;38(Web Server):W402–W406.

38. Kortemme, T., Kim, D.E., and Baker, D. Computational alanine scanning of protein–protein interfaces. *Sci Signal.* 2004 Feb 10; 2004(219):pl2.

39. Zhu, X., and Mitchell, J.C. KFC2: A knowledge-based hot spot prediction method based on interface solvation, atomic density, and plasticity features. *Proteins.* 2011 Sep;79(9):2671–2683.

40. Zhao, N., Han, J.G., Shyu, C.R., *et al.* Determining effects of non-synonymous SNPs on protein–protein interactions using supervised and semi-supervised learning. *PLoS Comput Biol.* 2014; 10(5):e1003592.

41. Capriotti, E., Calabrese, R., Fariselli, P., *et al.* WS-SNPs&GO: A web server for predicting the deleterious effect of human protein variants using functional annotation. *BMC Genomics.* 2013;14:S6.

42. Yates, C.M., Filippis, I., Kelley, L.A., *et al.* SuSPect: enhanced prediction of single amino acid variant (SAV) phenotype using network features. *J Mol Biol.* 2014 Jul 15;426(14):2692–2701.

43. Jankauskaitė, J., Jiménez-García, B., Dapkūnas, J., *et al.* SKEMPI 2.0: An updated benchmark of changes in protein–protein binding energy, kinetics and thermodynamics upon mutation. *Bioinformatics.* 2019 Feb 1;35(3):462–469;

44. Thorn, K.S., and Bogan, A.A. ASEdb: A database of alanine mutations and their effects on the free energy of binding in protein interactions. *Bioinformatics.* 2001 Mar 1;17(3):284–285.

45. Kumar, M.D.S., and Gromiha, M.M. PINT: Protein–protein interactions thermodynamic database. *Nucleic Acids Res.* 2006 Jan 1; 34(90001):D195–D198.

46. Liu, Z., Li, Y., Han, L., *et al.* PDB-wide collection of binding data: Current status of the PDBbind database. *Bioinformatics.* 2015 Feb 1;31(3):405–412.

47. Liu, Q,. Chen, P., Wang, B., *et al.* dbMPIKT: A database of kinetic and thermodynamic mutant protein interactions. *BMC Bioinformatics.* 2018 Dec 27;19(1):455.

48. Jemimah, S., Yugandhar, K., and Michael Gromiha, M. PROXiMATE: A database of mutant protein–protein complex thermodynamics and kinetics. *Bioinformatics.* 2017 Sep 1;33(17):2787–2788.

49. Moal, I.H., and Fernández-Recio, J. SKEMPI: A structural kinetic and energetic database of mutant protein interactions and its use in empirical models. *Bioinformatics.* 2012 Oct 15;28(20):2600–2607.

50. Yugandhar, K., and Gromiha, M.M. Protein–protein binding affinity prediction from amino acid sequence. *Bioinformatics.* 2014;30(24): 3583–3589.

51. Sacquin-Mora, S., Carbone, A., and Lavery, R. Identification of protein interaction partners and protein–protein interaction sites. *J Mol Biol.* 2008 Oct 24;382(5):1276–1289.

52. Fleishman, S.J., Whitehead, T.A., Strauch, E-M., *et al.* Community-wide assessment of protein-interface modeling suggests improvements to design methodology. *J Mol Biol.* 2011 Nov 25;414(2): 289–302.

53. Teyra, J., Sidhu, S.S., and Kim, P.M. Elucidation of the binding preferences of peptide recognition modules: SH3 and PDZ domains. *FEBS Lett.* 2012;586(17):2631–2637.

54. Sharabi, O., Yanover, C., Dekel, A., *et al.* Optimizing energy functions for protein–protein interface design. *J Comput Chem.* 2011 Jan 15; 32(1):23–32.

55. Horton, N., and Lewis, M. Calculation of the free energy of association for protein complexes. *Prot Sci.* 1992; Jan; 1(1):169–181.

56. Zhou, P., Wang, C., Tian, F., *et al.* Biomacromolecular quantitative structure–activity relationship (BioQSAR): A proof-of-concept study on the modeling, prediction and interpretation of protein–protein binding affinity. *J Comput Aided Mol Des.* 2013 Jan 10;27(1):67–78.

57. Barlow, K.A., Ó Conchúir, S., Thompson, S., *et al.* Flex ddG: Rosetta ensemble-based estimation of changes in protein–protein binding affinity upon mutation. *J Phys Chem B.* 2018 May 31;122(21): 5389–5399.

58. Benedix, A., Becker, C.M., de Groot, B.L., *et al.* Predicting free energy changes using structural ensembles. *Nat Methods.* 2009 Jan;6(1):3–4.

59. Sharabi, O., Dekel, A., and Shifman, J.M. Triathlon for energy functions: Who is the winner for design of protein–protein interactions? *Proteins.* 2011 May;79(5):1487–1498.

60. Vangone, A., and Bonvin, A.M. Contacts-based prediction of binding affinity in protein–protein complexes. *Elife.* 2015 Jul 20;4:e07454.

61. Raucci, R., Laine, E., and Carbone, A. Local interaction signal analysis predicts protein–protein binding affinity. *Structure.* 2018 Jun 5;26(6):905–915.e4.

62. Davis, I.W., Arendall, W.B., Richardson, D.C., *et al.* The backrub motion: How protein backbone shrugs when a sidechain dances. *Structure.* 2006 Feb;14(2):265–274.

63. Murciano-Calles, J., McLaughlin, M.E., Erijman. A., *et al.* Alteration of the C-terminal ligand specificity of the Erbin PDZ domain by allosteric mutational effects. *J Mol Biol.* 2014;426(21):3500–3508.

64. Kellogg, E.H., Leaver-Fay, A., and Baker, D. Role of conformational sampling in computing mutation-induced changes in protein structure and stability. *Proteins.* 2011 Mar;79(3):830–838.

65. Darnell, S.J., LeGault, L., and Mitchell, J.C. KFC server: Interactive forecasting of protein interaction hot spots. *Nucleic Acids Res.* 2008 May 19;36(Web Server):W265–W269.

66. Xia, J-F., Zhao, X-M., Song, J., *et al.* APIS: Accurate prediction of hot spots in protein interfaces by combining protrusion index with solvent accessibility. *BMC Bioinformatics.* 2010 Apr 8;11(1):174.

67. Lise, S., Archambeau, C., Pontil, M., *et al.* Prediction of hot spot residues at protein–protein interfaces by combining machine learning and energy-based methods. *BMC Bioinformatics.* 2009 Dec 30; 10(1):365.

68. Morrow, J.K., and Zhang, S. Computational prediction of protein hot spot residues. *Curr Pharm Des.* 2012;18(9):1255–1265.

69. Ofran, Y., and Rost, B. Protein–protein interaction hotspots carved into sequences. *PLoS Comput Biol.* 2007 Jul;3(7):e119.

70. Ardito, F., Giuliani, M., Perrone, D., *et al.* The crucial role of protein phosphorylation in cell signaling and its use as targeted therapy (Review). *Int J Mol Med.* 2017 Aug;40(2):271–80.

71. Woodsmith, J., and Stelzl, U. Studying post-translational modifications with protein interaction networks. *Curr Opin Struct Biol.* 2014 Feb;24:34–44.

72. Duan, G., and Walther, D. The roles of post-translational modifications in the context of protein interaction networks. *PLoS Comput Biol.* 2015;11(2):1–23.

73. Ren, J., Jiang, C., Gao, X., *et al.* PhosSNP for systematic analysis of genetic polymorphisms that influence protein phosphorylation. *Mol Cell Proteomics.* 2010 Apr;9(4):623–634.

74. Reimand, J., and Bader, G.D. Systematic analysis of somatic mutations in phosphorylation signaling predicts novel cancer drivers. *Mol Syst Biol.* 2013 Apr 28;9(1):637.

75. Tay, A.P., Liang, A., Wilkins, M.R., *et al.* Visualizing post-translational modifications in protein interaction networks using PTMOracle. *Curr Protoc Bioinformatics.* 2019;66(1):e71.

76. Tay, A.P., Pang, C.N.I., Winter, D.L., *et al.* PTMOracle: A cytoscape app for covisualizing and coanalyzing post-translational modifications in protein interaction networks. *J Proteome Res.* 2017;16(5): 1988–2003.

77. Audagnotto, M., and Dal Peraro, M. Protein post-translational modifications: In silico prediction tools and molecular modeling. *Comput Struct Biotechnol J.* 2017;15:307–319.

78. Grimes, M., Hall, B., Foltz, L., *et al.* Integration of protein phosphorylation, acetylation, and methylation data sets to outline lung cancer signaling networks. *Sci Signal.* 2018 May 22;11(531):eaaq1087.

79. Pazos, F., and Valencia, A. Protein co-evolution, co-adaptation and interactions. *EMBO J.* 2008;27(20):2648–2655.

80. Pazos, F., Helmer-Citterich, M., Ausiello, G., *et al.* Correlated mutations contain information about protein–protein interaction. *J Mol Biol.* 1997 Aug;271(4):511–523.

81. Guo, F., Ding, Y., Li, Z., *et al.* Identification of protein–protein interactions by detecting correlated mutation at the interface. *J Chem Inf Model.* 2015 Sep 28;55(9):2042–2049.

82. Hopf, T.A., Ingraham, J.B., Poelwijk, F.J., *et al.* Mutation effects predicted from sequence co-variation. *Nat Biotechnol.* 2017 Feb 16; 35(2):128–135.

83. Haidar, J.N., Pierce, B., Yu, Y., *et al.* Structure-based design of a T-cell receptor leads to nearly 100-fold improvement in binding affinity for pepMHC. *Proteins.* 2009 Mar;74(4):948–960.

84. Lippow, S.M., Wittrup, K.D., and Tidor, B. Computational design of antibody-affinity improvement beyond in vivo maturation. *Nat Biotechnol.* 2007 Oct 23;25(10):1171–1176.

85. Reynolds, K.A., Hanes, M.S., Thomson, J.M., *et al.* Computational redesign of the SHV-1 β-lactamase/β-lactamase inhibitor protein interface. *J Mol Biol.* 2008 Oct 24;382(5):1265–1275.

86. Sammond, D.W., Kimple, R.J., Siderovski, D.P., *et al.* Structure-based protocol for identifying mutations that enhance protein–protein binding affinities. *J Mol Biol.* 2007 Aug 31;371(5):1392–1404.

87. Schreiber, G., Selzer, T., and Albeck S. Rational design of faster associating and tighter binding protein complexes. *Nat Struct Biol.* 2000 Jul 1;7(7):537–541.

88. Sharabi, O., Peleg, Y., Mashiach, E., *et al.* Design, expression and characterization of mutants of fasciculin optimized for interaction with its target, acetylcholinesterase. *Protein Eng Des Sel.* 2009;22(10): 641–648.

89. Yosef, E., Politi, R., Choi, M.H., *et al.* Computational design of calmodulin mutants with up to 900-fold increase in binding specificity. *J Mol Biol.* 2009;385(5):1470–1480.

90. Shifman, J.M., and Mayo, S.L. Exploring the origins of binding specificity through the computational redesign of calmodulin. *Proc Natl Acad Sci U S A.* 2003 Nov 11;100(23):13274–13279.

91. Foight, G.W., Ryan, J.A., Gullá, S. V., *et al.* Designed BH3 peptides with high affinity and specificity for targeting Mcl-1 in cells. *ACS Chem Biol.* 2014 Sep 19;9(9):1962–1968.

92. Chen, T.S., Palacios, H., and Keating, A.E. Structure-based redesign of the binding specificity of anti-apoptotic Bcl-x(L). *J Mol Biol.* 2013 Jan 9;425(1):171–185.

93. Chen, T.S., and Keating, A.E. Designing specific protein–protein interactions using computation, experimental library screening, or integrated methods. *Protein Sci.* 2012 Jul;21(7):949–963.

94. Frappier, V., Jenson, J.M., Zhou, J., *et al.* Tertiary structural motif sequence statistics enable facile prediction and design of peptides that bind anti-apoptotic Bfl-1 and Mcl-1. *Structure.* 2019 Apr 2;27(4): 606–617.e5.

95. Zheng, F., and Grigoryan, G. *Design of Specific Peptide–Protein Recognition.* New York, NY: Humana Press; 2016:249–63.

96. Kortemme, T., Joachimiak, L.A., Bullock, A.N., *et al.* Computational redesign of protein–protein interaction specificity. *Nat Struct Mol Biol.* 2004 Apr 21;11(4):371–379.

97. Fromer, M., and Shifman, J.M. Tradeoff between stability and multi-specificity in the design of promiscuous proteins. *PLoS Comput Biol.* 2009;5(12): e1000627.

98. Humphris, E.L., and Kortemme, T. Design of multi-specificity in protein interfaces. *PLoS Comput Biol.* 2007 Aug;3(8):e164.

99. Filchtinski, D., Sharabi, O., Rüppel, A., *et al.* What makes Ras an efficient molecular switch: a computational, biophysical, and structural study of Ras-GDP interactions with mutants of Raf. *J Mol Biol.* 2010 Jun 11;399(3):422–435.

100. Qiao, C., Lv, M., Li, X., *et al.* Affinity maturation of antiHER2 monoclonal antibody MIL5 using an epitope-specific synthetic phage library by computational design. *J Biomol Struct Dyn.* 2013 May;31(5):511–521.

101. Guntas, G., Purbeck, C., and Kuhlman, B. Engineering a protein–protein interface using a computationally designed library. *Proc Natl Acad Sci U S A.* 2010 Nov 9;107(45):19296–192301.

102. Arkadash, V., Yosef, G., Shirian, J., *et al.* Development of high affinity and high specificity inhibitors of matrix metalloproteinase 14 through computational design and directed evolution. *J Biol Chem.* 2017; 292(8):3481–3495.

103. Dutta, S., Ryan, J., Chen, T.S., *et al.* Potent and specific peptide inhibitors of human pro-survival protein Bcl-x L. *J Mol Biol.* 2015 Mar 27;427(6):1241–1253.

104. Przulj, N., Wigle, D.A., and Jurisica, I. Functional topology in a network of protein interactions. *Bioinformatics.* 2004 Feb 12; 20(3):340–348.

105. Gulati, S., Cheng, T.M.K., and Bates, P.A. Cancer networks and beyond: Interpreting mutations using the human interactome and protein structure. *Semin Cancer Biol.* 2013 Aug 1;23(4):219–226.

106. Lamolet, B., Pulichino, A.M., Lamonerie, T., *et al.* A pituitary cell-restricted T box factor, Tpit, activates POMC transcription in cooperation with Pitx homeoproteins. *Cell.* 2001 Mar 23;104(6): 849–859.

107. Schuster-Böckler, B, and Bateman, A. Protein interactions in human genetic diseases. *Genome Biol.* 2008;9(1):1–12.

108. Seto, M.L., Lee, S.J., Sze, R.W., *et al.* Another TWIST on Baller-Gerold syndrome. Am J Med Genet. 2001 Dec 15;104(4):323–330.

109. Huang, Y.J., Hang, D., Lu, L.J., *et al.* Targeting the human cancer pathway protein interaction network by structural genomics. Mol Cell Proteomics. 2008 Oct 1;7(10):2048–2060.

110. Porta-Pardo, E., Garcia-Alonso, L., Hrabe, T., *et al.* A pan-cancer catalogue of cancer driver protein interaction interfaces. *PLOS Comput Biol.* 2015 Oct 20;11(10):e1004518.

111. Caleca, L., Catucci, I., Figlioli, G., *et al.* Two missense variants detected in breast cancer probands preventing BRCA2-PALB2 protein interaction. *Front Oncol.* 2018;8:480.

112. Rajalingam, K., Schreck, R., Rapp, U.R., *et al.* Ras oncogenes and their downstream targets. *Biochim Biophys Acta.* 2007 Aug;1773(8): 1177–1795.

113. Hao, Y., Wang, C., Cao, B., *et al.* Gain of interaction with IRS1 by p110α-helical domain mutants is crucial for their oncogenic functions. *Cancer Cell.* 2013 May 13;23(5):583–593.

114. Lashuel, H.A., Wurth, C., Woo, L., *et al.* The most pathogenic transthyretin variant, L55P, forms amyloid fibrils under acidic conditions and protofilaments under physiological conditions. *Biochemistry.* 1999 Oct 12;38(41):13560–13573.

115. Ingram, V.M. A specific chemical difference between the globins of normal human and sickle-cell anaemia haemoglobin. *Nature.* 1956 Oct;178(4537):792–794.

116. Adzhubei, I.A., Schmidt, S., Peshkin, L., *et al.* A method and server for predicting damaging missense mutations. *Nat Methods.* 2010 Apr 1;7(4):248–249.

117. Bao, L., Zhou, M., and Cui, Y. nsSNPAnalyzer: Identifying disease-associated nonsynonymous single nucleotide polymorphisms. *Nucleic Acids Res.* 2005 Jul 1;33(Web Server):W480–W482.

118. Pejaver, V., Urresti, J., Lugo-Martinez, J., *et al.* MutPred2: Inferring the molecular and phenotypic impact of amino acid variants. *bioRxiv.* 2017 May 9;134981.

119. Capriotti, E., Calabrese, R., and Casadio, R. Predicting the insurgence of human genetic diseases associated to single point protein mutations with support vector machines and evolutionary information. *Bioinformatics.* 2006 Nov 15;22(22):2729–2734.

120. Niroula, A., Urolagin, S., and Vihinen, M. PON-P2: Prediction method for fast and reliable identification of harmful variants. *PLOS ONE.* 2015 Feb 3;10(2):e0117380.

121. Heyne, H.M.M., Papo, P.N. N., and Shifman, S.J.J. Generating quantitative binding landscapes through fractional binding selections, deep sequencing and data normalization. *bioRxiv.* 2019 Jan 1:509984.

122. Jenson, J.M., Xue, V., Stretz, L., *et al.* Peptide design by optimization on a data-parameterized protein interaction landscape. *Proc Natl Acad Sci.* 2018 Oct 30;115(44):E10342–E10351.

Part II
Protein–Nucleic Acid Interactions

Chapter 7

Computational approaches for understanding the recognition mechanism of protein–nucleic acid complexes

Ambuj Srivastava[†], Dhanusha Yesudhas[†], A. Kulandaisamy,
Nisha Muralidharan, C. Ramakrishnan, R. Nagarajan and
M. Michael Gromiha*

*Department of Biotechnology, Bhupat and Jyoti Mehta
School of Biosciences, Indian Institute of Technology Madras,
Chennai 600036, Tamil Nadu, India*

Protein–nucleic acid (NA) interactions are involved in several cellular processes including gene regulation, replication, packaging and control. The functions of protein–NA complexes are studied with the interactions between proteins and NAs using experimental methods and computational techniques. In this chapter, we survey the databases reported in the literature based on structures, binding and affinity. In addition, the contribution of noncovalent interactions to the binding of protein–NA complexes and disorder-to-order transitions (DOTs) upon the complex formation has been reviewed. Further, we discuss various mechanisms proposed to understand the recognition of protein–NA complexes and the importance of molecular dynamics (MD) simulations to gain deep insights. The role

*gromiha@iitm.ac.in
[†]Contributed equally to this work.

of different features to relate the binding affinity of protein–DNA complexes has also been explored. In essence, this comprehensive review provides ample insights to understand the structure–function relationship in protein–NA complexes.

7.1. Introduction

Protein–NA interactions are crucial for various biological processes such as replication, transcription and translation (Si *et al.*, 2015; Van Nostrand *et al.*, 2016). These interactions also influence the gene expression, which in turn impacts the function of a cell (Poddar *et al.*, 2018). Impairment of the function of a protein–NA interaction leads to various diseases including neurological disorders and cancer (Tompa, 2012). Therefore, it is crucial to understand the molecular mechanisms and residue biases in them. Depending on the flexibility of protein and NA molecules, the recognition mechanisms are widely divided into two classes such as direct (intermolecular interactions) and indirect (intramolecular interactions) readouts (Gromiha *et al.*, 2004a). In the direct readout mechanism, proteins and NAs bind to each other due to noncovalent interactions (hydrogen bonding, electrostatic, van der Waals, etc.) between amino acid residues and nucleotides in proteins and NAs, respectively, whereas in the indirect readout mechanism, NA or protein changes its conformation to bind to its partner protein. These mechanisms are widely studied in protein–DNA complexes (Gromiha *et al.*, 2004a, 2005; Ahmad *et al.*, 2006) and few attempts have been made on protein–RNA complexes as well (Stefl *et al.*, 2005, 2010).

RNA-binding domains (RBDs) harbor several RNA recognition motifs (RRMs) such as K homology domains, zinc fingers, cold-shock domains and sterile alpha motif domains. In DNA-binding proteins (DBPs), commonly occurring motifs are helix-turn-helix, helix-loop-helix, zinc fingers and leucine zippers. DBPs mostly bind to the major groove of the DNA and there exist few minor groove binders also. Some proteins interact with both DNA and RNA, which perform various functions such as transcription, translation, gene silencing, microRNA biogenesis and telomere maintenance, also called as DNA–RNA binding proteins (DRBPs) (Hudson and Ortlund, 2014).

Various experimental techniques are used to determine protein–DNA and protein–RNA interactions (PRIs). Protein–DNA interactions are studied with electrophoretic mobility shift assay (EMSA), DNA footprinting, southwestern blotting, fluorescent resonance energy transfer (FRET), systematic evolution of ligands by exponential enrichment (SELEX) and yeast two-hybrid systems (Dey *et al.*, 2012). On the other hand, RNA pull down assays, RNA footprinting, EMSAs, RNA immunoprecipitation and yeast three-hybrid systems are used to identify PRIs (Marchese *et al.*, 2016; McHugh *et al.*, 2014; Ramanathan *et al.*, 2019). Further, three-dimensional (3D) structures of protein–NA complexes have been determined using X-ray crystallography, NMR spectroscopy and electron microscopy, and the structures are deposited in protein data bank (PDB) (Rose *et al.*, 2017; http://www.rcsb.org/) and NA database (NDB) (Narayanan *et al.*, 2014; http://ndbserver.rutgers.edu/).

The structures of protein–NA complexes have been extensively used to understand the importance of specific noncovalent interactions influencing the binding of complexes, recognition mechanism, binding affinity as well as initial conformations for MD simulations. In this chapter, we survey the available databases for protein–NA complexes and computational analyses to explore the importance of specific interactions along with the integration of folding and binding. Further, we review various recognition mechanisms, binding affinity and the DOTs upon binding.

7.2. Databases for protein–nucleic acid complexes/interactions

Several databases have been developed for studying protein–NA interactions on various perspectives: (1) protein–NA complex structures, (2) interactions between amino acid residues and nucleotides, (3) thermodynamic database for protein–NA interactions and (4) transcription and posttranscriptional regulation. Table 7.1 lists the databases on these categories.

Table 7.1.   List of databases on protein–NA complexes/interactions.

Name	Description	Database link	References
Databases for protein–NA complexes			
PDIdb	Protein–DNA Interface Database	http://melolab.org/pdidb/web/content/home	Norambuena and Melo (2010)
NPIDB	Nucleic Acid–Protein Interaction DataBase	http://npidb.belozersky.msu.ru/	Kirsanov et al. (2013)
RBPDB	RNA-Binding Protein DataBase	http://rbpdb.ccbr.utoronto.ca/	Cook et al. (2011)
PDB	Protein Data Bank contains protein–NA complex structures	http://rcsb.org	Rose et al. (2017)
NDB	Nucleic Acid Database contains protein–NA complex structures	http://ndbserver.rutgers.edu/	Narayanan et al. (2014)
Databases for interactions between amino acid residues and nucleotides in protein–NA complexes			
RAID	RNA Association Interaction Database	http://www.rnasociety.org/raid/	Yi et al. (2017)
RAIN	Database for RNA–protein Association and Interaction Networks	http://rth.dk/resources/rain	Junge et al. (2017)
3D-footprint	Database for the structural analysis of protein–DNA complexes	http://floresta.eead.csic.es/3dfootprint/	Contreras-Moreira (2010)
hPDI	Database of experimental human protein–DNA interactions	http://bioinfo.wilmer.jhu.edu/PDI/	Xie et al. (2010)
RBPmap	Database of RNA-binding motifs	http://rbpmap.technion.ac.il/	Paz et al. (2014)

PDBparam	Tool for providing binding site residues in protein–DNA and protein–RNA complexes	https://www.iitm.ac.in/bioinfo/pdbparam/	Nagarajan *et al.* (2016)

Databases for binding affinities of protein–NA interactions

KDBI	Kinetic data of biomolecular interactions database	http://bidd.nus.edu.sg/group/kdbi/kdbi.asp	Ji *et al.* (2003)
PDBbind	Experimentally measured binding affinities for all biomolecular complexes including protein–NA complexes	http://www.pdbbind.org.cn/	Liu *et al.* (2015)

Databases for transcription and posttranscriptional regulation factors and their NA binding sites

JASPER	Database of transcription factor-DNA-binding profiles	http://jaspar.genereg.net	Khan *et al.* (2018)
MeDReaders	Database for transcription factors that bind to methylated DNA	http://medreader.org/	Wang *et al.* (2018)
POSTAR2	Database on binding sites of RNA-binding proteins and functional annotations	http://lulab.life.tsinghua.edu.cn/postar	Zhu *et al.* (2019)
TRANSFAC	Database on transcription factors and their DNA-binding sites	http://genexplain.com/transfac/	Wingender *et al.* (1996)

Last accessed on May 20, 2019.

7.2.1. *Databases for protein–nucleic acid complexes*

Norambuena and Melo (2010) developed Protein–DNA Interface Database (PDIdb), which contains relevant structural information of protein–DNA complexes for the structures solved at a resolution of better than 2.5 Å. It has several potential applications such as predicting the transcription factor (TF)-binding sites in DNA, studying specificity determinants that mediate enzyme recognition events and designing new DBPs with distinct binding specificity and affinity. PDIdb is freely available at http://melolab.org/pdidb/web/content/home.

The Nucleic acid–Protein Interaction DataBase (NPIDB) (Kirsanov *et al.*, 2013) is an online resource that gives structural information about all the available protein–DNA and protein–RNA complexes. The server also provides a variety of tools for the analysis of the complexes such as identifying hydrophobic clusters on interfaces, detecting potential hydrogen bonds and water bridges and visualizing structures. NPIDB takes PDB code of the complex and biological units like GO terms, Pfam and SCOP as inputs. The output contains protein information, experimental type, kind of NA and organism, links to other databases, lists of Pfam and SCOP domains, secondary structure and contacts with NA. NPIDB is freely accessible at http://npidb.belozersky.msu.ru/.

RNA-Binding Protein DataBase (RBPDB) is a collection of RNA-binding proteins (RBPs), which are linked to other proteins through orthology relationships (Cook *et al.*, 2011). These proteins are linked to Ensembl, FlyBase and WormBase gene annotations and RNA-bound protein structures in PDB. The database allows users to browse using domain or organism, export records and download bulk data, and scan sequences for RBP-binding sites. RBPDB is freely available at http://rbpdb.ccbr.utoronto.ca/.

Further, both PDB and NDB contain more than 7000 protein–NA complex structures determined experimentally using X-ray crystallography, NMR spectroscopy and electron microscopy (Rose *et al.*, 2017; Narayanan *et al.*, 2014). These databases can be accessed at http://rcsb.org and http://ndbserver.rutgers.edu/.

7.2.2. *Databases for interactions between amino acid residues and nucleotides in protein–nucleic acid complexes*

Several databases are available for RNA-associated interactions, and interactions between amino acid residues and nucleotides at the interface of protein–NA complexes. Yi *et al.* (2017) developed RNA Association Interaction Database (RAID), which contains experimental and predicted interactions from the literature and other databases. It contains RNA-associated interactions, and evidence and reliability assessment for each RNA-associated interaction based on integrative confidence scores. The database contains over 1.2 million RNA–protein interactions across 60 species. RAID is available at http://www.rna-society.org/raid/.

RNA–Protein Association and Interaction Networks (RAINs) is a specific database (Junge *et al.*, 2017) for RNA and protein interactions, which contains data extracted from experimental reports, literature mining as well as computational predictions. RAIN can be accessible at http://rth.dk/resources/rain and it has an option to download the data on all interactions.

Contreras-Moreira (2010) developed 3D-footprint, which provides binding specificity for protein–DNA complexes. It allows users to search based on keywords, find proteins that recognize similar DNA motifs and perform BLAST search to identify similar DBPs. Further, it provides sequence alignments, which highlight interface residues responsible for DNA recognition. Moreover, the Web interface allows custom analyses of protein–DNA complexes provided by the user. 3D-footprint is available at http://floresta.eead.csic.es/3dfootprint.

Xie *et al.* (2010) developed Human Protein–DNA Interactome (hPDI), which contains experimental data on protein–DNA interactions. RBPmap is a database of RNA-binding motifs from Human, Mouse and *Drosophila melanogaster* using the data extracted from the literature (Paz *et al.*, 2014). Nagarajan *et al.* (2016) developed PDBparam, which computes several structure-based parameters including binding site residues in protein–NA complexes.

7.2.3. *Databases for binding affinities of protein– nucleic acid interactions*

The numerical data on binding specificity of protein–NA complexes provide information about the affinity of binding upon complex formation, strength of the interactions and the effects of mutations in amino acid or DNA or RNA for binding specificity. The binding affinity can be quantitatively known with the parameters, dissociation constant (K_d), association constant (K_a), Gibbs free energy change (ΔG), enthalpy change (ΔH) and heat capacity change (ΔC_p). Ji *et al.* (2003) developed Kinetic Data of Biomolecular Interactions database (KDBI), which provides experimentally determined kinetic data of protein–RNA, protein–DNA and other interaction events. It accepts the name of the protein, NA and ligand, SWISS-PROT code, ligand chemical abstracts services (CASs) number and full-text search of a binding or reaction event as input. The output contains the kinetic data such as association/dissociation (or on/off) constant, equilibrium and catalytic rates, as well as inhibition and binding affinity. KDBI is freely accessible at http://bidd.nus.edu.sg/group/kdbi/kdbi.asp.

PDBbind database (Liu *et al.*, 2015) contains PDB-wide collection of experimentally measured binding data for all kinds of biomolecular complexes such as protein–ligand, protein–NA, protein–protein and NA–ligand complexes. This database provides linkage between structural information and energetic properties of biomolecular complexes. The current version of the database contains binding affinity data for about 900 protein–NA complexes. PDBbind is freely accessible at http://www.pdbbind.org.cn/.

7.2.4. *Databases for transcription and posttranscriptional regulation factors and their nucleic acid-binding sites*

Khan *et al.* (2018) developed JASPER, which is a curated, nonredundant TF DNA-binding profile stored as position frequency matrices (PFMs) and TF flexible models (TFFMs) for TFs across multiple species in six taxonomic groups. These matrices can be converted into position weight matrices or position-specific scoring matrices for scanning genomic sequences. JASPER is available at http://jaspar.genereg.net.

Recently, Wang *et al.* (2018) created MeDReaders, which contains information on transcription factors from human and mouse that bind to methylated DNA. The data can be browsed by TF gene symbols and can be searched using TF names, Ensemble gene IDs, RefSeq gene ID or binding DNA sequences. It is available at http://medreader.org/.

Weingender *et al.* (1996) developed a commercial database called TRANSFAC, which contains data relevant for gene expression at the transcription level. The database is useful for finding out transcription factors binding to DNA and acting through them. TRANSFAC comprises of two domains: (1) TF-binding sites, usually in promoters or enhancers and (2) binding proteins. In this database, TFs are classified into classes based on the general properties of their DNA-binding domains.

A comprehensive database for exploring posttranscriptional regulation, POSTAR2 (Zhu *et al.*, 2019), consists of CLIP-seq data, Ribo-seq data, RNA-seq data and other high-throughput sequencing data from six species. POSTAR2 is the largest collection of RBP-binding sites and functional annotations. It provides three modules such as RBP module, RNA module and Translatome module (Open Reading Frames). RBP module provides annotations and functions of RBPs and their binding sites. RNA module annotates the RBP-binding sites using various regulatory events and genomic variants. Translatome module is for exploring the translation landscape of genes across different tissues and cell lines. POSTAR2 is freely available at http://lulab.life.tsinghua.edu.cn/postar.

7.3. Structural analysis of protein–nucleic acid complexes

The availability of 3D structures of protein–NA complexes (protein–DNA and protein–RNA) enabled researchers to explore their structure–function relationship as well as for understanding their recognition mechanisms. These protein–NA complexes are mediated by several noncovalent interactions at the interface, which assist to perform several biological functions.

7.3.1. *Noncovalent interactions*

7.3.1.1. *Hydrogen bonds*

Detailed studies of noncovalent interactions on protein–DNA complexes showed that hydrogen bonds are dominant (>50%) followed by van der Waals, hydrophobic and electrostatic interactions (Lejeune *et al.*, 2005). Mandel-Gutfreund *et al.* (1995) proposed that Asp, Asn, Glu, Gln, Ser and Thr form weak hydrogen bonds (CH...O) with C5 of cytosine, C5-Met of thymine, contributing to the specificity of recognition. Jones *et al.* (1999) accounted that DNA-binding sites are hydrophilic surfaces, which assist to form direct and water-mediated hydrogen bonds.

In contrast, van der Waals interactions are more common than hydrogen bonding in protein–RNA complexes (Ellis *et al.*, 2007). Recently, based on the percentage of base area buried in the RNA interface, Hu *et al.* (2018) classified protein–RNA complexes into high, medium and low, and analyzed the properties such as compositions, total area of the interface, contribution of different noncovalent interactions and secondary structures. They reported that the strength of H-bonding and hydrophobic interactions in high class is more than electrostatic interactions. In low-class complexes, electrostatic interactions frequently occur, whereas in medium-class complexes, there are no preferred interactions.

7.3.1.2. *π-Interactions*

Wilson *et al.* (2014) analyzed π–π stacking, π–π T-shaped and sugar–π interactions in a set of protein–DNA complexes and revealed that pyrimidines are more prevalent than purines for contributing to π–π interactions from the DNA side. On the protein side, Phe is highly preferred followed by Tyr, His and Trp to be involved in these interactions. Further, using quantum chemical calculations, Wilson *et al.* (2016) reported that the contribution of π interactions between the bound and unbounded forms of protein–DNA complexes is up to –50 kJ/mol for stabilizing the complexes. The analysis on sugar–π interactions revealed that the influence of Arg and Tyr is higher than other residues in these interactions in protein–DNA complexes (Wilson and Wetmore, 2015). Specifically, structural studies on RNAse T from *Escherichia coli* demonstrated that Phe residues are important for DNA binding by making π–π stacking interactions and the substitution of Phe into Tyr or Trp reduced the stability and DNA-binding activity (Duh *et al.*, 2015). Gromiha *et al.* (2004b) analyzed the cation–π interactions in protein–DNA complexes and reported that these interactions are mainly formed by long-range contacts and the preference of Arg is higher than Lys to form cation–π interactions.

7.3.1.3. *Contributions of different noncovalent interactions*

Jayaram *et al.* (2002) quantified the contributions of different noncovalent interactions as well as packing free energies in protein–DNA binding. They observed that the contribution of cavitation and van der Waals energies are favorable for complex formation, whereas electrostatics is slightly unfavorable. The positively charged residues bind preferably irrespective of the desolvation expense, whereas negatively charged and neutral residues prove unfavorable to the binding. Zhou *et al.* (2010) examined several physical and chemical properties of protein–DNA interactions and revealed the significance of conformational entropy for binding.

Ahmad and Sarai (2004) computed the net charge, electric dipole moment and quadrupole moment tensors for DBPs and observed that the electric charge distribution in DBPs varies significantly from nonbinding proteins. Further, analysis on thermodynamic and structural data of protein–DNA interactions. Ahmad *et al.* (2008) showed that most of the stabilizing or hotspot residues are evolutionarily conserved and these residues have high packing density. Nadassy *et al.* (1999) characterized the solvent accessibility of the protein–DNA interface and stated that the interface area varies between 1120 and 5800 Å^2, which depends on the protein–DNA complex. The binding sites are mainly occupied by positively charged residues such as Lys and Arg in proteins and negative charges from phosphate groups from DNA. Krüger *et al.* (2018) compared the interactions between PRIs and protein–protein interactions (PPIs) and revealed that the PRIs have more buried surface area than PPIs. Further, they compared the preference of hotspot residues and reported that Leu is more preferred than Ile in PPIs and on contrast, Ile is 10-fold more prevalent than Leu in PPIs (Bogan and Thorn, 1998).

7.3.2. *Propensity of amino acid residues at the interfaces of protein–DNA and protein–RNA complexes*

Gromiha and Fukui (2011) proposed a scoring function-based approach for identifying the binding site residues in protein–NA complexes and showed that positively charged, polar and aromatic residues are important for binding. These residues influence the formation of electrostatic, hydrogen bonding and stacking interactions. Figure 7.1 shows the binding propensity of all the 20 amino acid residues at the interfaces of protein–DNA, protein–RNA and protein–protein complexes.

The results revealed that the binding propensities of aromatic, negatively charged, sulfur-containing and hydrophobic residues are remarkably higher in protein–protein complexes than protein–DNA complexes. In protein–DNA complexes, positively charged residues dominate interactions with DNA and appreciable contribution is observed for polar and aromatic residues. Although binding propensity is similar for several residues in protein–DNA and protein–RNA complexes, notable difference is observed for Lys, Arg, Thr, Ser, Gln and Asn in protein–DNA complexes. Interestingly, these amino acid residues belong to positively charged and polar groups, which form electrostatic interactions/hydrogen bonds with DNA.

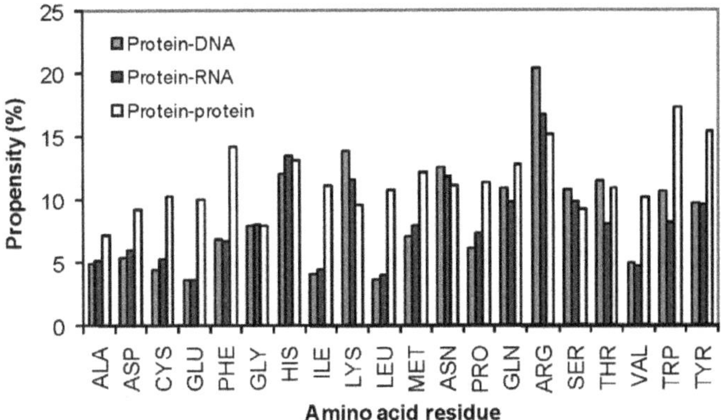

Figure 7.1. Binding propensity of amino acid residues in protein–DNA, protein–RNA and protein–protein complexes (Gromiha and Fukui, 2011).

7.3.3. *Conformational changes of DNA*

The DNA/RNA conformational change is an important factor in protein–DNA/RNA recognition. Olson *et al.* (1998) computed the structural features such as shift, slide, rise, tilt, roll and twist of DNA upon complex formation, which account for conformational changes of DNA. Gromiha *et al.* (2004a, 2005) quantified the contributions from DNA conformation and contacts between protein and DNA and proposed a mechanism based on inter- and intramolecular interactions. In order to account for the indirect readout mechanism, Yamasaki *et al.* (2012) reported a method based on MD simulations of DNAs that have all possible tetramer sequences. Rohs *et al.* (2009) studied a set of protein–DNA complexes and accounted that the local variations in DNA shape in the minor groove and electrostatic potential provide a general mechanism to achieve DNA-binding specificity. Bouvier *et al.* (2011) showed that protein–DNA recognition is induced by conformational changes in DNA.

Recently, Poddar *et al.* (2018) analyzed free and complex forms of DBPs and observed that binding residues (BRs) have significant changes in accessible surface area (ASA) values and RMSDs in protein–DNA interactions, whereas in PPIs these changes are minimal. In addition, 25% of proteins that have disordered residues are being changed to ordered residues upon complex formation.

Corona and Guo (2016) grouped protein–DNA complexes into highly specific, multi-specific and nonspecific classes, based on the binding specificity of DBPs. They showed that highly specific and multi-specific binding proteins mostly targeted the major groove of DNA, whereas nonspecific cases are majorly prevalent in the minor groove. In addition, highly specific and multi-specific DBPs influence conformational changes upon DNA binding and these proteins have high flexibility in free and complex forms. Further, Asp binds with cytosine preferably via a hydrogen bond in highly specific DBPs. Gardini *et al.* (2017) classified protein–DNA complexes into structural proteins, transcription factors and DNA-related enzymes, and reported that the presence of Arg at the protein–DNA interface is more abundant among other amino acids and the preference of Arg is different among proteins from these three classes.

7.3.4. *Protein–DNA/RNA interfacial properties and functional implications*

Prabakaran *et al.* (2006) categorized protein–DNA complexes into seven distinct clusters based on the contribution of different noncovalent interactions, conformation and structural properties of DNA, and buried surface area at the complex interface. They reported that proteins with identical motifs are classified into different clusters, whereas different proteins with distinct motifs are classified into the same cluster, suggesting that the conventional motif-based classification of DBPs may not necessarily correspond to structural and functional properties of protein–DNA complexes. Sathyapriya *et al.* (2008) carried out investigations on protein- and DNA-centric networks as well as protein–DNA graphs/networks. The results revealed that deoxyribose–amino acid clusters are dominant in β-sheet proteins and this approach could distinguish between interface clusters observed in helix-turn-helix, and zipper-type proteins. They have also proposed a classification scheme for protein–DNA complexes based on their interface clusters.

Wang *et al.* (2014) compared binding site properties in single-stranded (SSBs) and double-stranded DNA-binding proteins (DSBs) and reported that binding sites of DSBs and SSBs are preferred by α-helix and β-strand, respectively. Moreover, the interface of DSBs is wave shape, whereas in SSBs, the interface is smaller and irregular. Based on the charge at the interface, DSBs are more positively charged than SSBs. Hudson and Ortlund (2014) demonstrated that DRBPs composed of about 2% in human proteome and these proteins play important cellular roles such as transcription, translation, gene silencing and microRNA biogenesis. Cordeiro *et al.* (2014) reviewed the role of DNA and RNA in protein–NA interactions to the conformational change of proteins that aggregate in protein misfolding disorders, such as Alzheimer, Parkinson and prion diseases.

7.3.5. *Residues play dual roles on binding and stabilizing in protein–nucleic acid complexes*

Structural studies on protein–NA complexes revealed that residues at the interface are important for binding, whereas other residues contribute for stability. The analyses on residues, which play a dual role in both binding and stability, provide insights to understand the importance of these residues for simultaneously attaining stability and affinity.

Stabilizing residues (SRs) are identified using the concepts of hydrophobicity, long-range interactions and conservation of residues. Consequently, the structural parameters such as surrounding hydrophobicity (Hp > 16), long-range order (LRO > 0.01), stabilization center (SC > 0) and conservation score (CS > 6) are used to identify the SRs and are obtained from the SRide server, http://sride.enzim. hu/ (Gromiha *et al.*, 2004c; Magyar *et al.*, 2005).

BRs in protein–DNA/RNA complexes are obtained by using a distance cutoff ≤3.5 Å between any heavy atoms in protein and DNA/RNA. Residues that are important for both binding and stability of a protein are called as key residues (KRs) (Kulandaisamy *et al.*, 2017).

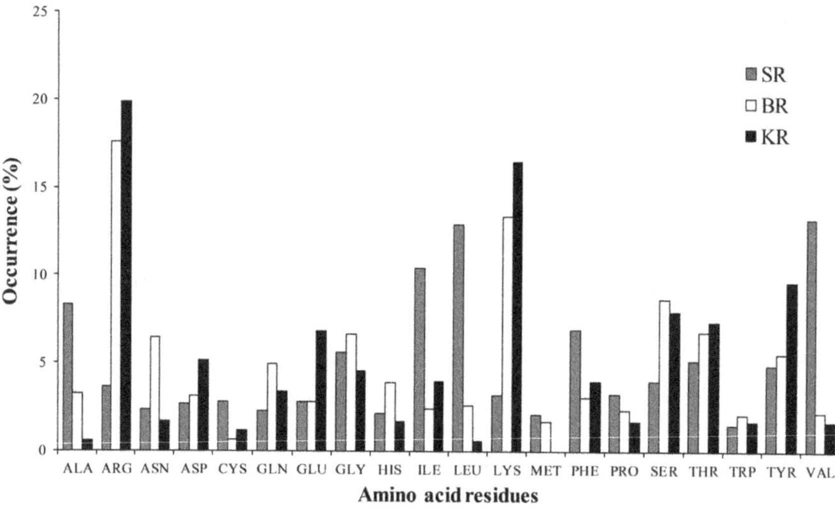

Figure 7.2.   Frequency of occurrence of 20 amino acid residues in stabilizing, binding and KRs in protein–DNA complexes (Kulandaisamy *et al.*, 2018b).

7.3.6. *Frequency of occurrence of stabilizing, binding and KRs in protein–DNA/RNA complexes*

In protein–DNA/RNA complexes, 4–5% of SRs and 3% of BRs are identified as KRs. Further analysis revealed that hydrophobic residues Val, Leu, Ile, Ala and Phe are enriched in SRs, which indicates that hydrophobic interactions play a major role in the stability of protein–DNA complexes (Figure 7.2). On the other hand, BRs are dominated by positively charged (Arg and Lys) and polar residues (Ser, Thr and Asn), and this might be due to the predominance of electrostatic and hydrogen bond interactions in protein–DNA complexes. The KRs are composed of charged, polar and nonpolar residues (Figure 7.2). Especially, Arg, Tyr, Ser, Gly are preferred over Lys, Trp, Asn, Leu, respectively. These trends have also been observed in protein–RNA and protein–protein complexes (Kulandaisamy *et al.*, 2017, 2018a), in which charged, polar and hydrophobic residues are important to both binding and stability.

7.3.7. *Preference of atomic contacts and DOT in KRs*

The comparison of atomic contacts in BRs and KRs between protein and DNA/RNA showed that BRs have marginally higher number of polar–polar and polar–nonpolar contacts than KRs; however, in both BRs and KRs, nonpolar–nonpolar contacts are minimal. In protein–DNA complexes, charged contacts are significantly higher in KRs than protein–RNA and protein–protein complexes (Kulandaisamy *et al.*, 2017, 2018a).

Further, the distribution of BRs, SRs and KRs in DOT regions in a set of protein–DNA complexes showed the presence of several BRs in DOT regions, whereas no SRs and KRs are observed in these regions. This suggests that disorderness is more important for binding than stability (Kulandaisamy *et al.*, 2018b).

7.3.8. *Functional importance and diseases associated with KRs*

Analysis on KRs revealed that they are involved in endocrine and metabolic disorders and various types of cancers (Kulandaisamy *et al.*, 2018b). In addition, KRs are involved in different functions, specifically phosphorylation. In human POTE1 protein, the mutation of a KR Gln 94 into Glu completely abolishes the POT1–DNA complex formation, which is a cause of melanoma, a cutaneous malignant 10 disease (Robles-Espinoza *et al.*, 2014). Moreover, a runt-related TF 1 protein strongly reduces the binding affinity due to mutations at R80, R135, K167 and T169 (Bravo *et al.*, 2001), and interestingly, these residues are observed to be KRs in protein–DNA complexes.

7.4. Disorder-to-order transition

Intrinsically disordered regions (IDRs)/proteins lack stable three dimension in solution and are involved in cell cycle, signaling and regulation (Tompa, 2012). Disordered proteins are highly populated in cells and help to regulate the function by binding multiple partners,

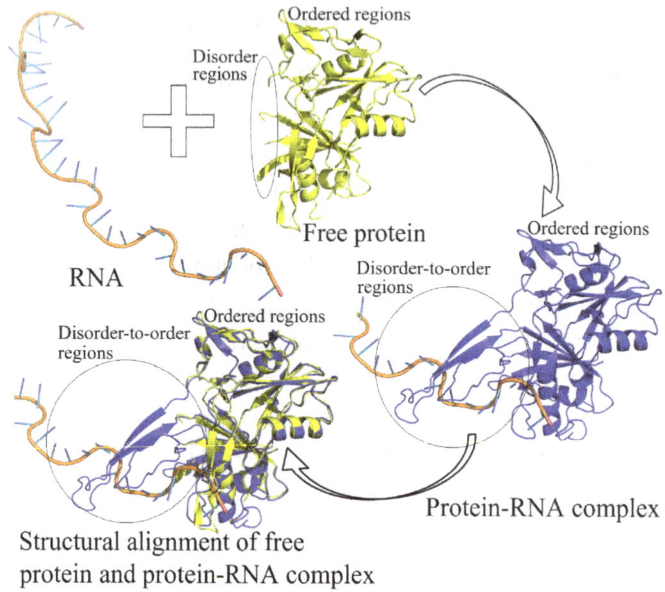

Figure 7.3. An example of DOT in a protein–RNA complex (Srivastava *et al.*, 2018).

fold upon binding, alter posttranslational modification and changing binding affinity (Toto *et al.*, 2016). The disordered regions of these proteins when bind to their substrates have a transition from disorder to order state, which are also known as DOT regions (Figure 7.3). These DOT regions have linear and low complexity domains and are also known as short linear motifs (SLIMs) (Tompa, 2012; Uversky, 2013; Bah and Forman-Kay, 2016; Srivastava *et al.*, 2018).

Protein–NA complexes often have the abundance of disordered regions. These regions provide essential flexibility to the protein to facilitate shape and chemical complementarity. The disordered proteins can tune their affinity of binding to the NA molecules (Vuzman and Levy, 2012; Järvelin *et al.*, 2016). In DOT regions of protein–RNA complexes, Arg, Lys and His are dominant, whereas Arg, Lys and Gly are enriched in those of protein–DNA complexes (Srivastava *et al.*, 2018; Poddar *et al.*, 2018).

IDRs are experimentally identified by biophysical and biochemical methods such as NMR, X-ray crystallography, circular dichroism and small angle X-ray scattering. The interaction of disordered regions with

NAs can also be identified by SELEX, RNA/DNA pull down assay, FRET and yeast two-/three-hybrid systems (Marasco and Scognamiglio, 2015). On the other hand, computational methods such as Hamiltonian mapping, PredictProtein, EPSLiM and IDEAL are used for predicting IDRs in proteins (Tompa *et al.*, 2015). ANCHOR and DisoRDPbind are two widely cited methods that are used to predict protein-, RNA- and DNA-binding residues in disordered regions of proteins (Dosztányi *et al.*, 2009; Peng and Kurgan, 2015).

The short DOT regions that are important in protein binding are also known as molecular recognition features (MoRFs). These MoRFs exist in all domains of life, in particular, the MoRF content of archaea and bacteria shows a good correlation with the disorder content. However, the amount of MoRF residues is poorly correlated with the disorder content in eukaryotes. In a proteome-wide study of organisms from three domains of life, it was observed that the disorder content in DRBPs is more than other proteins. The disorder enrichment correlation analysis showed that DBPs are more enriched in eukaryotic animals and plants having no biases in bacteria and archaea proteome; whereas RNA-binding proteins are enriched with disordered regions in all domains of life (Yan *et al.*, 2016; Varadi *et al.*, 2015).

7.4.1. *Examples*

Few examples showing the importance of IDRs are discussed below. HIV-1 Rev protein has a role in transportation of viral RNA and the C-terminal domain of this protein is disordered. This disordered region is shown to be important for both oligomerization and stability of the complex. A study on disordered part of c-Myb, a myeloblastosis oncoprotein and its ordered partner KIX showed that mutation in disordered regions impact binding affinity as well as helical stability of the protein (Poosapati *et al.*, 2018).

Wollenhaupt *et al.* (2018) showed that ntr2, yeast spliceosomal disassembly factor, is an intrinsically disordered protein, which down-regulates the Brr2 helicase activity by altering the helicase proteins bound to the RNA substrate. A study on the histone terminal domain showed the importance of IDRs in binding specificity (Hansen *et al.*, 2006). In the CRISPR-Cas system of prokaryotes, structures of

Cmr complexes have multiple disordered regions, which get fully structured upon binding with crRNA to perform their functions (Osawa *et al.*, 2015).

7.4.2. *Statistical analysis on DOT residues in protein–nucleic acid complexes*

Disordered regions are biased with specific amino acids such as Gly, Arg, His, Cys and Trp and generally have low complexity regions (Varadi *et al.*, 2015). The DOT residues, also termed as dual personality residues, have unique composition biases to ordered or IDRs (Zhang *et al.*, 2007) and most of them are short and dominated by Ala, Gly, Ser, Glu and Lys. These residues are more often targeted than ordered or disordered regions for posttranslation modifications and regulation activities (Zhang *et al.*, 2007). The preference of residues in disordered regions of DRBPs is different. In DNA-binding regions of proteins and transcription factors, the frequency of His, Lys, Glu, Ser and Pro are dominant (Guo *et al.*, 2012). The positively charged residues are found to be appreciable in protein–RNA complexes as shown in Figure 7.4 (Srivastava *et al.*,

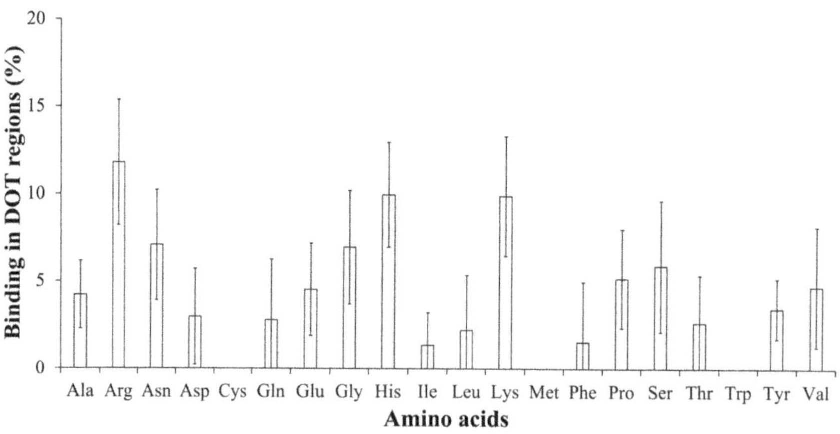

Figure 7.4. Composition of BRs in DOT regions of protein–RNA complexes (Srivastava *et al.*, 2018).

2018). Analysis on DRBP structures deposited in PDB showed that RNA-binding proteins are more disordered than DBPs and the disorderness is highly conserved in RNA-binding proteins (Varadi *et al.*, 2015). In protein–RNA complexes, disordered regions are generally small in size and less in number, which might be required to bring necessary shape complementarity during interactions (Srivastava *et al.*, 2018). Further, the secondary structure and solvent accessibility calculations suggested that disordered regions generally form loops, turns and coils, which mostly convert into β-strand upon binding.

The interaction energy between amino acid residues and nucleotides of DOT and non-DOT residues is computed using the following equation (Cornell *et al.*, 1995):

$$\text{Interaction energy} = \sum \left[\left(\frac{A_{ij}}{r_{ij}^{12}} - \frac{B_{ij}}{r_{ij}^{6}} \right) + \frac{q_i q_j}{\varepsilon r_{ij}} \right] \qquad (1)$$

where $A_{ij} = \varepsilon_{ij}^{*}(R_{ij}^{*})^{12}$ and $B_{ij} = 2\varepsilon_{ij}^{*}(R_{ij}^{*})^{6}$; $R_{ij}^{*} = (R_i^{*} + R_j^{*})$ and $\varepsilon_{ij}^{*} = (\varepsilon_i^{*} \varepsilon_j^{*})^{1/2}$; R^* and ε^* van der Waals radius and well depth, respectively; q_i and q_j are the charge on atoms i and j, respectively, and R_{ij} is the distance between atoms i and j.

Table 7.2 lists the interaction energy between amino acids and nucleotides in DOT regions along with ordered regions, and the results highlight specific strong and weak interactions of DOT and non-DOT residues in protein–RNA complexes. For example, His strongly interacts with G and C. Among other positively charged residues, Arg prefers to interact with G, whereas Lys has no such bias. Interestingly, due to the importance of electrostatic interaction for binding between proteins and NAs, Lys and Arg have higher preference in ordered regions than DOT regions. These differences in interaction energy could help to understand specific interactions between DOT regions in protein and RNA, which could also be used to distinguish between RNA BRs of proteins in DOT and other regions (Srivastava *et al.*, 2018).

Table 7.2. Interaction energy between amino acids and nucleotides in DOT regions.

Amino acids	A	G	C	U
Ala	−0.62 (−0.55)	−0.34 (−0.57)	−0.49 (−0.53)	−0.55 (−0.64)
Arg	−0.36 (−1.23)	**−1.15** (−0.83)	−0.89 (−0.95)	**−1.06** (−0.98)
Asn	−0.45 (−0.68)	−0.59 (−0.73)	−0.48 (−0.83)	**−1.85** (−0.82)
Asp	−0.75 (−0.74)	−0.39 (−0.79)	−0.19 (−0.56)	**−1.40** (−0.92)
Cys	0.00 (−0.87)	−0.01 (−0.03)	−0.03 (−1.10)	−0.63 (−1.13)
Gln	−0.15 (−0.87)	−0.57 (−0.74)	−0.08 (−0.84)	−0.36 (−0.71)
Glu	−0.72 (−0.80)	−0.41 (−0.64)	−0.43 (−0.62)	−0.68 (−0.59)
Gly	−0.28 (−0.47)	−0.37 (−0.69)	−0.58 (−0.57)	**−1.07** (−0.79)
His	−0.81 (−1.17)	**−2.13** (−1.41)	**−1.53** (−1.21)	−0.70 (−1.01)
Ile	−0.60 (−0.64)	**−1.63** (−0.80)	−0.54 (−0.50)	**−1.33** (−0.76)
Leu	−0.35 (−0.75)	**−1.19**(−0.50)	−0.42 (−0.49)	−0.54 (−0.41)
Lys	−0.74(−0.76)	−0.86 (−0.83)	−0.66 (−0.90)	−0.83 (−0.83)
Met	−0.64 (−1.05)	−0.07 (−0.75)	−0.16 (−1.03)	−0.83 (−1.19)
Phe	−0.81 (−1.03)	**−1.12** (−0.89)	−0.54 (−1.32)	−0.24 (−1.42)
Pro	−0.88 (−0.83)	−0.60 (−0.88)	−0.62 (−0.91)	−0.69 (−1.00)
Ser	−0.79 (−0.77)	−0.29 (−0.56)	**−1.24** (−0.71)	**−1.41** (−0.68)
Thr	−0.39 (−0.68)	−0.66 (−0.56)	−0.67 (−0.64)	−0.53 (−1.00)
Trp	**−1.15** (−1.10)	0.00 (−1.64)	0.00 (−0.99)	0.00 (−1.34)
Tyr	**−1.53**(−1.36)	−1.11 (−1.42)	−0.63 (−1.05)	−0.16 (−1.09)
Val	−0.38 (−0.70)	−0.66 (−0.53)	−0.64 (−0.53)	−0.52 (−0.76)

*Interaction energy for non-DOT residues is mentioned in the parenthesis. Amino acid–nucleotide pairs with favorable interaction energies in DOT regions are shown in bold. Data were taken from Srivastava *et al.* (2018).

7.5. Mechanisms for protein–DNA recognition

The structural data obtained from X-ray crystallography and NMR spectroscopy provide valuable information to understand the general features of DBPs. In addition, researchers have explored the interactions of TFs among each other and with their binding partners. Although significant progress has been achieved toward understanding the TF search process, the details of this mechanism remain controversial.

Several mechanisms have been proposed to describe how TFs find their target sites on DNA (Figure 7.5) and are reviewed in Yesudhas *et al.* (2017a). One of the main scenarios involves a "sliding" mechanism, in which the protein moves from its initial nonspecific site to its actual target site by sliding along the DNA (also known as one-dimensional sliding). The binding of the lactose (lac) repressor to nonoperator sequences is an ideal example of sliding, and its DNA binding entirely relies on electrostatic interactions thereby favoring the diffusion process (Komazin-Meredith *et al.*, 2008; Winter *et al.*, 1981; Kalodimos *et al.*, 2004). Only a few DBPs using the sliding mechanism from nonspecific to specific binding have been structurally solved, among them are BamHI (Viadiu and Aggarwal 2000), λ-repressor (Albright *et al.*, 1998) and the lactose repressor (Furini *et al.*, 2013).

The second scenario is a "hopping" mechanism, in which a TF might hop from one site to another in the 3D space by dissociating from its original site and subsequently binding to the new site.

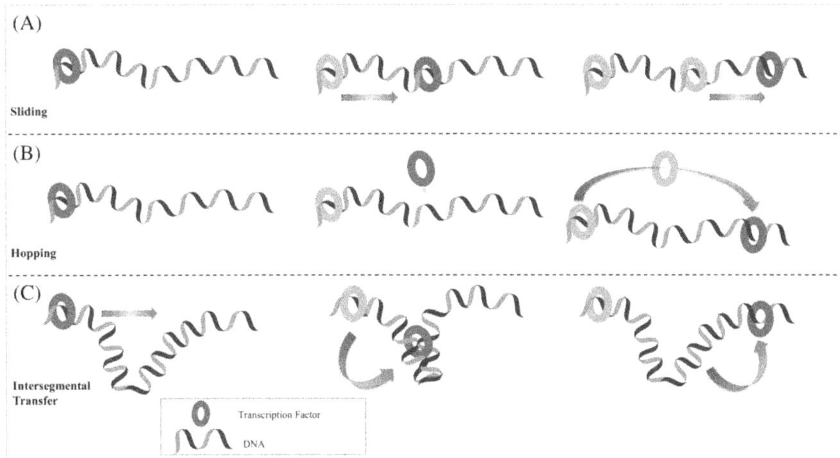

Figure 7.5. Protein–DNA recognition mechanisms: (A) sliding, (B) hopping and (C) intra-segmental transfer (Yesudhas *et al.*, 2017a).

This may happen within the same chain and reassociation occurs adjacent to the former dissociated site (Komazin-Meredith *et al.*, 2008). For example, in the case of Herpes Simplex Virus Type 1 UL42, the binding between protein and DNA is mostly governed by electrostatic interactions, which will be responsible for dissociation and reassociation of UL42 into its new position. Winter *et al.* (1981) suggested that hopping should be slower than sliding as its mechanism relies on dissociation followed by reassociation to a different site.

Third search mechanism, proposed by von Hippel and Berg (1989), is described as "inter-segmental transfer" in which, the protein moves between two sites via an intermediate "loop" formed by the DNA and subsequently bind at two different DNA sites. This mechanism is applicable to TFs with two DNA-binding sites (e.g., Lac repressor or SfiI endonuclease) (Halford *et al.*, 2000). Proteins with two DNA-binding sites can occasionally bind nonspecifically to two locations situated far apart within the DNA strand, which are brought into close contact through the formation of these loops. Such TFs transfer across a point of close contact without dissociating from the DNA (Lomholt *et al.*, 2009). This inter-segmental transfer accelerates the search for the DNA target site because this mechanism is based on constantly changing random configurations (von Hippel and Berg, 1989; Halford and Marko, 2004).

7.5.1. *Direct and indirect readout mechanisms*

7.5.1.1. *Direct readout*

Regulatory proteins and other TFs are known to regulate its target genes expression by recognizing its specific DNA sequences on the genome. Their interactions with DNA can be governed directly through atomic contacts (intermolecular readout) and/or indirectly through the conformational properties of the DNA (intramolecular readout). The intermolecular recognition occurs when the amino acid side chains of a protein interact with specific DNA and mainly through hydrogen bonding between the protein and the major/minor groove base pairs. However, the total number of interactions appears between the base pair and protein is also equally important to consider. Bidentate bonds (two hydrogen bonds with different donor and acceptor atoms) have higher degree of specificity than bifurcated hydrogen bonds (two hydrogen bonds sharing the same donor). The bidentate bonds tend to contribute more for specificity than the single hydrogen bond (Coulocheri *et al.*, 2007; Rohs *et al.*, 2010; Yesudhas *et al.*, 2017a).

7.5.1.2. *Computation of intermolecular interaction energy*

Kono and Sarai (1999) proposed a method for computing the intermolecular interaction energy based on the distribution of amino acid residues around DNA bases. In this method, each amino acid residue in a protein–DNA complex has been represented by its C_α atom and the distributions of C_α atoms of amino acid residues around bases have been transformed into statistical potentials using the following equations (Sippl, 1990):

$$\Delta E^{ab}(s) = -RT lnln \frac{f^{ab}(s)}{f(s)} \tag{2}$$

$$f^{ab}(s) = \frac{1}{1 + m_{ab}w} f(s) + \frac{m_{ab}w}{1 + m_{ab}w} g^{ab}(s) \tag{3}$$

where m_{ab} is the number of observed pairs (amino acid a and base b), w is the weight given to each observation, $f(s)$ is the relative frequency of occurrence of any amino acids at grid point s and $g^{ab}(s)$ is the equivalent relative frequency of occurrence of amino acid a against base b. R and T are gas constant and absolute temperature, respectively.

By threading random DNA sequences onto the protein–DNA framework, it is possible to calculate the Z-score of the specific sequences against the random sequences, which represents the specificity of the complex. The Z-score function is a quantitative measure of the specificity of protein–DNA recognition that helps to identify the relationship between structure and specificity. Assuming the additivity of potential energies, the sum of the potential energies $[E_{PD} = \Sigma_{ab,s} \Delta E^{ab}(s)]$ for a given DNA sequence in a complex form was defined as the energy for the sequences. The energy for a particular sequence, in a crystal structure, for example, was normalized to measure specificity by the Z-score against random sequences. The Z-score was defined as

$$Z\text{-score} = (X - m)/\sigma \tag{4}$$

where X is the energy of a particular sequence, m is the mean energy of a set of (e.g. 50,000) random DNA sequences and σ is the standard deviation (Kono and Sarai, 1999).

7.5.1.3. *Indirect readout*

Although direct readout exists in all protein–DNA complexes, structures of bound DNA frequently deviate from their standard ones. Such changes for protein–DNA recognition are also known as shape readout or indirect readout mechanisms, in which the binding relies on the base pairs create structural changes within the DNA to facilitate recognition. It accounts for structural rearrangements of DNA, which have been evaluated mainly with the average and deviations of six base-step parameters (shift, slide, rise, tilt, roll and twist).

7.5.1.4. *Computation of intramolecular interaction energy*

The intramolecular interaction energy of a protein–DNA complex structure was calculated from the conformational energy of DNA, proposed by Olson *et al.* (1998). It is given by

$$E_{DNA} = 1/2 \Sigma\Sigma f_{ij}\Delta\theta_i\Delta\theta j \tag{5}$$

where θ_i represents the base-step parameters and f_{ij} is the elastic force constant impeding deformation of the given base-step pair and $\Delta\theta_i = \theta_i - \theta_i^0$, in which θ_i^0 is the average base-step parameter. The base-step parameters used were shift, slide, rise, tilt, roll and twist. The unknown parameters f_{ij} and θ_i^0 were determined by statistical analysis of the same nonredundant protein–DNA complexes. Setting up a covariance matrix from observed distributions of θ_i thus refers to an effective inverse harmonic force-constant matrix. Inversion of this matrix transformed it to a force-constant matrix in the original coordinate basis. All parameters of a base step for which one parameter exceeded three standard deviations were removed in an iterative manner and the final force field was calculated. The conformational energy of DNA in a given complex structure was calculated as the sum of all the base steps. These potentials were used to quantify the specificity of intramolecular interactions of protein–DNA recognition, as a Z-score (Eq. 4), by using the same threading procedure as that of intermolecular interactions.

7.5.1.5. *Examples*

Most of the minor groove-binding proteins, which often distort the DNA geometry, fall under the indirect readout mechanism. Integration host factor, IHF–DNA complex (Rice *et al.*, 1996) (PDB code, 1IHF) DNA is severely bent (overall bending angle, 160°) and the conformation of the DNA is crucial for its recognition. TATA box-binding proteins (TBPs) bind to the minor groove and bend the DNA (Gromiha *et al.*, 2004a). DNA recognition by zinc finger proteins has been studied extensively and provides a good model system for the direct readout mechanism. Gromiha *et al.* (2004a) showed that intermolecular readout makes major contribution to its specificity of zinc finger proteins as well as restriction enzyme, EcoRV.

The elusive conformational shift occurring in biomolecules is the transition among B, A and Z conformations of DNA (Lee *et al.*, 2012). B-form DNA is the most favored conformation for a protein–DNA complex under physiological conditions, whereas A-form DNA is induced locally in some complexes. The important factors driving this transition are hydration and electrostatics; however, solvent conditions, counterions condensation and free energy contribution from phosphate–phosphate repulsion also contribute to this transition (Waters *et al.*, 2016; Yesudhas *et al.*, 2017a).

7.5.1.6. *Combination of inter- and intramolecular interactions*

The inter- and intramolecular energies obtained from statistical potentials are combined to calculate the total energy. Because the derivations of these empirical energies are based on different statistics, a weighting factor has been introduced such that

$$E_{tot} = cE_{PD} + (1 - c)\, E_{DNA} \qquad (6)$$

where E_{PD} and E_{DNA} are the energies of the inter- and intramolecular interactions, respectively, and c is a weighting coefficient ranging between 0 and 1, and it is determined by maximizing the total Z-score. Gromiha *et al.* (2004) reported that the combination of inter- and intramolecular energies enhanced the Z-score in most of the protein–DNA complexes, emphasizing the importance of both interactions for recognition.

7.5.2. *Role of water molecules in protein–DNA recognition*

Water-mediated contacts are shown to play major roles in the binding specificity of these complexes (Reddy *et al.*, 2001; VanSchouwen *et al.*, 2008). In addition, hydrogen bonding between proteins and DNA is facilitated by water molecules. Highly ordered water molecules also mediate the specific recognition between base pairs of major groove and protein by positioning hydrogen bond donors and acceptors at the base edges. For example, in the case of lac repressor, protein–DNA interface is enriched with water molecules when it binds nonspecifically; however, this interface is devoid of water when it binds to specific sites (Rohs *et al.*, 2010; Yesudhas *et al.*, 2017b). Most water molecules around DNA–protein interfacial areas are released to the bulk solvent environment upon formation of a protein–DNA complex, which is believed to play a role in the binding thermodynamics, but it is largely unknown whether the process also contributes to binding specificity (Sarai and Kono, 2005).

Protein side chains and water molecules in the binding interface create a flexible network of interactions that can readily adopt new conformations in response to changes in DNA or protein sequence.

Water molecules at protein–DNA interface help in electrostatic repulsion of like charges of protein and DNA, and also facilitate the hydrogen bonding between protein and DNA, thus the water-mediated interaction is also an important factor for the binding affinity. For example, in the case of wild-type and mutant zinc finger protein Zif268/Egr1 protein–DNA crystal structures, it was demonstrated that a single amino acid mutation (Zif268 D20A) could lead to altered protein side-chain conformations and DNA-binding preferences (Miller and Pabo, 2001). However, the side-chain rearrangement occurred without altering the protein-docking geometry, but resulted in a new, positioned water molecule in the binding interface for one of the DNA sequences (Siggers and Gordan, 2014). A comparative study on two Cre recombinase variants in complex with distinct DNA sequences revealed that both DNA and protein differences affect the contacts made in the binding interface (Baldwin *et al.*, 2003). These studies highlight the complex interplay of water molecules along with amino acid side chains and DNA bases in the binding interface.

7.6. Organism-specific recognition of protein–nucleic acid complexes

The availability of 3D structures of protein–NA complexes from different organisms enabled researchers to understand their recognition mechanisms in detail. Nagarajan *et al.* (2015) constructed 18 sets of protein–RNA complexes from different organisms and revealed the similarities and differences among them based on various sequence- and structure-based features such as root mean square deviation, sequence homology, propensity and conservation of binding site residues, binding motifs of both amino acid residues and nucleotides, preferred amino acid–nucleotide pairs and influence of neighboring residues for binding. They showed that proteins of mesophilic organisms have more number of binding sites than thermophiles in both protein and RNA, and the binding propensities of amino acid residues are distinct in *E. coli*, *Homo sapiens*, *Saccharomyces cerevisiae*, thermophiles and archaea. The residues Pro, Cys and Gln show high preference in *S. cerevisiae*, Lys, Arg and Phe are dominant in *E. coli*, whereas Gly and Trp are preferred in *H. sapiens*. The preference of amino acid residue–nucleotide pairs showed that amino acids show inclination toward pairing with cytosine in *E. coli*, whereas they prefer to interact with adenine in *H. sapiens* and *S. cerevisiae* (Nagarajan *et al.*, 2015).

Further, extension of the study with MD simulations of aspartyl tRNA synthetase complexed with aspartyl tRNA (AspRS-tRNA^{Asp}) form *E. coli*, *S. cerevisiae* and *T. thermophilus* revealed that the number of interface residues, hydrogen bonding and hydrophobic interactions and binding energy varies considerably among these organisms. Specifically, high variations observed in the anticodon binding domain of the AspRS signifies that the enzyme is evolved for highly specific recognition of the cognate tRNA, which substantiates the concept of codon bias in recognition. Similarly, variations in the catalytic and insertion domains revealed that the amino acid charging is also organism specific.

In addition, functional molecular motions associated with the charging of tRNA^{Asp} with the cognate aspartic acid are seen only in the AspRS-tRNA^{Asp} cognate complexes of *E. coli* (PDB-id: 1C0A) (Eiler *et al.*, 1999) and *S. cerevisae* (PDB-id: 1ASY) (Ruff *et al.*, 1991). It involves (1) an N-terminal domain rotation, (2) binding of D-arm of tRNA^{Asp} with the hinge domain of AspRS and (3) binding of 3′-OH acceptor arm of tRNA with the catalytic domain of AspRS. Simulations of crystal structures of the noncognate complexes, AspRS (*Thermus thermophilus*) — tRNA^{Asp} (*E. coli*) (PDB-id: 1EFW) (Briand *et al.*, 2000) and AspRS (*E. coli*) — tRNA^{Asp} (*S. cerevisiae*) (PDB-id: 1IL2) (Moulinier *et al.*, 2001) did not show these conformational changes and the result implies that the recognition is highly specific and it varies with the organism type.

7.7. Molecular dynamics simulations of protein–nucleic acid complexes

MD has been widely used to study a variety of normal and unusual forms of DNA and other NAs (Cheatham, 2004; Cheatham and

Young, 2000). The dynamic behavior of DNA structures mostly govern their binding properties, which can be understood through computational techniques. The monitored dynamic movements of atoms will shed light on the functional and structural phenomena undergone by proteins or DNA during the initial phase of complex formation and reveal details about the energetics and dynamics of the recognition process that are difficult to obtain through experiments. Application of MD simulations to calculate binding free energy landscapes for DBPs has several potential benefits especially; to obtain biophysical insights into the nature of the interactions governing specificity of the complexes would pave a way to build simplified methods for structure-based prediction for complex molecules (Khabiri and Freddolino, 2017).

7.7.1. *Force fields*

The MD simulations can provide the detailed atomic-level information, and several program packages such as AMBER, CHARMM, GROMOS, GROMACS and NAMD have been developed specifically for atomistic simulations of biomolecules including NAs. MD simulation of NAs was lagged behind protein simulation because of its structure, charge and availability of crystal structures to validate the NA simulation methodologies. About 80–90% of the computational cost comes from solvent–solvent interactions, and it is necessary to treat the solvent and electrostatics more carefully for NA structures and care should be taken for choosing force field parameters for NA structures.

The AMBER force field for NAs has been used extensively and multiple artifacts have been discovered, corrected and reassessed recently (Galindo-Murillo *et al.*, 2016). AMBER has good NA-specific force field corrections (Svozil *et al.*, 2008; Perez *et al.*, 2007; Ivani *et al.*, 2016), whereas CHARMM is commonly used for lipid and protein simulations. In addition, substantial refinements and implementations have been made in AMBER (Ivani *et al.*, 2016), OL15 (Zgarbova *et al.*, 2015) and CHARMM (Hart *et al.*, 2012) DNA parameters (Galindo-Murillo *et al.*, 2015, 2016), which yielded significant improvements in the reproduction of physically realistic conformations of duplex DNA (Khabiri and Freddolino, 2017). Recently, computational facilities made researchers not only to focus on conformational sampling but also to discover deficiencies and overcome problems with the force fields.

7.7.2. *Conformational changes*

The deformability of DNA plays an important role in the biological processes. Then, it is necessary to analyze the flexibility of DNA at the base-pair level, including its classical transitions, such as A–B and B–Z (Wan *et al.*, 2013). The sequence-dependent kinks in the DNA molecule or minor deviations in the DNA groove width that can lead to preferential binding to the cognate protein can also be analyzed from the MD trajectories. Curves is a widely used NA conformational analysis program, which provides a full analysis of DNA structure, including base pair–axis parameters, intra-base and inter-base pair parameters, backbone and groove parameters.

The elastic nature of the DNA can be obtained from analyzing the MD trajectories and can be approached in many other ways. One of them is based on techniques such as principal component analysis (PCA), which captures the deformation profile of the molecule (eigenvectors) and the relative importance of the eigenvectors (associated eigenvalues). Eigenvalues can easily be converted into force constants, which provides an energetic measure of deformation. Similar analysis can be performed on the DNA structure parameters such as shift, slide, rise (translations parameters) and tilt, roll, twist (rotation parameters) relating the two base pairs of a base-pair step.

7.7.3. *MD simulation studies on protein–DNA complexes*

In addition, MD simulation studies have illustrated the complexity of the intercalation of small drugs to DNA, suggesting a mechanism that explains the drug favors the DNA unwinding. Mukherjee *et al.* (2008) reported the complexity of drug binding to DNA minor groove using MD simulations and explained the view of drug traveling paths among many conformations leading to nonspecific contacts before reaching the high-affinity spot.

Generally, protein–DNA complexes are categorized into three groups based on their recognition: (i) specific hydrogen bond interactions between protein and nucleobases, where the DNA structure is not altered, (ii) electrostatic interactions mediated by grooves, whose geometry is sequence dependent, in which DNA is largely distorted but maintains the duplex integrity and (iii) sequence-dependent DNA deformability where the duplex integrity is lost, either locally or globally (Perez *et al.*, 2012). The exact mechanism by which DNA deforms its geometry upon protein binding is still being debated, and simulation results support either the induced fit or the conformational selection paradigms (Yesudhas *et al.*, 2016, 2017b). The mechanism of proteins scanning the genome to find the target sequence is still unclear; however, MD simulation studies revealed that the DNA deformability of SRY protein is responsible for its target sequence identification, which suggests the possibility of similar mechanism by other TFs also to recognize their substrates (Banerjee *et al.*, 2005; Qi *et al.*, 2009; Perez *et al.*, 2012; Yesudhas *et al.*, 2017a). Similarly, specific and nonspecific binding nature of TFs are also studied from MD simulation, for example, the lactose repressor first binds nonspecifically to DNA, and subsequently it rapidly searches for its actual target binding site. The detailed description of the steps involved in the transformation of nonspecific complex into the specific one was explained with MD simulations (Furini *et al.*, 2013). All these studies illustrate the importance of MD simulations to clarify long-standing issues, such as the thermodynamics and kinetics of drug DNA binding, and to explain specificity and selectivity of TFs (Perez *et al.*, 2012).

7.7.4. *Free energy calculations*

For each trajectory in MD simulations, the binding energy (ΔG_{bind}) between protein and DNA in a protein–DNA complex is computed using the following equation:

$$\Delta G_{\text{bind}} = G_{p+D} - \left(G_p + G_D\right) \tag{7}$$

where G_{p+D}, G_p and G_D are the free energies of the complex, the isolated protein and DNA, respectively. In the Molecular Mechanics Poisson Boltzmann Surface Area approach (MMPBSA), each free energy term in Eq. (1) is calculated as

$$G = E_{\text{bond}} + E_{vdw} + E_{\text{elec}} + G_{PB} + G_{SA} - TS_s \tag{8}$$

where E_{bond} is the contribution from bond energy, that is, the sum of the bond length, angle and dihedral energies; E_{vdw}, E_{elec}, G_{PB} and G_{SA} are van der Waals, electrostatic, polar and nonpolar contributions to the solvation energy, respectively; T is the absolute temperature and S_s is the solute entropy (Furini *et al.*, 2013). Polar and nonpolar contributions to the solvation energy were calculated using the Adaptive Poisson–Boltzmann Solver (APBS) software.

MD simulations is a valuable tool for free energy calculations, and often the computed values agree with experimental free energies. The partition of free energy into entropic and enthalpic contributions, and the use of force field approaches in this context, was discussed earlier. It has been shown that entropies are very difficult to estimate, and the errors in calculations of this quantity are often high in magnitude even larger than those for the free energy. Unlike free energies, the entropies cannot be calculated as a simple average from an MD simulation and extensive sampling of all degrees of freedom is required. Schlitter (1993) proposed an ad hoc approximation for estimating an upper limit to the absolute entropy of a macromolecule from MD simulations using the covariance matrix of atomic fluctuations. This method has been used to estimate the absolute translational, rotational, and conformational entropy, as well as the change in entropy upon folding of certain biomolecules. However, Levy *et al.* (1984) introduced a method based on the quasi-harmonic approximation, which also connects the absolute entropy of the atoms to the covariance matrix. This was initially used to calculate vibrational entropies using internal coordinates, and the method can be extended to estimate the total entropy using the covariance matrix of Cartesian coordinates (Carlsson and Aqvist, 2005; Andricioaei and Karplus, 2001). In recent years, because of the advancement in the computational power, the already existing methods have been tested and compared on the basis of reproducing translational, rotational, and vibrational entropies. In addition, new approaches are being discovered to estimate the absolute and relative entropies of biomolecules.

7.8. Binding affinity

The binding affinities of protein–DNA complexes are experimentally determined using SELEX, chromatin immunoprecipitation (ChIP), EMSA, DNA pull-down assays and the reporter assays. Binding affinity is generally expressed in terms of shape and chemical complementarities (Strauch, 2001; Rawat and Biswas, 2011; Parisien *et al.*, 2012), which depend on the area of binding and physicochemical properties of interacting atoms, respectively. Specifically, noncovalent interactions such as electrostatics, van der Waals, hydrophobic and dipole attractions are important for the chemical complementarity (Hu *et al.*, 2018). These interactions are successfully used to compute the binding affinity of protein–NA complexes. Yan and Wang (2013) proposed an optimized scoring function to estimate the binding affinity. Setny *et al.* (2012) showed that absolute free energy of protein–DNA complexes can be reproduced with the help of force field–based calculations. In addition, Rosetta-Vienna predicts the free energy change of binding upon protein–RNA complex formation (Kappel *et al.*, 2019).

Hogan and Austin (1987) reported that the elastic properties of DNA may be an important factor for protein–DNA-binding specificity by performing the studies on 434 repressor proteins. Further, the significance of these elastic properties of DNA has been extensively reviewed with an application to protein–DNA recognition. The results, especially from the Cro protein–DNA complexes, revealed that the free energy of complex formation increases with stiffness for the nonspecific interactions, and an opposite trend was observed for specific ones (Takeda *et al.*, 1989). The decomposition energy terms suggest that the binding energy in the nonspecific case is used mainly to compensate the free energy changes due to entropy lost by DNA, while the energy of specific interactions provide enough energy to bend the DNA molecule and to change the conformations of the Cro proteins upon binding (Gromiha *et al.*, 1997).

The influence of DNA stiffness to the affinity of protein–DNA complexes has been studied by Gromiha (2000) and developed a sequence-dependent stiffness scale (Table 7.3) to calculate the correlation between DNA stiffness and binding affinity. The stiffness values for each trinucleotide was assigned appropriately from Table 7.3 and the average Young's modulus was computed using the following equation:

Table 7.3.   Structure-based DNA stiffness scale for trinucleotides*.

Trinucleotide	E (10^8 N/m^2)	Trinucleotide	E (10^8 N/m^2)
AAA/TTT	4.80	CAG/CTG	2.40
AAC/GTT	3.90	CCA/TGG	3.25
AAG/CTT	1.91	CCC/GGG	6.07
AAT/ATT	2.96	CCG/CGG	2.40
ACA/TGT	4.70	CGA/TCG	2.82
ACC/GGT	1.57	CGC/GCG	3.33
ACG/CGT	7.09	CTA/TAG	4.75
ACT/AGT	3.63	CTC/GAG	4.03
AGA/TCT	4.03	GAA/TTC	2.70
AGC/GCT	4.58	GAC/GTC	7.83
AGG/CCT	4.34	GCA/TGC	3.75
ATA/TAT	2.36	GCC/GGC	3.16
ATC/GAT	1.83	GGA/TCC	3.69
ATG/CAT	3.19	GTA/TAC	2.19
CAA/TTG	2.53	TAA/TTA	2.72
CAC/CTG	3.36	TCA/TGA	2.97

*Data were taken from Gromiha (2000).

$$E = \frac{\sum E_i}{n} \tag{9}$$

where E_i is the Young's modulus for the ith trinucleotide and n is the total number of trinucleotide units.

The correlation coefficient between average stiffness (E) and protein–DNA-binding affinity was computed using the familiar formula:

$$r = \frac{N \sum XY - (\sum X \sum Y)}{\{[N \sum X^2 - (\sum X)^2][N \sum Y^2 - (\sum Y)^2]\}^{1/2}} \tag{10}$$

where r is the correlation coefficient, N, X and Y are, respectively, the number of data, average stiffness values and experimental protein–DNA-binding affinity.

The average stiffness values obtained with the target sequences of protein–DNA complexes have been related with experimental protein–DNA-binding specificity. The observed correlations were in the range of 0.65–0.97 between DNA stiffness and binding free energy change in several protein–DNA complexes (Gromiha, 2005). These results reveal the influence of DNA stiffness to protein–DNA-binding specificity. In addition, the direct contacts between protein and DNA through hydrogen bonds and electrostatic and other interactions are also important for understanding the mechanism of protein–DNA recognition.

7.9. Conclusion

Protein–NA interactions are influenced by various noncovalent interactions and subsequently various mechanisms have been proposed to understand the recognition. In this chapter, we reviewed broad computational approaches for understanding the recognition mechanism on different perspectives such as development of databases, structural analysis on interface residues, DOTs, direct and indirect readout mechanisms, organism-specific recognitions, MD simulations and binding affinity. This comprehensive review would be helpful to gain deep insights to understand the structure–function relationship of protein–NA complexes.

Acknowledgment

The work is partially supported by the Council of Scientific and Industrial Research, Government of India, to MMG (37(1694)/17/EMR-II).

References

Ahmad, S., Kono, H., Arauzo-Bravo, M.J., *et al.* ReadOut: structure-based calculation of direct and indirect readout energies and specificities for protein-DNA recognition. *Nucleic Acids Res.* 2006;34:W124–W127.

Ahmad, S., Keskin O., Sarai A., *et al.* Protein-DNA interactions: Structural, thermodynamic and clustering patterns of conserved residues in DNA-binding proteins. *Nucleic Acids Res.* 2008;36:5922–5932.

Ahmad, S., and Sarai, A. Moment-based prediction of DNA-binding proteins. *J Mol Biol.* 2004;341:65–71.

Albright, R.A., Mossing, M.C., and Matthews, B.W. Crystal structure of an engineered Cro monomer bound nonspecifically to DNA: Possible implications for nonspecific binding by the wild-type protein. *Protein Sci.* 1998;7:1485–1494.

Andricioaei, I., and Karplus, M. On the calculation of entropy from covariance matrices of the atomic fluctuations. *J Chem Phys.* 2001; 115:6289.

Bah, A., and Forman-Kay, J.D. Modulation of intrinsically disordered protein function by post-translational modifications. *J Biol Chem.* 2016;291:6696–6705.

Baldwin, E.P., Martin S.S., Abel J., *et al.* A specificity switch in selected cre recombinase variants is mediated by macromolecular plasticity and water. *Chem Biol.* 2003;10:1085–1094.

Banerjee, A., Yang W., Karplus M., *et al.* Structure of a repair enzyme interrogating undamaged DNA elucidates recognition of damaged DNA. *Nature.* 2005;434:612–618.

Bouvier, B., Zakrzewska, K., and Lavery R. Protein-DNA recognition triggered by a DNA conformational switch. *Angew Chem Int Ed Engl.* 2011;50:6516–6518.

Bogan, A.A., and Thorn, K.S. Anatomy of hot spots in protein interfaces. *J Mol Biol.* 1998;280:1–9.

Briand, C., Poterszman A., Eiler S., *et al.* An intermediate step in the recognition of tRNA(Asp) by aspartyl-tRNA synthetase." *J Mol Biol.* 299: 1051–60.

Bravo, J., Li Z., Speck N.A., *et al.* The leukemia-associated AML1 (Runx1) — CBF beta complex functions as a DNA-induced molecular clamp. *Nat Struct Biol.* 2001;8:371–378.

Carlsson, J., and Aqvist, J. Absolute and relative entropies from computer simulation with applications to ligand binding. *J Phys Chem B.* 2005;109:6448–6456.

Cheatham, T.E., 3rd, and Young, M.A. Molecular dynamics simulation of nucleic acids: successes, limitations, and promise. *Biopolymers.* 2000;56:232–256.

Cheatham, T.E., 3rd. Simulation and modeling of nucleic acid structure, dynamics and interactions. *Curr Opin Struct Biol.* 2004;14:360–367.

Contreras-Moreira, B. 3D-footprint: A database for the structural analysis of protein-DNA complexes. *Nucleic Acids Res.* 2010;38:D91-D97.

Cook, K.B., Kazan H., Zuberi K., *et al.* RBPDB: A database of RNA-binding specificities. *Nucleic Acids Res.* 2011;39:D301–D308.

Cordeiro, Y., Macedo, B., Silva J.L., *et al.* Pathological implications of nucleic acid interactions with proteins associated with neurodegenerative diseases. *Biophys Rev.* 2014;6:97–110.

Corona, R.I., and Guo, J.T. Statistical analysis of structural determinants for protein-DNA-binding specificity. *Proteins.* 2016;84:1147–1161.

Cornell, W.D., Cieplak, P., Bayly, C.I., *et al.* A second generation force field for the simulation of proteins, nucleic acids, and organic molecules. *J Am Chem Soc.* 1995;117:5179–5197.

Coulocheri, S.A., Pigis D.G., Papavassiliou K.A., *et al.* Hydrogen bonds in protein-DNA complexes: where geometry meets plasticity. *Biochimie.* 2007;89:1291–1303.

Dey, B., Thukral S., Krishnan S., *et al.* DNA-protein interactions: methods for detection and analysis. *Mol Cell Biochem.* 2012;365:279–299.

Dosztányi, Z., Meszaros B., and Simon I. ANCHOR: Web server for predicting protein binding regions in disordered proteins. *Bioinformatics.* 2009;25:2745–2746.

Duh, Y., Hsiao, Y.Y., Li, C.L., *et al.* Aromatic residues in RNase T stack with nucleobases to guide the sequence-specific recognition and cleavage of nucleic acids. *Protein Sci.* 2015;24:1934–1941.

Eiler, S., Dock-Bregeon A., Moulinier L., *et al.* Synthesis of aspartyl-tRNA(Asp) in Escherichia coli — A snapshot of the second step. *EMBO J.* 1999;18:6532–6541.

Ellis, J.J., Broom, M., and Jones, S. Protein-RNA interactions: Structural analysis and functional classes. *Proteins.* 2007;66:903–911.

Furini, S., Barbini, P., and Domene, C. DNA-recognition process described by MD simulations of the lactose repressor protein on a specific and a non-specific DNA sequence. *Nucleic Acids Res.* 2013;41:3963–3972.

Galindo-Murillo, R., Roe, D.R., and Cheatham, T.E., 3rd. Convergence and reproducibility in molecular dynamics simulations of the DNA duplex d(GCACGAACGAACGAACGC). *Biochim Biophys Acta.* 2015;1850: 1041–1058.

Galindo-Murillo, R., Robertson, J.C., Zgarbova, M., *et al.* Assessing the current state of amber force field modifications for DNA. *J Chem Theory Comput.* 2016;12:4114–4127.

Gardini, S., Furini S., Santucci A., *et al.* A structural bioinformatics investigation on protein-DNA complexes delineates their modes of interaction. *Mol Biosyst.* 2017;13:1010–1017.

Gromiha, M.M. Structure based sequence dependent stiffness scale for trinucleotides: A direct method. *J Biol Phys.* 2000;26:43–50.

Gromiha, M.M., Munteanu M.G., Simon I., *et al.* The role of DNA bending in Cro protein-DNA interactions. *Biophys Chem.* 1997;69:153–160.

Gromiha, M.M., Siebers J.G., Selvaraj S., *et al* Intermolecular and intramolecular readout mechanisms in protein-DNA recognition. *J Mol Biol.* 2004a;337:285–294.

Gromiha, M.M., Santhosh, C., and Ahmad, S. Structural analysis of cation-pi interactions in DNA binding proteins. *Int J Biol Macromol.* 2004b;34:203–211.

Gromiha, M.M., Pujadas G., Magyar C., *et al.* Locating the stabilizing residues in (alpha/beta)8 barrel proteins based on hydrophobicity, long-range interactions, and sequence conservation. *Proteins.* 2004c;55:316–329.

Gromiha, M.M. Influence of DNA stiffness in protein-DNA recognition. *J Biotechnol.* 2005;117:137–145.

Gromiha, M.M., Siebers J.G., Selvaraj S., *et al.* Role of inter and intramolecular interactions in protein-DNA recognition. *Gene.* 2005; 364:108–13.

Gromiha, M.M., Yokota, K., and Fukui, K. Sequence and structural analysis of binding site residues in protein-protein complexes. *Int J Biol Macromol.* 2010;46:187–192.

Gromiha, M.M., and Fukui, K. Scoring function based approach for locating binding sites and understanding recognition mechanism of protein-DNA complexes. *J Chem Inf Model.* 2011;51:721–729.

Guo, X., Bulyk, M.L., and Hartemink, A.J. Intrinsic disorder within and flanking the DNA-binding domains of human transcription factors. *Pac Symp Biocomput.* 2012;2012:104–115.

Halford, S.E., Gowers, D.M., and Sessions, R.B. Two are better than one. *Nat Struct Biol.* 2000;7:705–707.

Halford, S.E., and Marko, J.F. How do site-specific DNA-binding proteins find their targets? *Nucleic Acids Res.* 2004;32:3040–3052.

Hansen, J.C., Lu X., Ross E.D., *et al.* Intrinsic protein disorder, amino acid composition, and histone terminal domains. *J Biol Chem.* 2006;281:1853–1856.

Hart, K., Foloppe, N., Baker, C.M., *et al.* Optimization of the CHARMM additive force field for DNA: Improved treatment of the BI/BII conformational equilibrium. *J Chem Theory Comput.* 2012;8: 348–362.

Hogan, M.E., and Austin, R.H. Importance of DNA stiffness in protein-DNA binding specificity. *Nature.* 1987;329:263–266.

Hu, W., Qin L., Li M., *et al.* A structural dissection of protein–RNA interactions based on different RNA base areas of interfaces. *RSC Advances.* 2018;8:10582–10592.

Hudson, W.H., and Ortlund, E.A. The structure, function and evolution of proteins that bind DNA and RNA. *Nat Rev Mol Cell Biol.* 2014;15: 749–760.

Ivani, I., Dans, P.D., Noy, A., *et al.* Parmbsc1: A refined force field for DNA simulations. *Nat Methods.* 2016;13:55–58.

Järvelin, A.I., Noerenberg M., Davis I., *et al.* The new (dis)order in RNA regulation. *Cell Commun Signal.* 2016;14:9.

Jayaram, B., McConnell K., Dixit S.B., *et al.* Free-energy component analysis of 40 protein-DNA complexes: a consensus view on the thermodynamics of binding at the molecular level. *J Comput Chem.* 2002;23:1–14.

Ji, Z.L., Chen, X., Zhen, C.J., *et al.* KDBI: Kinetic data of bio-molecular interactions database. *Nucleic Acids Res.* 2003;31:255–257.

Jones, S., van Heyningen P., Berman H.M., *et al.* Protein-DNA interactions: A structural analysis. *J Mol Biol.* 1999;287:877–896.

Junge, A., Refsgaard, J.C., Garde, C., *et al.* RAIN: RNA-protein Association and Interaction Networks. *Database.* 2017;2017:1–7.

Kalodimos, C.G., Biris, N., Bonvin, A.M., *et al.* Structure and flexibility adaptation in nonspecific and specific protein-DNA complexes. *Science.* 2004;305:386–389.

Kappel, K., Jarmoskaite, I., Vaidyanathan, P.P., *et al.* Blind tests of RNA-protein binding affinity prediction. *Proc Natl Acad Sci USA.* 2019;116: 8336–8341.

Khabiri, M., and Freddolino, P.L. Deficiencies in molecular dynamics simulation-based prediction of protein-DNA binding free energy landscapes. *J Phys Chem B.* 2017;121:5151–5161.

Khan, A., Fornes, O., Stigliani, A., *et al.* JASPAR 2018: Update of the open-access database of transcription factor binding profiles and its web framework. *Nucleic Acids Res.* 2018;46:D260–D266.

Kirsanov, D.D., Zanegina, O.N., Aksianov, E.A., *et al.* NPIDB: Nucleic acid-Protein Interaction DataBase. *Nucleic Acids Res.* 2013;41:D517–D523.

Komazin-Meredith, G., Mirchev R., Golan D.E., *et al.* Hopping of a processivity factor on DNA revealed by single-molecule assays of diffusion. *Proc Natl Acad Sci U S A.* 2008;105:10721–10726.

Kono, H., and Sarai, A. Structure-based prediction of DNA target sites by regulatory proteins. *Proteins.* 1999;35:114–131.

Krüger, D.M., Neubacher, S., and Grossmann, T.N. Protein-RNA interactions: Structural characteristics and hotspot amino acids. *RNA.* 2018;24:1457–1465.

Kulandaisamy, A., Lathi V., ViswaPoorani K., *et al.* Important amino acid residues involved in folding and binding of protein-protein complexes. *Int J Biol Macromol.* 2017;94:438–444.

Kulandaisamy, A., Srivastava, A., Kumar, P., *et al.* Identification and analysis of key residues in protein-RNA complexes. *IEEE/ACM Trans Comput Biol Bioinform.* 2018a;15:1436–1444.

Kulandaisamy, A., Srivastava A., Nagarajan R., *et al.* Dissecting and analyzing key residues in protein-DNA complexes. *J Mol Recognit.* 2018b;31:e2692.

Lee, O.S., Cho, V.Y., and Schatz, G.C. A- to B-form transition in DNA between gold surfaces. *J Phys Chem B.* 2012;116:7000–7005.

Lejeune, D., Delsaux N., Charloteaux B., *et al.* Protein-nucleic acid recognition: statistical analysis of atomic interactions and influence of DNA structure. *Proteins.* 2005;61:258–271.

Levitt, M. Computer simulation of DNA double-helix dynamics. *Cold Spring Harb Symp Quant Biol.* 1983;47:251–262.

Levy, R.M., Karplus, M., Kushick, J., *et al* Evaluation of the configurational entropy for proteins: Application to molecular dynamics simulations of an α-helix. *Macromolecules.* 1984;17:1370–1374.

Liu, Z., Li, Y., Han, L., *et al.* PDB-wide collection of binding data: current status of the PDBbind database. *Bioinformatics.* 2015;31:405–412.

Lomholt, M.A., van den Broek, B., Kalisch S.M., *et al.* Facilitated diffusion with DNA coiling. *Proc Natl Acad Sci U S A.* 2009;106:8204–8208.

Magyar, C., Gromiha M.M., Pujadas G., *et al.* SRide: A server for identifying stabilizing residues in proteins. *Nucleic Acids Res.* 2005;33: W303–W305.

Mandel-Gutfreund, Y., Schueler, O., and Margalit, H. Comprehensive analysis of hydrogen bonds in regulatory protein DNA-complexes: In search of common principles. *J Mol Biol.* 1995;253:370–382.

Marasco, D., and Scognamiglio, P.L. Identification of inhibitors of biological interactions involving intrinsically disordered proteins. *Int J Mol Sci.* 2015;16:7394–7412.

Marchese, D., de Groot N.S., Lorenzo Gotor N., *et al.* Advances in the characterization of RNA-binding proteins. *Wiley Interdiscip Rev RNA.* 2016;7:793–810.

McHugh, C.A., Russell, P., and Guttman, M. Methods for comprehensive experimental identification of RNA-protein interactions. *Genome Biol.* 2014;15:203.

Miller, J.C., and Pabo, C.O. Rearrangement of side-chains in a Zif268 mutant highlights the complexities of zinc finger-DNA recognition. *J Mol Biol.* 2001;313:309–315.

Moulinier, L., Eiler, S., Eriani, G., *et al.* The structure of an AspRS-tRNA(Asp) complex reveals a tRNA-dependent control mechanism. *EMBO J.* 2001;20:5290–5301.

Mukherjee, A., Lavery, R., Bagchi, B., *et al.* On the molecular mechanism of drug intercalation into DNA: a simulation study of the intercalation pathway, free energy, and DNA structural changes. *J Am Chem Soc.* 2008;130:9747–9755.

Nadassy, K., Wodak, S.J., and Janin, J. Structural features of protein-nucleic acid recognition sites. *Biochemistry.* 1999;38:1999–2017.

Narayanan, C.B., *et al.* The nucleic acid database: new features and capabilities. *Nucleic Acids Res.* 2014;42:D114–D122.

Nagarajan, R., Chothani S.P., Ramakrishnan C., *et al.* Structure based approach for understanding organism specific recognition of protein-RNA complexes. *Biol Direct.* 2015;10:8.

Nagarajan, R., Archana, A., Thangakani, A.M., *et al.* PDBparam: Online resource for computing structural parameters of proteins. *Bioinform Biol Insights.* 2016;10:73–80.

Norambuena, T., and Melo, F. The protein-DNA interface database. *BMC Bioinformatics.* 2016;11:262.

Olson, W.K., Gorin A.A., Lu X.J., *et al.* DNA sequence-dependent deformability deduced from protein-DNA crystal complexes. *Proc Natl Acad Sci USA.* 1998;95:11163–11638.

Osawa, T., Inanaga H., Sato C., *et al.* Crystal structure of the CRISPR-Cas RNA silencing Cmr complex bound to a target analog. *Mol Cell.* 2015;58:418–430.

Paz, I., Kosti I., Ares M., Jr., *et al.* RBPmap: A web server for mapping binding sites of RNA-binding proteins. *Nucleic Acids Res.* 2014;42: W361–W367.

Parisien, M., Freed, K.F., and Sosnick, T.R. On docking, scoring and assessing protein-DNA complexes in a rigid-body framework. *PLOS ONE.* 2012;7:e32647.

Peng, Z., and Kurgan, L. High-throughput prediction of RNA, DNA and protein binding regions mediated by intrinsic disorder. *Nucleic Acids Res.* 2015;43:e121.

Perez, A., Marchan, I., Svozil, D., *et al.* Refinement of the AMBER force field for nucleic acids: improving the description of alpha/gamma conformers. *Biophys J.* 2007;92:3817–3829.

Perez, A., Luque, F.J., and Orozco, M. Frontiers in molecular dynamics simulations of DNA. *Acc Chem Res.* 2012;45:196–205.

Poddar, S., Chakravarty, D., and Chakrabarti, P. Structural changes in DNA-binding proteins on complexation. *Nucleic Acids Res.* 2018;46: 3298–3308.

Poosapati, A., Gregory E., Borcherds W.M., *et al.* Uncoupling the folding and binding of an intrinsically disordered protein. *J Mol Biol.* 2018;430:2389–2402.

Prabakaran, P., Siebers, J.G., Ahmad, S., *et al.* Classification of protein-DNA complexes based on structural descriptors. *Structure.* 2006;14: 1355–1367.

Qi, Y., Spong, M.C., Nam, K., *et al.* Encounter and extrusion of an intrahelical lesion by a DNA repair enzyme. *Nature.* 2009;462:762–766.

Ramanathan, M., Porter, D.F., and Khavari, P.A. Methods to study RNA-protein interactions. *Nat Methods.* 2019;16:225–234.

Rawat, N., and Biswas, P. Shape, flexibility and packing of proteins and nucleic acids in complexes. *Phys Chem Chem Phys.* 2011;13:9632–9643.

Reddy, C.K., Das, A., and Jayaram, B. Do water molecules mediate protein-DNA recognition? *J Mol Biol.* 2001;314:619–632.

Rice, P.A., Yang S., Mizuuchi K., *et al.* Crystal structure of an IHF-DNA complex: A protein-induced DNA U-turn. *Cell.* 1996;87:1295–1306.

Robles-Espinoza, C.D., Harland, M., Ramsay, A.J., *et al.* POT1 loss-of-function variants predispose to familial melanoma. *Nat Genet.* 2014;46:478–481.

Rohs, R., West, S.M., Sosinsky, A., *et al.* The role of DNA shape in protein-DNA recognition. *Nature.* 2009;461:1248–1253.

Rohs, R., Jin, X., West, S.M., *et al.* Origins of specificity in protein-DNA recognition. *Annu Rev Biochem.* 2010;79:233–269.

Rose, P.W., Prlic, A., Altunkaya, A., *et al.* The RCSB protein data bank: integrative view of protein, gene and 3D structural information. *Nucleic Acids Res.* 2017;45:D271–D281.

Ruff, M., Krishnaswamy, S., Boeglin, M., *et al.* Class II aminoacyl transfer RNA synthetases: Crystal structure of yeast aspartyl-tRNA synthetase complexed with tRNA(Asp). *Science.* 1991;252:1682–1689.

Sarai, A., and Kono, H. Protein-DNA recognition patterns and predictions. *Annu Rev Biophys Biomol Struct.* 2005;34:379–398.

Sathyapriya, R., Vijayabaskar, M.S., and Vishveshwara, S. Insights into protein-DNA interactions through structure network analysis. *PLoS Comput Biol.* 2008;4:e1000170.

Schlitter, J. Estimation of absolute and relative entropies of macromolecules using the covariance matrix. *Chem Phys Lett.* 1993;215:617–621.

Setny, P., Bahadur, R.P., and Zacharias, M. Protein-DNA docking with a coarse-grained force field. *BMC Bioinformatics.* 2012;13:228.

Si, J., Zhao, R., and Wu, R. An overview of the prediction of protein DNA-binding sites. *Int J Mol Sci.* 2015;16:194–215.

Sippl, M.J. Calculation of conformational ensembles from potentials of mena force: an approach to the knowledge-based prediction of local structures in globular proteins. *J Mol Biol.* 1990;213:859–883.

Siggers, T., and Gordan, R. Protein-DNA binding: Complexities and multi-protein codes. *Nucleic Acids Res.* 2014;42:2099–2111.

Srivastava, A., Ahmad, S., and Gromiha, M.M. Deciphering RNA-recognition patterns of intrinsically disordered proteins. *Int J Mol Sci.* 2018;19:1595.

Stefl, R., Skrisovska, L., and Allain, F.H. RNA sequence- and shape-dependent recognition by proteins in the ribonucleoprotein particle. *EMBO Rep.* 2005;6:33–38.

Stefl, R., Oberstrass, F.C., Hood, J.L., *et al.* The solution structure of the ADAR2 dsRBM-RNA complex reveals a sequence-specific readout of the minor groove. *Cell.* 2010;143:225–237.

Strauch, M.A. *Protein–DNA Complexes: Specific.* In: *Encyclopedia of Life Sciences.* Nature Publishing group; 2001:1–7.

Svozil, D., Sponer, J.E., Marchan, I., *et al.* Geometrical and electronic structure variability of the sugar-phosphate backbone in nucleic acids. *J Phys Chem B.* 2008;112:8188–8197.

Takeda, Y., Sarai, A. and Rivera, V.M. Analysis of the sequence-specific interactions between Cro repressor and operator DNA by systematic base substitution experiments. *Proc Natl Acad Sci USA.* 1989;86:439–443.

Tompa, P. Intrinsically disordered proteins: A 10-year recap. *Trends Biochem Sci.* 2012;37:509–516.

Tompa, P., Schad E., Tantos A., *et al.* Intrinsically disordered proteins: Emerging interaction specialists. *Curr Opin Struct Biol.* 2015;35:49–59.

Toto, A., Camilloni, C., Giri, R., *et al.* Molecular recognition by templated folding of an intrinsically disordered protein. *Sci Rep.* 2016;6:21994.

Uversky, V.N. The alphabet of intrinsic disorder: II. Various roles of glutamic acid in ordered and intrinsically disordered proteins. *Intrinsically Disord Proteins.* 2013;1:e24684.

Van Nostrand, E.L., Huelga, S.C., and Yeo, G.W. Experimental and computational considerations in the study of RNA-binding protein-RNA interactions. *Adv Exp Med Biol.* 2016;907:1–28.

Varadi, M., Zsolyomi F., Guharoy M., *et al.* Functional advantages of conserved intrinsic disorder in RNA-binding proteins. *PLOS ONE.* 2015;10:e0139731.

VanSchouwen, B.M., Gordon H.L., Rothstein S.M., *et al.* Water-mediated interactions in the CRP-cAMP-DNA complex: does water mediate sequence-specific binding at the DNA primary-kink site? *Comput Biol Chem.* 2008;32:149–158.

Viadiu, H., and Aggarwal, A.K. Structure of BamHI bound to nonspecific DNA: A model for DNA sliding. *Mol Cell.* 2000;5:889–895.

von Hippel, P.H., and Berg, O.G. Facilitated target location in biological systems. *J Biol Chem.* 1989;264:675–678.

Vuzman, D., and Levy, Y. Intrinsically disordered regions as affinity tuners in protein-DNA interactions. *Mol Biosyst.* 2012;8:47–57.

Wan, H., Hu, J.P., Li, K.S., *et al.* Molecular dynamics simulations of DNA-free and DNA-bound TAL effectors. *PLOS ONE.* 2013;8:e76045.

Wang, G., Luo, X., Wang, J., *et al.* MeDReaders: A database for transcription factors that bind to methylated DNA. *Nucleic Acids Res.* 2018;46: D146–D151.

Wang, W., Liu, J., Xiong, Y., *et al.* Analysis and classification of DNA-binding sites in single-stranded and double-stranded DNA-binding proteins using protein information. *IET Syst Biol.* 2014;8:176–183.

Waters, J.T., Lu, X.J., Galindo-Murillo, R., *et al.* Transitions of double-stranded DNA between the A- and B-forms. *J Phys Chem B.* 2016; 120:8449–8456.

Wollenhaupt, J., Henning, L.M., Sticht, J., *et al.* Intrinsically disordered protein Ntr2 modulates the spliceosomal RNA helicase Brr2. *Biophys J.* 2018;114:788–799.

Wilson, K.A., Kellie, J.L., and Wetmore, S.D. DNA-protein pi-interactions in nature: abundance, structure, composition and strength of contacts between aromatic amino acids and DNA nucleobases or deoxyribose sugar. *Nucleic Acids Res.* 2014;42:6726–6741.

Wilson, K.A., and Wetmore, S.D. A survey of DNA–protein π–interactions: A comparison of natural occurrences and structures, and computationally predicted structures and strengths. *Challeng Adv Comput Chem Phys.* 2015;19:501–532.

Wilson, K.A., Wells, R.A., Abendong, M.N., *et al.* Landscape of pi-pi and sugar-pi contacts in DNA-protein interactions. *J Biomol Struct Dyn.* 2016;34:184–200.

Wingender, E., Dietze P., Karas H., *et al.* TRANSFAC: A database on transcription factors and their DNA binding sites. *Nucleic Acids Res.* 1996;24:238–241.

Winter, R.B., Berg, O.G., and von Hippel, P.H. Diffusion-driven mechanisms of protein translocation on nucleic acids. 3. The Escherichia coli lac repressor — Operator interaction: Kinetic measurements and conclusions. *Biochemistry.* 1981;20:6961–6977.

Xie, Z., Hu, S., Blackshaw, S., *et al.* hPDI: A database of experimental human protein-DNA interactions. *Bioinformatics.* 2010;26:287–289.

Yamasaki, S., Terada, T., Kono, H., *et al.* A new method for evaluating the specificity of indirect readout in protein-DNA recognition. *Nucleic Acids Res.* 2012;40:e129.

Yan, J., Dunker, A.K., Uversky, V.N., *et al.* Molecular recognition features (MoRFs) in three domains of life. *Mol Biosyst.* 2016;12:697–710.

Yan, Z., and Wang, J. Optimizing scoring function of protein-nucleic acid interactions with both affinity and specificity. *PLOS ONE.* 2013;8:e74443.

Yesudhas, D., Anwar, M.A., Panneerselvam, S., *et al.* Structural mechanism behind distinct efficiency of Oct4/Sox2 proteins in differentially spaced DNA complexes. *PLOS ONE.* 2016;11:e0147240.

Yesudhas, D., Batool, M., Anwar, M.A., *et al.* Proteins recognizing DNA: Structural uniqueness and versatility of DNA-binding domains in stem cell transcription factors. *Genes (Basel).* 2017a;8:E192.

Yesudhas, D., Anwar, M.A., Panneerselvam, S., *et al.* Evaluation of Sox2 binding affinities for distinct DNA patterns using steered molecular dynamics simulation. *FEBS Open Bio.* 2017b;7:1750–1767.

Yi, Y., Zhao, Y., Li, C., *et al.* RAID v2.0: An updated resource of RNA-associated interactions across organisms." *Nucleic Acids Res.* 2017;45:D115–D118.

Zgarbova, M., Sponer, J., Otyepka, M., *et al.* Refinement of the sugar-phosphate backbone torsion beta for AMBER force fields improves the description of Z- and B-DNA. *J Chem Theory Comput.* 2015;11: 5723–5736.

Zhang, Y., Stec, B., and Godzik, A. Between order and disorder in protein structures: analysis of "dual personality" fragments in proteins. *Structure.* 2007;15:1141–7.

Zhou, P., Tian, F., Ren, Y., *et al.* Systematic classification and analysis of themes in protein-DNA recognition. *J Chem Inf Model.* 2010;50: 1476–1488.

Zhu, Y., Xu, G., Yang, Y.T., *et al.* POSTAR2: Deciphering the post-transcriptional regulatory logics. *Nucleic Acids Res.* 2019;47: D203–D211.

Chapter 8
Prediction of nucleic acid binding proteins and their binding sites

Dhanusha Yesudhas[†], Ambuj Srivastava[†], Nisha Muralidharan,
A. Kulandaisamy, R. Nagarajan and M. Michael Gromiha[*]

*Department of Biotechnology, Bhupat and Jyoti Mehta
School of Biosciences, Indian Institute of Technology Madras,
Chennai 600036, Tamil Nadu, India*

The information on interactions between proteins and nucleic acids in protein–nucleic acid complexes provides deep insights for understanding their functions. Experimentally, several techniques including X-ray crystallography, isothermal titration calorimetry and DNA/RNA footprinting are used to identify nucleic acid binding proteins and their binding sites. On the other hand, computational tools have been developed for annotating nucleic acid-binding proteins to compensate the requirements of manpower, resources and time for experiments. In this chapter, the methods for identifying DNA- or RNA-binding proteins and features used in these algorithms have been reviewed. Different methods for identifying the binding sites in protein–nucleic acid complex structures are outlined along with computational methods for predicting the residues at the interface. We have also listed a set of tools available in the literature for predicting DNA/RNA-binding proteins and their binding sites along with annotating those proteins in genomic sequences.

*gromiha@iitm.ac.in
[†]Contributed equally to this work.

8.1. Introduction

Nucleic acid binding proteins are important for their functions in various biological processes including replication, transcription and translation. Binding sites in protein–nucleic acid complexes are vital for understanding their specificity and affinity. Alteration in binding sites leads to the loss of binding and/or change in the affinity, which may impact the function of the complex. Even a single amino acid mutation in protein–nucleic acid binding site can cause deadly diseases, including cancer (Taniguchi *et al.*, 2018). Hence, it is important to study the binding sites of the protein–nucleic acid complexes.

On the experimental side, several techniques such as systematic evolution of ligands by exponential enrichment (SELEX), chromatin immunoprecipitation (ChIP), electrophoretic mobility shift assay (EMSA), DNA/RNA Pull-down assays, DNA/RNA footprinting, southwestern blotting, fluorescent resonance energy transfer (FRET), X-ray crystallography and nuclear magnetic resonance are used to identify the nucleic acid binding proteins, binding sites and binding affinity. However, high-throughput identification and large-scale analysis require huge man power, resources and time (McHugh *et al.*, 2014; Carey *et al.*, 2012). Hence, computational methods have been proposed and shown to be effective for identifying nucleic acid binding proteins and their binding sites.

Computational prediction methods and algorithms are classified into two groups: (i) identification of nucleic acid interacting proteins and (ii) prediction of their binding site residues with DNA/RNA. Different methods have been proposed specifically for identifying DNA-binding proteins (Kumar *et al.*, 2007; Kumar *et al.*, 2009; Zaman *et al.*, 2017; Ali *et al.*, 2018; Mishra *et al.*, 2019) and RNA-binding proteins (Kumar *et al.*, 2011; Zheng *et al.*, 2018; Paz *et al.*, 2014; Livi *et al.*, 2016; Bressin *et al.*, 2019). Most of the methods for identifying DNA- or RNA-binding proteins are based on evolutionary information and machine learning techniques. Evolutionary-based methods use multiple sequence alignment and position-specific scoring matrices (PSSM), whereas machine learning methods utilize sequence- and/or structure-based features to identify the DNA- or RNA-interacting proteins.

The prediction of binding site residues requires detailed information of the query protein to derive suitable parameters. In addition, binding site information along with its partner nucleic acid helps to model the three-dimensional structures of protein–nucleic acid complexes (Tuszynska *et al.*, 2015; Yan *et al.*, 2017; de Vries *et al.*, 2010). Binding site prediction methods are developed using structural information or just from amino acid sequence alone, and these methods utilize the features hydrophobicity, charge, free energy, dipole moment, amino acid–nucleotide interaction propensity, conservation score, secondary structure and solvent accessibility for predicting the binding sites in DNA- and RNA-binding proteins (Chen *et al.*, 2012a; Hwang *et al.*, 2007; Kumar *et al.*, 2007; Wang and Brown, 2006a, 2006b; Wang *et al.*, 2010; Yan and Kurgan, 2017).

Further, prediction tools are also used to annotate genomic sequences to identify novel proteins interacting with RNA or DNA. Gao and Skolnick (2009) developed a method for predicting DNA-binding proteins from sequence using threading approach and applied to identify DNA-binding proteins in human proteome. Langlois and Lu (2010) used sequence-based features for predicting DNA-binding proteins using machine learning approaches and applied their model on large-scale annotations. Ahmad *et al.* (2018) developed a sequence and gene expression-based model to predict DNA-binding proteins in multiple data sets constructed from human, mouse and Arabidopsis proteomes.

In this chapter, we survey various prediction methods developed for identifying DNA- and RNA-binding proteins. Further, widely used approaches for identifying binding site residues from protein–nucleic acid complex structures are outlined along with computational tools for predicting the binding sites. In addition, binding site prediction in disordered regions will be discussed. Flowchart in Figure 8.1 summarizes the overall contents of this chapter.

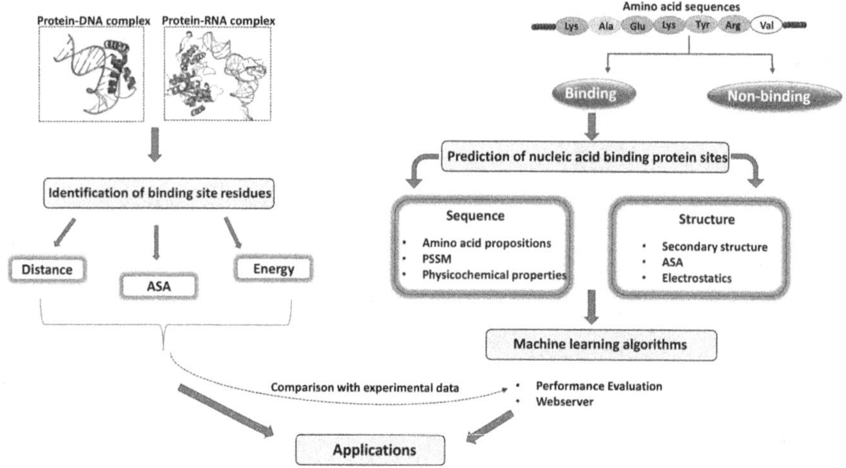

Figure 8.1.   Flowchart illustrating the overall contents of this chapter.

8.2. Identification of nucleic acid binding proteins

Proteome-wide annotations of nucleic acid binding proteins are important for understanding their functions. Hence, several computational tools have been developed for identifying DNA- or RNA-binding proteins just from amino acid sequences as well as using structural information. These methods generally follow a series of steps: construction of data sets, derivation of features, feature reduction, development of methods, assessment of performance and establishment of Web servers. Construction of data sets involves generating a set of nucleic acid binding and non-binding proteins from Uniprot (UniProt Consortium, 2018) or Protein Data Bank (Rose *et al.*, 2017) and culled with sequence redundancy. Features are derived from amino acid sequences and/or three-dimensional structures from non-redundant data sets followed by feature reduction. The major sequence-based features are PSSM, amino acid composition, physico-chemical properties, conservation score, secondary structure and so on. Developing methods utilize statistical analysis and machine learning algorithms such as logistic regression, random forest, neural network, support vector machine (SVM), etc. Recently, deep learning techniques are also widely used for developing prediction methods. The performance of the prediction methods is evaluated with sensitivity, specificity, accuracy and area under the curve (AUC) and are defined as follows:

$$\text{Sensitivity} = \frac{TP}{TP + FN} \tag{1}$$

$$\text{Specificity} = \frac{TN}{TN + FP} \tag{2}$$

$$\text{Accuracy} = \frac{TP + TN}{TP + TN + FP + FN} \tag{3}$$

where, TP, TN, FP and FN are true positives (binding predicted as binding), true negatives (non-binding predicted as non-binding), false positives (non-binding predicted as binding) and false negatives (binding predicted as non-binding), respectively. The AUC is also utilized to evaluate the performance of a prediction method, which is a trade-off between true positive rate and false positive rate. The AUC is 0.5 and 1 for random and perfect predictions, respectively.

8.3. Specific methods for identifying DNA-binding proteins

Computational methods are developed for distinguishing DNA-binding and non-DNA-binding proteins using (i) single descriptors such as amino acid composition, residue pair preference, motifs and amino acid properties (Xu *et al.*, 2014; Adilina *et al.*, 2019), (ii) evolutionary information through PSSM (Xu *et al.*, 2015; Wei *et al.*, 2017), (iii) structure-based properties (Gao and Skolnick, 2009; Paz *et al.*, 2016; Asghari and Abdolmaleki, 2019) and (iv) combinations of them (Zhang and Liu, 2017a; Wang *et al.*, 2017; Ali *et al.*, 2018; Mishra *et al.*, 2019).

Ahmad *et al.* (2004) analyzed several sequence-based features such as amino acid composition, solvent accessibility and secondary structure, and reported that these features have the ability to identify DNA-binding proteins. Xu *et al.* (2014) developed a sequence-based predictor based on different physicochemical properties, such as hydrophobicity, polarity, charge and solvent accessibility for distinguishing DNA-binding and non-DNA-binding proteins. In addition, the compositions of mono-, di- and tripeptides were effectively utilized to predict the DNA-binding proteins (Lin *et al.*, 2011; Rahman *et al.*, 2018; Adilina *et al.*, 2019).

Wei *et al.* (2017) derived PSSM profiles of sequences and trained with random forest for predicting DNA-binding proteins. Zaman *et al.* (2017) used monogram and bigram features of Hidden Markov Models (HMM) and used HMM profiles for developing a prediction method with SVM.

On the other hand, methods have also been proposed to identify DNA binding proteins using structure-based features such as statistical pair potentials, DNA-binding propensity, electrostatic potential, electrostatic patches of protein surfaces and other electrostatic features including charge, dipole and quadrupole moments (Bhardwaj *et al.*, 2005; Nimrod *et al.*, 2010; Gao and Skolnick, 2008, 2009; Zhao *et al.*, 2010; Zhao *et al.*, 2014; Paz *et al.*, 2016; Zhang *et al.*, 2016; Asghari and Abdolmaleki, 2019).

Mishra *et al.* (2019) developed a method for predicting DNA-binding proteins using the combination of PSSM and residue-specific contact energies. Ma *et al.* (2016) incorporated PSSM profiles with different physicochemical features as well as binding propensity of both DNA-binding and non-binding residues for developing a method. Further, amino acid and pseudo-amino acid compositions and different physiochemical features are combined with PSSM profiles for developing several prediction methods (Fujishima *et al.*, 2007; Ho *et al.*, 2007; Fang *et al.*, 2008; Kumar *et al.*, 2007; Langlois and Lu, 2010; Ma *et al.*, 2013; Niu *et al.*, 2014; Song *et al.*, 2014; Xu *et al.*, 2014; Guan *et al.*, 2015; Liu *et al.*, 2015a; Zhang and Liu, 2017a; Zou *et al.*, 2013; Chowdhury *et al.*, 2017; Wang *et al.*, 2017; Ali *et al.*, 2018; Hu *et al.*, 2019).

Motion *et al.* (2015) developed a species-specific SVM model by incorporating the features of amino acid composition and applied to plant genome annotations. They reported that the performance of species-specific methods is better than generic methods. Qu *et al.* (2017a) constructed a large data set of DNA-binding and non-DNA-binding proteins from UniProt database (UniProt Consortium, 2018) and used deep learning approach for developing a sequence-based model for the discrimination of the DNA and non-DNA-binding proteins. Further, they developed species-specific models for human, mouse and rice. The available computational prediction methods on DNA-binding proteins are listed in Table 8.1.

8.4. Methods for identifying RNA-binding proteins

Several methods have been developed specifically for identifying RNA-binding proteins and annotating them in different organisms. These computational tools utilized amino acid compositions of different lengths (mono-, di- and tripeptides), amino acid properties, PSSM profiles and structure-based features. Table 8.2 lists the computational methods for the prediction of RNA-binding proteins.

Kumar *et al.* (2011) used mono- and dipeptide amino acid compositions along with PSSM and trained the data with SVM for developing a method. Sharan *et al.* (2017) proposed a method by incorporating composition of dipeptides, tripeptides, physicochemical properties, secondary structure and different functional annotations. These features are commonly used in other methods also (Yu *et al.*, 2006; Fujishima *et al.*, 2007; Ma *et al.*, 2015; Zhang and Liu, 2017b).

On the other hand, structure-based features such as structural alignment, protein surface accessibility, secondary structure and electrostatic features are used for identifying RNA-binding proteins (Shazman and Mandel Gutfreund, 2008; Zhao *et al.*, 2011; Ahmad and Sarai, 2011; Paz *et al.*, 2016; Asghari and Abdolmaleki, 2019). Zhao *et al.* (2011) computed the interaction energy between protein and RNA and used specified thresholds for predicting RNA-binding proteins.

Livi *et al.* (2016) established an SVM-based method using a data set on human proteins and tested with seven different organisms.

Table 8.1. Computational methods for discriminating DNA-binding proteins.

Name	Algorithm	URL	Reference
Sequence based			
iDNA-Prot	RF	http://www.jci-bioinfo.cn/iDNA-Prot	Lin *et al.* (2011)
iDNA-Prot_dis	SVM	http://bioinformatics.hitsz.edu.cn/iDNA-Prot_dis/	Liu *et al.* (2014)
SVM-PSSM-DT	SVM	http://bioinformatics.hitsz.edu.cn/PSSM-DT/	Xu *et al.* (2015)
HMMBinder	SVM	http://brl.uiu.ac.bd/HMMBinder/	Zaman *et al.* (2017)
Local-DPP	RF	http://server.malab.cn/Local-DPP/Index.html	Wei *et al.* (2017)
Wang *et al.*, 2017	SVM	Source code available at https://doi.org/10.6084/m9.figshare.5104084	Wang *et al.* (2017)
DPP-PseAAC	SVM	http://77.68.43.135:8080/DPP-PseAAC/	Rahman *et al.* (2018)
Adilina *et al.*, 2019	RF	Source code available at https://github.com/SkAdilina/DNA_Binding	Adilina *et al.* (2019)
DNAbinder	SVM	https://webs.iiitd.edu.in/raghava/dnabinder/index.html	Kumar *et al.* (2007)
DNA-Prot	RF	Standalone version of software available at http://www3.ntu.edu.sg/home/EPNSugan/index_files/dnaprot.htm	Kumar *et al.* (2009)
newDNA-Prot	SVM	https://sourceforge.net/projects/newdnaprot/	Zhang *et al* (2014)
StackDPPred	Stacking	https://bmll.cs.uno.edu/add and standalone available	Mishra *et al.* (2019)
IKP-DBPPred	SVM	http://server.malab.cn/IKP-DBPPred/index.jsp	Qu *et al.* (2017b)
PSFM-DBT	SVM	http://bioinformatics.hitsz.edu.cn/PSFM-DBT/	Zhang and Liu (2017a)
PseDNA-Pro	SVM	http://bioinformatics.hitsz.edu.cn/PseDNA-Pro/	Liu *et al* (2015b)
iDNAPro-PseAAC	SVM	http://bioinformatics.hitsz.edu.cn/iDNAPro-PseAAC/	Liu *et al.* (2015a)

enDNA-Prot	Ensemble	http://bioinformatics.hitsz.edu.cn/Ensemble-DNA-Prot/	Xu *et al.* (2014)
Structure based			
DNABIND	LogReg	http://dnabind.szialab.org/	Szilágyi and Skolnick (2006)
SPOT-Struct-DNA	Template based	http://sparks-lab.org/yueyang/server/SPOT-Struct-DNA/	Zhao *et al.* (2010)
iDBPs	RF	http://idbps.tau.ac.il/	Nimrod *et al.* (2010)
BindUP	SVM	http://bindup.technion.ac.il/	Paz *et al.* (2016)
iDNAprot-ES	SVM	http://brl.uiu.ac.bd/iDNAProt-ES/	Chowdhury *et al.* (2017)

Notes: SVM: Support vector machine; LogReg: logistic regression; KNN: K-nearest neighbor; RF: random forest; RDF: random decision forest; NN: neural network; NB: Naive Bayes

Last accessed on 28/6/2019

Servers that are not working are omitted.

Table 8.2.   Computational methods for discriminating RNA-binding proteins.

Name	Algorithm	URL	Reference
Sequence based			
TriPepSVM	SVM	Source code available at https://github.com/marsicoLab/TriPepSVM	Bressin *et al.* (2019)
catRAPID signature	SVM	http://s.tartaglialab.com/new_submission/signature	Livi *et al.* (2016)
RBPPred	SVM	Standalone available at http://rnabinding.com/RBPPred.html	Zhang and Liu (2017b)
Deep-RBPPred	Deep learning	Source code available at http://www.rnabinding.com/Deep_RBPPred/Deep-RBPPred.html	Zheng *et al.* (2018)
RNApred	SVM	http://www.imtech.res.in/raghava/rnapred/	Kumar *et al.* (2011)
SPOT-Seq-RNA	Template based	http://sparks-lab.org/yueyang/server/SPOT-Seq-RNA/	Yang *et al.* (2014)
APRICOT	Bayesian	command-line tool available at https://pypi.python.org/pypi/bio-apricot	Sharan *et al.* (2017)
Structure based			
SPOT-Struct-RNA	Template based	http://sparks-lab.org/yueyang/server/SPOT-Struct-RNA/	Zhao *et al.* (2011)
BindUP	SVM	http://bindup.technion.ac.il/	Paz *et al.* (2016)

Notes: SVM: Support vector machine; LogReg : logistic regression; KNN: K-nearest neighbor; RF: random forest; RDF: random decision forest; NN: neural network; NB: Naive Bayes.

Last accessed on 28/6/2019

Servers that are not working on this date are omitted.

Bressin *et al.* (2019) derived tripeptide frequencies from protein sequences belonging to three different organisms such as *Homo sapiens*, *Salmonella clade* and *Escherichia coli*, and developed a SVM-based classification model. Zheng *et al.* (2018) developed a fast- and high-performance deep-learning-based classification model based on hydrophobicity, normalized van der Waals volume, polarity and polarizability, solvent accessibility, charge and polarity of the side chain, and identified RNA-binding proteins from different species: *Arabidopsis thaliana*, *Saccharomyces cerevisiae* and *H. sapiens*.

8.5. Identification of binding site residues from protein–nucleic acid complex structures

Different methods have been proposed for identifying the binding sites in protein–DNA complex structures. They are based on (i) distance between atoms in protein and DNA, (ii) reduction in solvent accessibility upon binding and (iii) interaction energy between residues in protein and nucleotides in DNA. Figure 8.2 illustrates the details for identifying the binding sites.

In distance-based approach, an amino acid residue in a protein–nucleic acid complex structure is designated as binding if any of its heavy atoms falls within a cut-off distance of 3.5 Å from any atom in

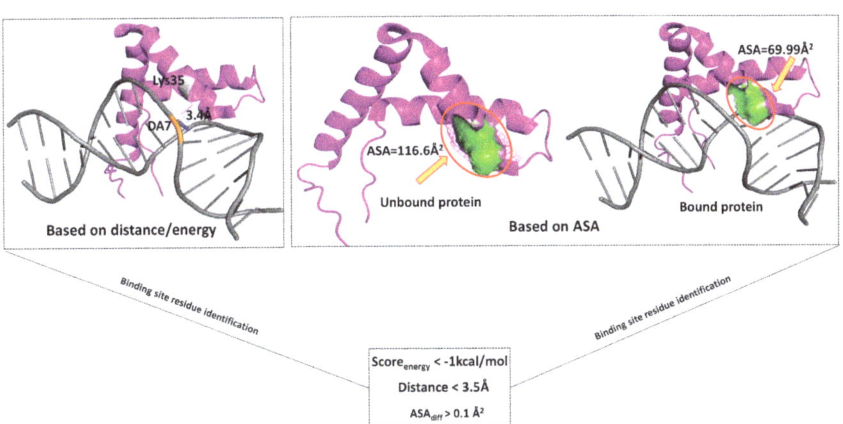

Figure 8.2.   Identification of binding site residues.

DNA (Ahmad *et al.*, 2004; Gromiha and Nagarajan, 2013). Other cut-off values (3–6 Å) have also been used to assign the binding sites for predicting the binding sites in the literature (Kuznetsov *et al.*, 2006; Ofran *et al.*, 2007).

The concept of solvent accessibility is another criterion for identifying the binding site residues. In this procedure, binding site residues are identified using the following steps: (i) compute the accessible surface area (ASA) of each residue in a protein–nucleic acid complex structure [$ASA_{complex}(i)$], where, i stands for any amino acid residue in a protein, (ii) compute the ASA of the same residue in its unbound form [$ASA_{unbound}(i)$], (iii) compute the difference between them [$ASA_{diff}(i)$]:

$$ASA_{diff}(i) = |ASA_{complex}(i) - ASA_{unbound}(i)| \qquad (4)$$

(iv) if $ASA_{diff}(i)$ is more than 0.1 Å^2, then the respective residue is considered to be binding (Bahadur *et al.*, 2003).

Gromiha and Fukui (2011) developed an energy-based approach for identifying the binding sites in protein–DNA complexes. In a protein–DNA complex, the interaction energy ($Score_{energy}$) for each residue in a protein is computed using the contribution of its heavy atoms with all heavy atoms in the partner DNA using the non-bonded interactions in AMBER force field (Cornell *et al.*, 1995). It is given by

$$Score_{energy} = \sum [(A_{ij}/r_{ij}^{12} - B_{ij}/r_{ij}^{6}) + q_i q_j / r_{ij}] \qquad (5)$$

where, $A_{ij} = \varepsilon_{ij}^*(R_{ij}^*)^{12}$ and $B_{ij} = 2\varepsilon_{ij}^*(R_{ij}^*)^6$; $R_{ij}^* = (R_i^* + R_j^*)$ and $\varepsilon_{ij}^* = (\varepsilon_i^* \varepsilon_j^*)^{1/2}$; R^* and ε^* are, respectively, the van der Waals radius and well depth, and these parameters are obtained from Cornell *et al.* (1995), and q_i and q_j are, respectively, the charges for the atoms i and j; and r_{ij} is the distance between them. The residues that have the $Score_{energy}$ of less than −1 kcal/mol are identified as binding site residues.

8.6. Prediction of binding site residues in DNA-binding proteins

Several methods have been proposed to identify the binding site residues in DNA-binding proteins (Ahmad *et al.*, 2004; Zhang *et al.*,

2019; Zhu *et al.*, 2019; Amirkhani *et al.*, 2019; Deng *et al.*, 2018; Zhou *et al.*, 2017; Hu *et al.*, 2017). These methods are mainly based on sequence- and structure-based features such as physiochemical properties, solvent accessibility, secondary structures, evolutionary information and so on. Nagarajan *et al.* (2013) analyzed a set of 11 methods and showed that performance of methods using PSSM are better than other methods in predicting the binding site residues in DNA-binding proteins. Consequently, several methods have been proposed recently using PSSM combined with other features.

Zhang *et al.* (2019) developed a method, PrPDH (Prediction of Protein–DNA-binding Hotspots) exploiting four feature groups such as sequence, structure, network and solvent accessible area, to predict the hotspots in protein–DNA binding interfaces. These features include PSSM, secondary structure, disordered regions, residue wise contact energy matrix (RCEM), depth index (DPX) and protrusion index (CX).

DNAPred developed by Zhu *et al.* (2019) utilizes amino acid frequency difference between binding and non-binding residues for improving the accuracy in DNA-binding site prediction, along with other features such as PSSM, secondary structure and relative solvent accessibility. Hu *et al.* (2017) developed a method based on SVM (Target DNA) for identifying DNA-binding sites using only two typical features such as PSSM and predicted solvent accessibility.

Zhou *et al.* (2017) developed a residue-encoding method, referred to as the PSSM Relation Transformation (PSSM-RT), to encode residues by utilizing the relationships of evolutionary information between residues. Along with PSSM-RT, sequence features such as amino acid composition, predicted secondary structure, solvent accessible area, pKa of carboxyl group and side chain, and hydrophobicity index are used in this method.

Further, several machine learning methods have been developed for predicting the binding sites in DNA-binding proteins using evolutionary information (Chai *et al.*, 2016; Wang *et al.*, 2014; Amirkhani *et al.*, 2019). On the other hand, structure-based methods have also been proposed for identifying the binding sites in DNA-binding proteins (Li *et al.*, 2014; Sukumar *et al.*, 2016). Table 8.3 lists some of

Table 8.3.   Online tools for predicting binding sites in DNA-binding proteins.

Name	Methods used	URL	Reference
Sequence			
DP-Bind	SVM, Logistic regression	http://lcg.rit.albany.edu/dp-bind/	Hwang et al. (2007)
DBS-PSSM	NN	http://ccbb.jnu.ac.in/shandar/servers/dbs-pssm/	Ahmad and Sarai (2005)
DBS-Pred	NN	http://ccbb.jnu.ac.in/shandar/servers/dbs-pred/	Ahmad et al. (2004)
SDCPred	NN	http://sciwhylab.jnu.ac.in/shandar/servers/sdcpred/index.html	Andrabi et al. (2009)
DRNApred	HMM and logistic regression	http://biomine.cs.vcu.edu/servers/DRNApred/	Yan and Kurgan (2017)
DQPred-DBR	SVM	http://www.inforstation.com/webservers/DQPred-DBR/predict.html	Chai et al. (2016)
funDNApred	SVM, NB and KNN	http://biomine.cs.vcu.edu/servers/funDNApred/#References	Amirkhani et al. (2019)
Structure			
PrPDH	SVM	http://bioinfo.ahu.edu.cn:8080/PrPDH/Server.jsp	Zhang et al. (2019)
DBSI	SVM	https://mitchell-lab.biochem.wisc.edu/DBSI_Server/index.php	Sukumar et al. (2016)
DNABINDPROT	Gaussian network model	http://www.prc.boun.edu.tr/appserv/prc/dnabindprot/	Ozbek et al. (2010)
DR_bind	Empirical	http://dnasite.limlab.ibms.sinica.edu.tw/	Chen et al. (2012b)
DISPLAR	NN	http://pipe.scs.fsu.edu/displar.html	Tjong and Zhou (2007)

Notes: SVM = Support vector machine; NN = neural network; RF = random forest; KNN = k-nearest neighbor; NB = Naive Bayes; HMM = hidden Markov model

Last accessed on 10/07/2019.

Servers that are not working are omitted.

the available online tools for predicting binding sites in DNA-binding proteins.

8.7. Prediction of binding site residues in RNA-binding proteins

Similar to DNA-binding proteins, several computational methods have been proposed for identifying protein–RNA-binding sites using machine learning techniques with sequence- and/or structure-based features. The commonly used features are amino acid composition, conservation, solvent accessibility, evolutionary information, ASA, secondary structures and hydrophobicity (Si *et al.*, 2015; Mukherjee and Bahadur, 2018). Nagarajan and Gromiha (2014) developed a Web server (https://www.iitm.ac.in/bioinfo/RNA-protein/) for relating the correspondence between the type of protein and RNA and the best method for identifying the binding sites in RNA-binding proteins.

Jung *et al.* (2019) developed sequence- (PSPRInt-Seq) and structure-based predictors (PSPRInt-Str) using five types of features: (a) PSSM, (b) secondary structure and relative ASA, (c) RNA tri-nucleotide composition for RNA sequence, (d) RNA secondary structure and (e) pairs of interacting amino acids and nucleotides.

Sun *et al.* (2016) developed a method, RNAProSite, by incorporating different features such as evolutionary information in the form of PSSMs, physicochemical properties and geometrical features along with residue electrostatic surface potential and triplet interface propensity.

Tang *et al.* (2017) proposed PredRBR, which uses a feature set selected from a large number of sequence and structure characteristics and two categories of structural neighborhood properties. Sequence- and structure-based features include physiochemical properties, pKa, PSSM, solvent accessible area, secondary structure, atomic contacts and residue contacts. Kumar *et al.* (2008) developed an SVM-based method, Pprint, using PSSM for predicting the binding site residues. Yasser *et al.* (2016) developed FastRNABindR, which uses PSSM on three machine learning techniques such as Naive Bayes, random forest

Table 8.4.   Online tools for predicting binding sites in RNA-binding proteins.

Name	Methods used	Server link	References
Sequence			
RBind	Network	https://zhaolab.com.cn/RBind	Wang *et al.* (2018)
PredRBR	Gradient tree boosting	http://denglab.org/PredRBR/	Tang *et al.* (2017)
RBPmap	Weighted-rank approach	http://rbpmap.technion.ac.il/	Paz *et al.* (2014)
Structure			
Pprint	SVM	http://crdd.osdd.net/raghava/pprint/	Kumar *et al.* (2008)
RPISeq	SVM	http://pridb.gdcb.iastate.edu/RPISeq/	Muppirala *et al.* (2011)
DRNApred	HMM and logistic regression	http://biomine.cs.vcu.edu/servers/DRNApred/	Yan and Kurgan (2017)

Notes: SVM: Support vector machine; HMM: hidden Markov model
Last accessed on 10/07/2019
Servers that are not working are omitted.

and SVM. Walia *et al.* (2014) developed a method RNABindRPlus which combines two approaches called HomPRIP, a sequence homology-based method and PSSM profile of SVM classifier for predicting RNA-binding sites in proteins. Table 8.4 lists the available online tools for predicting the binding sites in RNA-binding proteins.

8.8. Best predictor for identifying the binding sites in protein–nucleic acid complexes

Several computational methods have been developed for predicting the interacting residues in DNA/RNA-binding proteins using sequence and/or structural information. These methods showed different levels of accuracies, and the accuracy varies in the range of

20–90%. However, there is no single existing method uniformly predicts the binding sites at high accuracy for all the available data sets. This leads confusions to experimentalists for selecting the best method to identify the binding sites for designing their experiments. Hence, it is essential to choose the best method based on the query protein/nucleic acid. Nagarajan *et al.* (2013) tackled the problem and developed Web servers https://www.iitm.ac.in/bioinfo/DNA-protein/and https://www.iitm.ac.in/bioinfo/RNA-protein/ (Nagarajan and Gromiha, 2014) for selecting the best method for identifying the binding sites in DNA- and RNA-binding proteins, respectively, based on the structural class, fold, family, superfamily, motif and function of the query protein single/double-stranded DNA, RNA type and DNA/RNA conformation. This correspondence is reported to enhance the prediction accuracy of 5–10%.

8.9. Binding site prediction in disordered regions

Intrinsically disordered proteins/regions (IDPs/IDRs) lack unique three-dimensional structure in solution and are highly flexible. The flexibility allows these proteins to bind to multiple partners including protein, DNA and RNA molecules. Although, protein–protein interactions through IDRs are very well studied, DNA– and RNA–protein interactions through IDRs are poorly understood. Consequently, several computational tools are available for predicting disorder-to-order regions in protein–protein interactions (Dosztányi *et al.*, 2009; Perovic *et al.*, 2018; Xue *et al*, 2010), and only very few methods are developed to predict the binding sites in disordered regions of protein–DNA and protein–RNA complexes. Peng and Kurgan (2015) developed a method for predicting RNA-, DNA- and protein-binding sites in the disordered regions of a protein using sequence-based features, hydrophobicity, solvent accessibility, low-complexity regions, charge and free energy. However, the performance of predicting the binding sites in disordered regions of DNA- and RNA-binding proteins is lesser than ordered regions (Nagarajan *et al.*, 2013; Nagarajan and Gromiha, 2014), and it is possible to improve the performance upon the availability of large amounts of data in future.

8.10. Conclusion

Information on binding site residues in protein–DNA and protein–RNA complexes is important to understand their binding affinity and functions. We have reviewed various tools and methods for identifying DNA- and RNA-binding proteins and their binding sites. Methods developed with PSSM profiles are shown to have better performance than other methods. Further, specific methods and refinements are necessary to predict the binding site residues in disordered regions with high accuracy.

Acknowledgment

The work is partially supported by the Council of Scientific and Industrial Research, Government of India, to MMG (37(1694)/17/EMR-II).

References

Adilina, S., Farid, D.M., and Shatabda, S. Effective DNA binding protein prediction by using key features via Chou's general PseAAC. *J Theor Biol.* 2019;460:64–78.

Ahmad, S., and Sarai, A. PSSM-based prediction of DNA binding sites in proteins. *BMC Bioinformatics.* 2005;6:33.

Ahmad, S. and Sarai, A. Analysis of electric moments of RNA-binding proteins: implications for mechanism and prediction. *BMC Struct Biol.* 2011;11(1):8.

Ahmad, S., Gromiha, M.M., and Sarai, A. Analysis and prediction of DNA-binding proteins and their binding residues based on composition, sequence and structural information. *Bioinformatics.* 2004;20:477–486.

Ahmad, S., Prathipati, P., Tripathi, L.P., *et al.* Integrating sequence and gene expression information predicts genome-wide DNA-binding proteins and suggests a cooperative mechanism. *Nucleic Acids Res.* 2018;46: 54–70.

Ali, F., Kabir, M., Arif, M., *et al.* DBPPred-PDSD: Machine learning approach for prediction of DNA-binding proteins using Discrete Wavelet Transform and optimized integrated features space. *Chemometr Intell Lab Syst.* 2018;182:21–30.

Amirkhani, A., Kolahdoozi, M., Wang, C., *et al.* Prediction of DNA-binding residues in local segments of protein sequences with Fuzzy Cognitive Maps. *IIEEE/ACM Trans Comput Biol Bioinform.* 2019. (In press).

Andrabi, M., Mizuguchi, K., Sarai, A., *et al.* Prediction of mono-and di-nucleotide-specific DNA-binding sites in proteins using neural networks. *BMC Struct Biol.* 2009;9:30.

Asghari, M.P., and Abdolmaleki, P. Prediction of RNA-and DNA-binding proteins using various machine learning classifiers. *Avicenna J Med Biotechnol.* 2019;11:104–111.

Bahadur, R.P., Chakrabarti, P., Rodier, F., *et al.* Dissecting subunit interfaces in homodimeric proteins. *Proteins.* 2003;53:708–719.

Bhardwaj, N., Langlois, R.E., Zhao, G., *et al.* Kernel-based machine learning protocol for predicting DNA-binding proteins. *Nucleic Acids Res.* 2005;33:6486–6493.

Bressin, A., Schulte-Sasse, R., Figini, D., *et al.* TriPepSVM: De novo prediction of RNA-binding proteins based on short amino acid motifs. *Nucleic Acids Res.* 2019;47:4406–4417.

Carey, M.F., Peterson, C.L., and Smale, S.T. Experimental strategies for the identification of DNA-binding proteins. *Cold Spring Harb Protoc.* 2012;2012:18–33.

Chai, H., Zhang, J., Yang, G., *et al.* An evolution-based DNA-binding residue predictor using a dynamic query-driven learning scheme. *Mol Biosyst.* 2016;12:3643–3650.

Chakravarty, D., Janin, J., Robert, C.H., *et al.* Changes in protein structure at the interface accompanying complex formation. *IUCrJ.* 2015;2: 643–652.

Chen, C.Y., Chien, T.Y., Lin, C.K., *et al*. Predicting target DNA sequences of DNA-binding proteins based on unbound structures. *PLOS ONE.* 2012a;7:e30446.

Chen, Y.C., Wright, J.D., and Lim, C. . DR_bind: A web server for predicting DNA-binding residues from the protein structure based on electrostatics, evolution and geometry. *Nucleic Acids Res.* 2012b;40:W249–W256.

Chowdhury, S.Y., Shatabda, S., and Dehzangi, A. iDNAprot-es: Identification of DNA-binding proteins using evolutionary and structural features. *Sci Rep.* 2017;7:14938.

Cornell, W.D., Cieplak, P., Bayly, C.I., *et al*. A second generation force field for the simulation of proteins, nucleic acids, and organic molecules. *J Am Chem Soc*. 1995;117:5179–5197.

Deng, L., Pan, J., Xu, X., *et al*. PDRLGB: Precise DNA-binding residue prediction using a light gradient boosting machine. *BMC Bioinformatics*. 2018;19:522.

de Vries, S.J., van Dijk, M., and Bonvin, A.M. The HADDOCK web server for data-driven biomolecular docking. *Nat Protoc*. 2010;5:883–897.

Dosztányi, Z., Mészáros, B., and Simon, I. ANCHOR: Web server for predicting protein binding regions in disordered proteins. *Bioinformatics*. 2009;25:2745–2746.

Fang, Y., Guo, Y., Feng, Y., *et al*. Predicting DNA-binding proteins: approached from Chou's pseudo amino acid composition and other specific sequence features. *Amino Acids*. 2008;34:103–109.

Fujishima, K., Komasa, M., Kitamura, S., *et al*. Proteome-wide prediction of novel DNA/RNA-binding proteins using amino acid composition and periodicity in the hyperthermophilic archaeon Pyrococcus furiosus. *DNA Res*. 2007;14:91–102.

Gao, M., and Skolnick, J. DBD-Hunter: A knowledge-based method for the prediction of DNA–protein interactions. *Nucleic Acids Res*. 2008;36: 3978–3992.

Gao, M., and Skolnick, J. A threading-based method for the prediction of DNA-binding proteins with application to the human genome. *PLoS Comput Biol*. 2009;5:e1000567.

Gromiha, M.M., and Fukui, K. Scoring function based approach for locating binding sites and understanding recognition mechanism of protein–DNA complexes. *J Chem Inf Model*. 2011;51:721–729.

Gromiha, M.M., and Nagarajan, R. Computational approaches for predicting the binding sites and understanding the recognition mechanism of protein–DNA complexes. *Adv Protein Chem Struct Biol*. 2013;91: 65–99.

Guan, C., Niu, X., Shi, F., *et al*. Predicting a DNA-binding protein using random forest with multiple mathematical features. *Biomed Mater Eng*. 2015;26:S1883–S1889.

Ho, S.Y., Yu, F. C., Chang, C.Y., *et al*. Design of accurate predictors for DNA-binding sites in proteins using hybrid SVM–PSSM method. *Biosystems*. 2007;90:234–241.

Hu, J., Li, Y., Zhang, M., *et al*. Predicting protein-DNA binding residues by weightedly combining sequence-based features and boosting multiple SVMs. *IEEE/ACM Trans Comput Biol Bioinform*. 2017;14:1389–1398.

Hu, J., Zhou, X., Zhu, Y.H., *et al.* TargetDBP: Accurate DNA-binding protein prediction via sequence-based multi-view feature learning. *IEEE/ACM Trans Comput Biol Bioinform.* 2019 (In press).

Hwang, S., Gou, Z., and Kuznetsov, I.B. DP-Bind: A web server for sequence-based prediction of DNA-binding residues in DNA-binding proteins. *Bioinformatics.* 2007;23:634–636.

Jung, Y., EL-Manzalawy, Y., Dobbs, D., *et al.* Partner-specific prediction of RNA-binding residues in proteins: A critical assessment. *Proteins.* 2019;87:198–211.

Kumar, K.K., Pugalenthi, G., and Suganthan, P.N. DNA-Prot: Identification of DNA binding proteins from protein sequence information using random forest. *J Biomol Struct Dyn.* 2009;26:679–686.

Kumar, M., Gromiha, M.M., and Raghava, G.P.S. Identification of DNA-binding proteins using support vector machines and evolutionary profiles. *BMC Bioinformatics.* 2007;8:463.

Kumar, M., Gromiha, M.M., and Raghava, G.P.S. Prediction of RNA binding sites in a protein using SVM and PSSM profile. *Proteins.* 2008;71:189–194.

Kumar, M., Gromiha, M.M., and Raghava, G.P.S. SVM based prediction of RNA-binding proteins using binding residues and evolutionary information. *J Mol Recognit.* 2011;24:303–313.

Kuznetsov, I.B., Gou, Z., Li, R., *et al.* Using evolutionary and structural information to predict DNA-binding sites on DNA-binding proteins. *Proteins.* 2006;64:19–27.

Langlois, R.E., and Lu, H. Boosting the prediction and understanding of DNA-binding domains from sequence. *Nucleic Acids Res.* 2010;38:3149–3158.

Li, B.Q., Feng, K.Y., Ding, J., *et al.* Predicting DNA-binding sites of proteins based on sequential and 3D structural information. *Mol Genet Genomics.* 2014;289:489–499.

Lin, W.Z., Fang, J.A., Xiao, X., *et al.* iDNA-Prot: Identification of DNA binding proteins using random forest with grey model. *PLOS ONE.* 2011;6:e24756.

Liu, B., Xu, J., Lan, X., *et al.* iDNA-Prot| dis: Identifying DNA-binding proteins by incorporating amino acid distance-pairs and reduced alphabet profile into the general pseudo amino acid composition. *PLOS ONE.* 2014;9:e106691.

Liu, B., Wang, S., and Wang, X. DNA binding protein identification by combining pseudo amino acid composition and profile-based protein representation. *Sci Rep.* 2015a;5:15479.

Liu, B., Xu, J., Fan, S., *et al.* PseDNA-Pro: DNA-binding protein identification by combining Chou's PseAAC and physicochemical distance transformation. *Molecular Informatics.* 2015b;34:8–17.

Livi, C.M., Klus, P., Delli Ponti, R., *et al.* Cat Rapid signature: Identification of ribonucleoproteins and RNA-binding regions. *Bioinformatics.* 2016;32:773–775.

Ma, X., Wu, J., and Xue, X. Identification of DNA-binding proteins using support vector machine with sequence information. *Comput Math Methods Med.* 2013;2013:524502.

Ma, X., Guo, J., and Sun, X. Sequence-based prediction of RNA-binding proteins using random forest with minimum redundancy maximum relevance feature selection. *Biomed Res Int.* 2015;425810.

Ma, X., Guo, J., and Sun, X. DNABP: Identification of DNA-binding proteins based on feature selection using a random forest and predicting binding residues. *PLOS ONE.* 2016;11, e0167345.

McHugh, C.A., Russell, P., and Guttman, M.Methods for comprehensive experimental identification of RNA-protein interactions. *Genome Biol.* 2014;15: 203.

Mishra, A., Pokhrel, P., and Hoque, M.T. StackDPPred: A stacking based prediction of DNA-binding protein from sequence. *Bioinformatics.* 2019;35:433–441.

Motion, G.B., Howden, A.J., Huitema, E., *et al.* DNA-binding protein prediction using plant specific support vector machines: Validation and application of a new genome annotation tool. *Nucleic Acids Res.* 2015;43:e158–e158.

Mukherjee, S., and Bahadur, R.P. An account of solvent accessibility in protein-RNA recognition. *Sci Rep.* 2018;8:10546.

Muppirala, U.K., Honavar, V.G., and Dobbs, D. Predicting RNA-protein interactions using only sequence information. *BMC Bioinformatics.* 2011;12:489.

Nagarajan, R., Ahmad, S., and Gromiha, M.M. Novel approach for selecting the best predictor for identifying the binding sites in DNA binding proteins. *Nucleic Acids Res.* 2013;41(16): 7606–7614.

Nagarajan, R., and Gromiha, M.M. Prediction of RNA binding residues: An extensive analysis based on structure and function to select the best predictor. *PLOS ONE.* 2014;9:e91140.

Nimrod, G., Schushan, M., Szilágyi, A., *et al.* iDBPs: A web server for the identification of DNA binding proteins. *Bioinformatics.* 2010;26: 692–693.

Niu, X.H., Hu, X.H., Shi, F., *et al.* Predicting DNA binding proteins using support vector machine with hybrid fractal features. *J Theor Biol.* 2014;343:186–192.

Ofran, Y., Mysore, V., and Rost, B. Prediction of DNA-binding residues from sequence. *Bioinformatics.* 2007;23:i347–i353.

Ozbek, P., Soner, S., Erman, B., *et al.* DNABINDPROT: Fluctuation-based predictor of DNA-binding residues within a network of interacting residues. *Nucleic Acids Res.* 2010;38:W417–W423.

Paz, I., Kligun, E., Bengad, B., *et al.* BindUP: A web server for non-homology-based prediction of DNA and RNA binding proteins. *Nucleic Acids Res.* 2016;44:W568–W574.

Paz, I., Kosti, I., Ares Jr, M., *et al.* RBPmap: A web server for mapping binding sites of RNA-binding proteins. *Nucleic Acids Res.* 2014;42: W361–W367.

Peng, Z., and Kurgan, L. High-throughput prediction of RNA, DNA and protein binding regions mediated by intrinsic disorder. *Nucleic Acids Res.* 2015;43:e121–e121.

Perovic, V., Sumonja, N., Marsh, L.A., *et al.* IDPpi: Protein–protein interaction analyses of human intrinsically disordered proteins. *Sci Rep.* 2018;8:10563.

Qu, K., Han, K., Wu, S., *et al.* Identification of DNA-binding proteins using mixed feature representation methods. *Molecules.* 2017b ;22:1602.

Qu, Y.H., Yu, H., Gong, X.J., *et al.* On the prediction of DNA-binding proteins only from primary sequences: A deep learning approach. *PLOS ONE.* 2017a;12:e0188129.

Rahman, M.S., Shatabda, S., Saha, S., *et al.* DPP-PseAAC: A DNA-binding protein prediction model using Chou's general PseAAC. *J Theor Biol.* 2018;452:22–34.

Rose, P.W., Prlić, A., Altunkaya, A., *et al.* The RCSB protein data bank: Integrative view of protein, gene and 3D structural information. *Nucleic Acids Res.* 2017;45: D271–D281.

Sharan, M., Förstner, K.U., Eulalio, A., *et al.* APRICOT: An integrated computational pipeline for the sequence-based identification and characterization of RNA-binding proteins. *Nucleic Acids Res.* 2017;45: e96–e96.

Shazman, S., and Mandel-Gutfreund, Y. Classifying RNA-binding proteins based on electrostatic properties. *PLoS Comput Biol.* 2008;4: e1000146.

Si, J., Cui, J., Cheng, J., *et al.* Computational prediction of RNA-binding proteins and binding sites. *Int J Mol Sci.* 2015;16: 26303–26317.

Song, L., Li, D., Zeng, X., *et al.* nDNA-prot: Identification of DNA-binding proteins based on unbalanced classification. *BMC Bioinformatics.* 2014;15: 298.

Sukumar, S., Zhu, X., Ericksen, S.S., *et al.* DBSI server: DNA binding site identifier. *Bioinformatics.* 2016;32:2853–2855.

Sun, M., Wang, X., Zou, C., *et al.* Accurate prediction of RNA-binding protein residues with two discriminative structural descriptors. *BMC Bioinformatics.* 2016;17: 231.

Szilágyi, A., and Skolnick, J. Efficient prediction of nucleic acid binding function from low-resolution protein structures. *J Mol Bio.* 2006;358:922–933.

Tang, Y., Liu, D., Wang, Z., *et al.* A boosting approach for prediction of protein-RNA binding residues. *BMC Bioinformatics.* 2017;18:465.

Taniguchi, H., Fujimoto, A., Kono, H., *et al.* Loss-of-function mutations in Zn-finger DNA-binding domain of HNF4A cause aberrant transcriptional regulation in liver cancer. *Oncotarget.* 2018;9:26144–26156.

Tjong, H., and Zhou, H.X. DISPLAR: An accurate method for predicting DNA-binding sites on protein surfaces. *Nucleic Acids Res.* 2007;35: 1465–1477.

Tuszynska, I., Magnus, M., Jonak, K., *et al.* NPDock: A web server for protein–nucleic acid docking. *Nucleic Acids Res.* 2015;43:W425–W430.

UniProt Consortium. UniProt: The universal protein knowledgebase. *Nucleic Acids Res.* 2018;46:2699.

Walia, R.R., Xue, L.C., Wilkins, K., *et al.* RNABindRPlus: A predictor that combines machine learning and sequence homology-based methods to improve the reliability of predicted RNA-binding residues in proteins. *PLOS ONE.* 2014;9:e97725.

Wang, L., and Brown, S.J. BindN: A web-based tool for efficient prediction of DNA and RNA binding sites in amino acid sequences. *Nucleic Acids Res.* 2006a;34:W243–W248.

Wang, L., and Brown, S.J. Prediction of DNA-binding residues from sequence features. *J Bioinform Comput Biol.* 2006b;4:1141–1158.

Wang, Y., Ding, Y., Guo, F., *et al.* Improved detection of DNA-binding proteins via compression technology on PSSM information. *PLOS ONE.* 2017;12:e0185587.

Wang, L., Huang, C., Yang, M.Q., *et al.* BindN+ for accurate prediction of DNA and RNA-binding residues from protein sequence features. *BMC Syst Biol.* 2010;4:S3.

Wang, K., Jian, Y., Wang, H., *et al*. RBind: Computational network method to predict RNA binding sites. *Bioinformatics*. 2018;34:3131–3136.

Wang, W., Liu, J., Xiong, Y., *et al*. Analysis and classification of DNA-binding sites in single-stranded and double-stranded DNA-binding proteins using protein information. *IET Syst Biol*. 2014;8:176–183.

Wei, L., Tang, J., and Zou, Q. Local-DPP: An improved DNA-binding protein prediction method by exploring local evolutionary information. *Inf Sci*. 2017;384:135–144.

Xu, R., Zhou, J., Liu, B., *et al*. enDNA-Prot: Identification of DNA-binding proteins by applying ensemble learning. *Biomed Res Int*. 2014;2014: 294279.

Xu, R., Zhou, J., Wang, H., *et al*. Identifying DNA-binding proteins by combining support vector machine and PSSM distance transformation. *BMC Syst Biol*. 2015;9: S10.

Xue, B., Dunbrack, R.L., Williams, R.W., *et al*. PONDR-FIT: A meta-predictor of intrinsically disordered amino acids. *Biochim Biophys Acta*. 2010;1804:996–1010.

Yan, J., and Kurgan, L. DRNApred, fast sequence-based method that accurately predicts and discriminates DNA-and RNA-binding residues. *Nucleic Acids Res*. 2017;45:e84.

Yan, Y., Zhang, D., Zhou, P., *et al*. HDOCK: A web server for protein–protein and protein–DNA/RNA docking based on a hybrid strategy. *Nucleic Acids Res*. 2017;45:W365–W373.

Yang, Y., Zhao, H., Wang, J., *et al*. SPOT-Seq-RNA: Predicting protein–RNA complex structure and RNA-binding function by fold recognition and binding affinity prediction. *Methods Mol Biol*. 2014;1137:119–130.

Yasser, E. M., Abbas, M., Malluhi, Q., *et al*. Fastrnabindr: Fast and accurate prediction of protein-RNA interface residues. *PLOS ONE.*; 2016: 11:e0158445.

Yu, X., Cao, J., Cai, Y., Shi, T., *et al*. Predicting rRNA-, RNA-, and DNA-binding proteins from primary structure with support vector machines. *J Theor Biol*. 2006;240:175–184.

Zaman, R., Chowdhury, S.Y., Rashid, M.A., *et al*. Hmmbinder: Dna-binding protein prediction using hmm profile based features. *Biomed Res Int*. 2017;2017:4590609.

Zhang, J., and Liu, B. PSFM-DBT: Identifying DNA-binding proteins by combing position specific frequency matrix and distance-bigram transformation. *Int J Mol Sci*. 2017a;18:e1856.

Zhang, X., and Liu, S. RBPPred: Predicting RNA-binding proteins from sequence using SVM. *Bioinformatics.* 2017b;33:854–862.

Zhang, J., Gao, B., Chai, H., *et al.* Identification of DNA-binding proteins using multi-features fusion and binary firefly optimization algorithm. *BMC Bioinformatics.* 2016;17:323.

Zhang, Y., Xu, J., Zheng, W., *et al.* newDNA-Prot: Prediction of DNA-binding proteins by employing support vector machine and a comprehensive sequence representation. *Comput Biol Chem.* 2014;52: 51–59.

Zhang, S., Zhao, L., Zheng, C.H., *et al.* A feature-based approach to predict hot spots in protein–DNA binding interfaces. *Brief Bioinform.* 2019 (in press).

Zhao, H., Wang, J., Zhou, Y., *et al.* Predicting DNA-binding proteins and binding residues by complex structure prediction and application to human proteome. *PLOS ONE.* 2014;9:e96694.

Zhao, H., Yang, Y., and Zhou, Y. Structure-based prediction of DNA-binding proteins by structural alignment and a volume-fraction corrected DFIRE-based energy function. *Bioinformatics.* 2010;26:1857–1863.

Zhao, H., Yang, Y., and Zhou, Y. Structure-based prediction of RNA-binding domains and RNA-binding sites and application to structural genomics targets. *Nucleic Acids Res.* 2011;39:3017–3025.

Zheng, J., Zhang, X., Zhao, X., *et al.* Deep-RBPPred: Predicting RNA binding proteins in the proteome scale based on deep learning. *Sci Rep.* 2018;8:15264.

Zhou, J., Lu, Q., Xu, R., *et al.* EL_PSSM-RT: DNA-binding residue prediction by integrating ensemble learning with PSSM Relation Transformation. *BMC Bioinformatics.* 2017;18(1):379.

Zhu, Y. H., Hu, J., Song, X.N., *et al.* DNAPred: Accurate identification of DNA-binding sites from protein sequence by ensembled hyperplane-distance-based support vector machines. *J Chem Inf Model.* 2019;59: 3057–3071.

Zou, C., Gong, J., and Li, H. An improved sequence based prediction protocol for DNA-binding proteins using SVM and comprehensive feature analysis. *BMC Bioinformatics.* 2013;14:90.

Chapter 9
Predicting protein-binding sites in nucleic acids

Kyungsook Han*

*Department of Computer Engineering
Inha University, Incheon, Korea*

Interactions between protein and nucleic acid (DNA or RNA) molecules are essential for a variety of cellular processes. A large amount of interaction data generated by experimental technologies have triggered the development of several computational methods. Most computational methods have focused on predicting binding sites in a sequence or on determining whether an interaction exists between a pair of protein and nucleic acid sequences. Regarding the problem of predicting binding sites between proteins and nucleic acids, many existing computational methods are limited to finding nucleic acid binding sites in proteins instead of protein-binding sites in nucleic acids. Compared to the problem of predicting nucleic acid binding sites in proteins, predicting protein-binding sites in nucleic acids is more challenging for several reasons. In this article, the existing computational methods for predicting protein-binding sites in nucleic acids and future directions in predicting protein-binding sites in nucleic acids have been reviewed.

*khan@inha.ac.kr

9.1. Introduction

Proteins interact with nucleic acids (DNA or RNA) in a variety of biological activities within a cell. For instance, a regulatory protein called transcription factor (TF) binds to a specific region of DNA to activate or repress the synthesis of RNA (Latchman, 1997). Interactions between RNA-binding proteins (RBPs) and RNA are involved in many cellular processes such as transcriptional and post-transcriptional regulation of gene expression, translation, and alternative splicing (König *et al.*, 2012).

Due to recent advances in high-throughput experimental technologies such as chromatin immunoprecipitation sequencing (ChIP-seq) and cross-linking and immunoprecipitation (CLIP), a large amount of data on the interaction between proteins and nucleic acids have been generated. Despite the increased data on protein-nucleic acid interactions, the mechanism of of the interaction between proteins and nucleic acids is not fully uncovered and a large number of nucleic acid binding proteins and their target nucleic acids remain to be uncovered. For example, for about 500 protein-coding genes in humans, only 1,542 RNA-binding proteins (RBPs) and their target RNAs have been identified so far (Gerstberger *et al.*, 2014).

As a complement to experimental methods, several computational methods have been proposed, which are largely motivated by the increased amount of data on protein-nucleic acid interactions and the availability of complete genome sequences of several organisms. Most computational methods attempted to solve one of the two problems: predicting binding sites in a sequence (Walia *et al.*, 2014; Tuvshinjargal *et al.*, 2016; Choi *et al.*, 2017) and determining whether an interaction exists between a pair of protein and nucleic acid sequences (Alipanahi *et al.*, 2015; Akbaripour-Elahabad *et al.*, 2016; Zhang and Liu, 2017).

Regarding the problem of predicting binding sites between proteins and nucleic acids, many existing computational methods are limited to finding nucleic acid-binding sites in proteins instead of protein-binding sites in nucleic acids. For instance, BindN+ (Wang *et al.*, 2010), which is an upgraded version of BindN (Wang and

Brown, 2006), uses a support vector machine (SVM) to predict the RNA- or DNA-binding residues from biochemical features and evolutionary information of protein sequences. RNABindRPlus (Walia *et al.*, 2014) also predicts RNA-binding residues in a protein sequence by combining predictions from an optimized SVM and those from a sequence homology method. aaRNA (Li *et al.*, 2014) predicts RNA binding residues in protein using sequence- and structure-based features.

Compared to the problem of predicting nucleic acid-binding sites in proteins, predicting protein-binding sites in nucleic acids is more challenging for several reasons (Choi and Han, 2013). One reason for this is that there are only four types of nucleotides in nucleic acids whereas there are 20 types of amino acids in proteins. For a sequence of length w, there are 4^w possible nucleic acid sequences whereas a protein has 20^w possible sequences, which is 5^w fold larger. As nucleic acid sequences show a much lower diversity than protein sequences, predicting the protein-binding nucleotides is more difficult than predicting the nucleic acid binding amino acids using sequence data alone. Another reason is that nucleotides show much less diverse interaction propensities than amino acids in protein-nucleic acid interactions. In proteins interacting with nucleic acids, the side chain contacts are more dominant than the main chain contacts, so amino acids reveal diverse interaction propensities. In contrast, in nucleic acids interacting with proteins, the backbone contacts are more frequent than the base contacts, so the four types of nucleotides have a similar interaction propensity from each other (Choi and Han, 2013). Thus, it is difficult to distinguish binding nucleotides from non-binding nucleotides.

Recently a small number of computational methods have been developed that can predict protein-binding sites in nucleic acids. For example, catRAPID estimates the binding propensity of RNA and protein molecules by combining secondary structure, hydrogen bonding and van der Waals contributions (Bellucci *et al.*, 2011). It often predicts an entire RNA sequence as a binding site even for an RNA sequence of 50 or more nucleotides. DeepBind (Alipanahi *et al.*, 2015) is known to outperform state-of-the-art experimental and computational methods for identification of protein binding sites in DNA or RNA sequences. It uses deep convolutional neural networks, trained on a huge amount of data from high-throughput experiments. For the problem of predicting RBP-binding sites in RNA sequences, DeepBind was trained on a large amount of data from RNAcompete, CLIP-seq and RIP-seq (Ray *et al.*, 2013). As DeepBind contains about 200 distinct prediction models for RBP-binding and about 500 distinct models for TF-binding, each for different target proteins, the user should try all of them in the absence of prior information on a target protein. As output, it provides a predictive binding score without protein-binding sites in the input nucleic acid sequence. PRIdictor (Tuvshinjargal *et al.*, 2015; Tuvshinjargal *et al.*, 2016) predicts not only protein-binding sites in RNA but also RNA-binding sites in protein at the nucleotide and residue level using sequence data alone. A computational method developed by Wong *et al.* (2015) predicts interacting nucleotides and residues between DNA and proteins.

In this article, we review existing computational methods for predicting protein-binding sites in nucleic acids and future directions in predicting protein-binding sites in nucleic acids. We also propose implications for novel prediction methods and practical applications.

9.2. Definition of a binding site between proteins and nucleic acids

Finding binding sites in the interactions between proteins and nucleic acids requires a binding site to be defined, and different definitions of a binding site have been used in previous studies on the interactions between proteins and nucleic acids (Wang and Brown, 2006; Hwang *et al.*, 2007; Yan *et al.*, 2006). The most common definition of a binding site is based on the distance between proteins and nucleic acids. For example, in BindN (Wang and Brown, 2006) and RNABindR (Terribilini *et al.*, 2007), an amino acid with an atom within a distance of 5Å from any atom of a nucleotide was considered to be an RNA-binding or DNA-binding amino acid. Likewise, a nucleotide with an atom within distance of 5Å from any atom of an amino acid was considered to be a protein-binding nucleotide. Some studies used 3.5Å instead of 5Å as the distance criteria to define a binding site.

Several types of interactions are involved in interactions between proteins and nucleic acids. Hydrogen bonds (H bonds), water bridges and hydrophobic interactions are the most typical interactions between proteins and nucleic acids. The nucleic acid-protein interaction database (NPIDB) (Kirsanov *et al.*, 2013) provides protein-DNA pairs and protein-RNA pairs involved in the three types of interactions. In the study by Im *et al.* (2015), a nucleotide participating in any of the interactions of the three types was considered as a protein-binding site.

Among the three types of the interactions, the hydrogen bonding interaction is the strongest. Some studies [references] use hydrogen bond interaction only to define a binding site. A protein-binding nucleotide should satisfy the following geometric criteria for hydrogen bonding interactions with a protein: the contacts with the donor-acceptor (D-A) distance < 3.9Å, the hydrogen-acceptor (H-A) distance < 2.5Å, the donor-hydrogen-acceptor (D-H-A) angle > 90° and the hydrogen-acceptor-acceptor antecedent (H-A-AA) angle > 90°. The geometric criteria for the hydrogen bond interaction can be modified by a slight change in the D-A distance and H-A distance.

9.3. Data of binding sites between proteins and nucleic acids

Several data sources are available for obtaining binding sites between proteins and nucleic acids. The best known data source is the protein data bank (PDB) (Berman *et al.*, 2000), which provides structure data of protein-DNA complexes, protein-RNA complexes, and other molecules as well. From the structure data of complexes of proteins with nucleic acids, one can derive binding sites at various levels, from the atomic level to the sequence level.

In addition to PDB, there are other data sources that are specialized in nucleic acid–protein interaction such as NPIDB (Kirsanov *et al.*, 2013). In NPIDB, one can extract binding sites involved in one of the three types of interactions between proteins and nucleic acids: hydrogen bonds, water bridges and hydrophobic interactions.

Recent advances in high throughput experiments such as ChIP-seq, eCLIP, iCLIP, RIP-chip, and RIP-seq resulted in an exponential growth of data on protein-binding sites in nucleic acid sequences. These experiments do not provide information on binding sites at the atomic level, but at the sequence or nucleotide level. Figure 9.1 shows the amount of data on protein-binding nucleic acids in the human genome at ENCODE (https://www.encodeproject.org/). As of July 26, 2018, there are 2,238 samples from ChIP-seq assays for protein-DNA binding, and 201 samples from eCLIP, iCLIP, RIP-chip, RIP-seq assays for protein-RNA binding. Storing the data from the 2,439 (= 2,238 + 201) samples requires about 80 TB.

The JASPAR CORE database (http://jaspar.genereg.net/collection/core/) contains a curated, non-redundant set of DNA-binding sites of TFs in eukaryotes, which were experimentally identified by ChIP-seq, microarray, or SELEX. The data set of binding sites can be converted into a position weight matrix (PWM; also known as PSSM) for use in scanning genomic sequences.

CLIPdb (Yang *et al.*, 2015) provides RBP-binding sites in RNA in four organisms (human, mouse, worm and yeast). POSTAR (Hu *et al.*, 2017) is an extended version of CLIPdb, and provides RBP-binding sites in RNA in six organisms (human, mouse, fly, worm, Arabidopsis and yeast). In addition to RBP-binding sites, POSTAR provides the confidence level of RBP-binding sites in the range [0, 1].

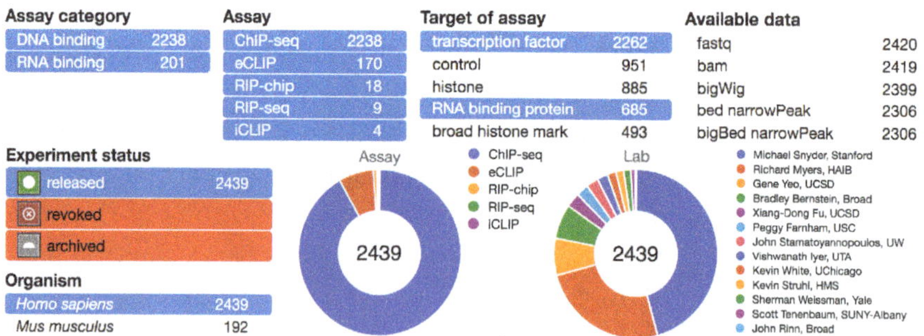

Figure 9.1. The amount of data on protein-binding nucleic acids in the human genome at ENCODE. As of July 26, 2018, there are 2,238 samples from ChIP-seq assays for protein-DNA binding, and 201 samples from eCLIP, iCLIP, RIP-chip, RIP-seq assays for protein-RNA binding.

9.4. Predicting protein-binding sites in DNA

Understanding the specificity of protein-DNA interactions has important consequence for the analysis and prediction of the gene regulatory relations that are associated with several biological processes. Regulatory proteins called TFs bind to specific regions of genomic DNA to activate or repress the synthesis of RNA (Latchman, 1997).

Recent advances in experimental techniques have led to the identification of many previously unknown transcription factor-binding sites (TFBSs) in DNA sequences *in vitro* or *in vivo* in many species and tissues. Different TFs have different preferences for their binding regions in DNA, and TF–DNA binding specificity is influenced by several factors, including nucleotide sequence, 3D structure and flexibility of TFs, presence of cofactors, chromatin accessibility and nucleosome occupancy, pioneer TFs that bind to nucleosomal DNA and DNA methylation (Slattery *et al.*, 2014). The mechanism by which TFs select specific binding regions is complex, and there are still a large number of TFBSs to be determined at various levels.

Several computational methods have been developed to study TF–DNA binding. For example, PAnDA (Cirillo *et al.*, 2015) predicts TF interactions with human DNA regions using expression profiles, protein–protein interactions and recognition motifs. GERV (Zeng *et al.*, 2016) predicts regulatory variants for TF binding using k-mers learned from ChIP-seq and DNase-seq data.

However, many computational methods are either tissue- or species-specific methods and are restricted to short DNA sequences only. Thus, they cannot be used to find potential TFBSs in long DNA sequences in the absence of information on tissue or species. For example, PROMO (Messeguer *et al.*, 2002) finds TFBSs in DNA sequences using species-tailored searches, and PIPES (Zhong *et al.*, 2013) finds tissue-specific binding targets of mouse TFs. For the problem of predicting TFBSs in DNA, DeepBind (Alipanahi *et al.*, 2015) contains about 500 distinct prediction models, each for different TFs. As output, it provides a predictive binding score without protein-binding sites in the input DNA sequence. However, DeepBind cannot be used to find TFBS in long DNA sequences with more than 1000 nucleotides.

Prediction methods that utilize biological contexts of TFBSs such as tissue types, species or cell development stages perform better than those that use sequence data alone without biological context information. However, context-based approaches cannot be used when context information is not available. The primary focus of this study was to predict potential TFBSs in DNA sequences in the absence of biological context information. Im *et al.* (2015) developed a prediction model known as PNImodeler to predict protein-binding nucleotides in DNA and DNA-binding amino acids in protein. PNImodeler predicts binding sites at the nucleotide level and residue level.

MotEvo (Arnold *et al.*, 2012) predicts TFBS on a genome-wide scale using a Bayesian probabilistic method and was used to find TFBSs from ChIP-seq data of several TFs (CTCF, GABP, NRSF, SRF and STAT1). Unipro UGENE (Okonechnikov *et al.*, 2012) has many features, and one of them is to search for TFBS in genomic regions of several organisms. A recent program called Contra v3 (Kreft *et al.*, 2017) is known to identify TFBS across species (Table 9.1).

9.5. Predicting protein-binding sites in RNA

Until very recently, there were few computational methods that can predict protein-binding sites in RNA. catRAPID estimates the binding propensity of RNA and protein molecules by combining secondary structure, hydrogen bonding and van der Waals contributions (Bellucci *et al.*, 2011). It often predicts an entire RNA sequence as a binding site even for an RNA sequence of 50 or more nucleotides.

Table 9.1.   A selection of programs that predict protein-binding sites in DNA.

Program	URL address	References
MotEvo	http://swissregulon.unibas.ch/sr/software	(Arnold *et al.*, 2012)
Unipro UGENE	http://ugene.net/key_features.html	(Okonechnikov *et al.*, 2012)
Contra v3	http://bioit2.irc.ugent.be/contra/v3	(Kreft *et al.*, 2017)
DeepBind	http://http://tools.genes.toronto.edu/deepbind/	(Alipanahi *et al.*, 2015)
PNImodeler	http://http://bclab.inha.ac.kr/pnimodeler/	(Im *et al.*, 2015)

Table 9.2.   A selection of programs that predict protein-binding sites in RNA.

Program	URL address	References
catRAPID	http://s.tartaglialab.com/page/catrapid_group	(Bellucci *et al.*, 2011)
DeepBind	http://http://tools.genes.toronto.edu/deepbind/	(Alipanahi *et al.*, 2015)
PRIdictor	http://bclab.inha.ac.kr/pridictor	(Tuvshinjargal *et al.*, 2016)

DeepBind (Alipanahi *et al.*, 2015) uses deep convolutional neural networks, trained on a huge amount of data from high-throughput experiments, including RNAcompete, CLIP-seq and RIP-seq. DeepBind contains about 200 distinct models, each for different RBPs, so the user should try all of them in the absence of prior information on RBP. As output, it only provides a predictive binding score without protein-binding sites in the input RNA sequence.

A more recent model called DeeperBind (Hassanzadeh and Wang, 2016) predicts the protein-binding specificity of DNA sequences using a long short-term recurrent convolutional network. By employing more complex and deeper layers, DeeperBind showed a better performance than DeepBind for some proteins, but its use is limited to the data sets from protein-binding microarrays. Both DeepBind and DeeperBind are classification models rather generative models, so cannot be used to construct nucleic acid sequences that potentially bind to a target protein.

A prediction model called PRIdictor (http://bclab.inha.ac.kr/pridictor) predicts mutual binding sites in RNA and protein at the nucleotide- and residue-level resolutions from their sequences (Tuvshinjargal *et al.*, 2016). In independent testing, PRIdictor showed an accuracy over 70% (Table 9.2).

9.6. Conclusion and future perspective

The problem of predicting the interactions between nucleic acids and proteins is becoming increasingly important, and several studies have been conducted to predict the interactions by computational approaches. So far, many computational approaches have attempted to solve one of the two problems: predicting binding sites in a sequence and determining whether an interaction exists between a pair of protein and nucleic acid sequences of sequences. Regarding the problem of predicting binding sites, many existing computational methods are limited to finding nucleic acid binding sites in proteins instead of protein-binding sites in nucleic acids.

In this article, we reviewed existing computational methods for predicting protein-binding sites in nucleic acids. These methods will help efficient *in vitro* or *in vivo* experiments for finding protein-binding sites in nucleic acids. The methods will also help understand the mechanism of the interactions between proteins and nucleic acids.

However, further studies are required to improve the accuracy and applicability of the existing computational methods. Another direction of research is to develop a computational method for finding nucleic acid sequences that contains binding sites to a target protein. Such a method will help design efficient *in vitro* experiments by constructing potential aptamers that bind to a target protein with high affinity and specificity.

Acknowledgments

This work was supported by the National Research Foundation of Korea (NRF) grant funded by the Ministry of Science and ICT (2017R1E1A1A03069921, NRF-2018K2A9A2A11080914).

References

Akbaripour-Elahabad, M., Zahiri, J., Rafeh, R., *et al.* rpiCOOL: A tool for In Silico RNA-protein interaction detection using random forest. *J Theor Biol.* 2016;402:1–8.

Alipanahi, B., Delong, A., Weirauch, M.T., *et al.* Predicting the sequence specificities of DNA-and RNA-binding proteins by deep learning. *Nat Biotechnol.* 2015;33:831–838.

Arnold, P., Erb, I., Pachkov, M., *et al.* MotEvo: Integrated Bayesian probabilistic methods for inferring regulatory sites and motifs on multiple alignments of DNA sequences. *Bioinformatics.* 2012;28(4):487–494.

Bellucci, M., Agostini, F., Masin, M., *et al.* Predicting protein associations with long noncoding RNAs. *Nat Methods.* 2011;8(6):444–446.

Berman, H.M., Westbrook, J., Feng, Z., *et al.* The protein data bank. *Nucleic Acids Res.* 2000;28:235–242.

Choi, D., Park, B., Chae, H., *et al.* Predicting protein-binding regions in RNA using nucleotide profiles and compositions. *BMC Syst Biol.* 2017;11:16.

Choi, S., and Han, K. Prediction of RNA-binding amino acids from protein and RNA sequences. *BMC Bioinformatics.* 2011;12: S7.

Choi, S., and Han, K. Predicting protein-binding RNA nucleotides using the feature-based removal of data redundancy and the interaction propensity of nucleotide triplets. *Comput Biol Med.* 2013;43:1687–1697.

Cirillo, D., Botta-Orfila, T., and Tartaglia, G.G. By the company they keep: Interaction networks define the binding ability of transcription factors. *Nucleic Acids Res.* 2015;43:e125.

Corcoran, D.L., Georgiev, S., Mukherjee, N., *et al.* PARalyzer: Definition of RNA binding sites from PAR-CLIP short-read sequence data. *Genome Biol.* 2011;12(8):R79.

Gerstberger, S., Hafner, M., and Tuschl, T. A census of human RNA-binding proteins. *Nat Rev Genet.* 2014;15:829–845.

Hassanzadeh, H.R., and Wang, M.D. DeeperBind: Enhancing prediction of sequence specificities of DNA binding proteins. *Proceedings (IEEE Int Conf Bioinformatics Biomed).* 2016;178–183.

Hu, B., Yang, Y.T., Huang, Y.M., *et al.* POSTAR: A platform for exploring post-transcriptional regulation coordinated by RNA-binding proteins. *Nucleic Acids Res.* 2017;45:D104–D114.

Hwang, S., Gou, Z., and Kuznetsov, I.B. DP-Bind: A web server for sequence-based prediction of DNA-binding residues in DNA-binding proteins. *Bioinformatics.* 2007;23(5):634–636.

Im, J., Tuvshinjargal, N., Park, P., *et al.* PNImodeler: Web server for inferring protein-binding nucleotides from sequence data. *BMC Genomics.* 2015;16:S6. doi:10.1186/1471-2164-16-S3-S6.

Kirsanov, D.D., Zanegina, O.N., Aksianov, E.A., *et al.* NPIDB: Nucleic acid–protein interaction database. *Nucleic Acids Res.* 2013;41(D1): D517–D523.

König, J., Zarnack, K., Luscombe, N.M., *et al.* Protein-RNA interactions: New genomic technologies and perspectives. *Nat Rev Genet.* 2012;13:77–83.

Kreft, L., Soete, A., Hulpiau, P., *et al.* ConTra v3: A tool to identify transcription factor binding sites across species, update 2017. *Nucleic Acids Res.* 2017;45(W1):W490–W494.

Latchman, D.S. Transcription factors: An overview. *Int J Biochem Cell Biol.* 1997;29(12):1305–1312.

Lee, N.K., Azizan, F.L., Wong, Y.S., *et al.* DeepFinder: An integration of feature-based and deep learning approach for DNA motif discovery. *Biotechnol Biotech EQ.* 2018;32(3):759–768.

Li, S., Yamashita, K., Amada, K. M., *et al.* Quantifying sequence and structural features of protein-RNA interactions. *Nucleic Acids Res.* 2014;42: 10086–10098.

Messeguer, X., Escudero, R., Farre, D., *et al.* PROMO: Detection of known transcription regulatory elements using species-tailored searches. *Bioinformatics.* 2002;18(2):333–334.

Okonechnikov, K., Golosova, O., Fursov, M., *et al.* Unipro UGENE: A unified bioinformatics toolkit. *Bioinformatics.* 2012;28(8):1166–1167.

Pan, X., and Shen, H.B. Predicting RNA–protein binding sites and motifs through combining local and global deep convolutional neural networks. *Bioinformatics.* 2018;34(20):3427–3436.

Ray, D., Kazan, H., Cook, K.B., *et al.* A compendium of RNA-binding motifs for decoding gene regulation. *Nature.* 499;172–177.

Si, J., Zhang, Z., Lin, B., *et al.* MetaDBSite: A meta approach to improve protein DNA-binding sites prediction. *BMC Syst Biol.* 2011;5:S7.

Slattery, M., Zhou, T.Y., Yang, L., *et al.* Absence of a simple code: How transcription factors read the genome. *Trends Biochem Sci.* 2014;39: 381–399.

Terribilini, M., Sander, J.D., Lee, J.H., *et al.* RNABindR: A server for analyzing and predicting RNA-binding sites in proteins. *Nucleic Acids Res.* 2007;35:W578–W584.

Tuvshinjargal, N., Lee, W., Park, B., *et al.* (2016). PRIdictor: Protein-RNA interaction predictor. *Biosystems.* 2016;139:17–22.

Walia, R.R., Xue, L.C., Wilkins K., *et al.* (2014). RNABindRPlus: A predictor that combines machine learning and sequence homology-based methods to improve the reliability of predicted RNA-binding residues in proteins. *PLOS ONE.* 2014;9:e97725.

Wang, L., and Brown, S.J. BindN: A web-based tool for efficient prediction of DNA and RNA binding sites in amino acid sequences. *Nucleic Acids Res.* 2006;34:243–248.

Wang, L., Huang, C., Yang, M.Q., *et al.* BindN+ for accurate prediction of DNA and RNA-binding residues from protein sequence features. *BMC Syst Biol.* 2010;4:S3.

Yan, C., Terribilini, M., Wu, F., *et al.* Predicting DNA-binding sites of proteins from amino acid sequence. *BMC Bioinformatics.* 2006;7(1):262.

Yang, Y.-C., Di, C., Hu, B., *et al.* CLIPdb: A CLIP-seq database for protein-RNA interactions. *BMC Genomics.* 2015;16:51.

Zeng, H.Y., Hashimoto, T., Kang, D.D., *et al.* GERV: A statistical method for generative evaluation of regulatory variants for transcription factor binding. *Bioinformatics.* 2016;32:490–496.

Zhang, X., and Liu, S. RBPPred: Predicting RNA-binding proteins from sequence using SVM. *Bioinformatics.* 2017;33:854–862.

Zhong, S., He, X., and Bar-Joseph, Z. Predicting tissue specific transcription factor binding sites. *BMC Genomics.* 2013;14:796.

Chapter 10
Docking algorithms and scoring functions

Arina Afanasyeva, Chioko Nagao and Kenji Mizuguchi*

Bioinfomatics Project, National Institutes of Biomedical Innovation, Health and Nutrition, Ibaraki City, Osaka, Japan

Molecular docking is a traditional *in silico* method to obtain the structure of multi-molecule complexes, allowing to get a deeper insight into the functions of the biologically important molecules through their interactions in the complex structure. Although there are a number of highly developed docking methods and algorithms, which were exhaustively tested for the docking of the small molecules into protein, in many cases they cannot be directly applied for docking of oligonucleotides due to the specificity of oligonucleotide molecule construction, flexibility and other aspects. For that reason, a number of original docking algorithms and scoring functions were developed specifically for DNA/RNA docking recently.

10.1. Introduction

DNA/RNA-binding proteins (DRBP) are involved in many critical cellular processes associated with RNA and DNA regulation and are also associated with various disorders; therefore, DRBPs have

---

*kenji@nibiohn.go.jp

gathered increasing attention as promising drug targets (Hudson and Ortlund, 2014).

In many cases the structure of a protein–oligonucleotide complex is difficult to obtain due to the problems with crystallization, and in such cases, structure may be obtained only by methods such as cryo-electron microscopy (cryo-EM) or solution nuclear magnetic resonance (NMR) spectroscopy, even though these methods have some limitations, for example, NMR can be applied only for relatively small molecules (<30 kDa) (Zhao *et al.*, 2014). For that reason, protein–DNA/RNA complexes are notably underrepresented in Protein Data Bank (PDB) (Berman *et al.*, 2000). The insufficient availability of crystal structures of protein–DNA/RNA complexes has led to an increasing interest *in silico* methods for structure reconstruction, such as docking and molecular dynamics. Nevertheless, docking of oligonucleotides in proteins has many complexities and is, therefore, a non-trivial task. In this review, we will discuss the primary challenges associated with the protein–DNA/RNA docking. First, we briefly explain protein–DNA/RNA recognition and interaction specificity. Next, we review a selection of algorithms and scoring functions that are commonly used or developed specifically for protein–DNA/RNA docking. Finally, we discuss the vexing issue of structural flexibility and the solutions that have been developed to overcome it.

10.2. Specificity of protein–RNA/DNA interaction

One of the main difficulties associated with the docking of oligonucleotides with a protein is the problem of identification of the probable binding interface for both the protein and the DNA/RNA; this situation is unlike the protein–small molecule docking when the binding pocket is well characterized or can be identified relatively easily by geometrical and biophysical properties.

DNA/RNA-binding sites are usually enriched in positively charged amino acid residues, such as Lys and Arg, which is explained by the necessity to compensate for the negatively charged DNA/RNA phosphate backbone (Kumar *et al.*, 2008). Some patterns and structural motifs for RNA/DNA recognition are known as well based on the findings from known DRBPs, such as OB-fold (oligonucleotide/oligosaccharide-binding fold) or zinc fingers. Consequently, many computational methods have been developed that variously employ sequence and motif information, or structure comparison to the known DRBPs for the identification of nucleic acid (NA)-binding sites. However, such a task is often non-trivial, chiefly because in some instances proteins with non-classic DNA/RNA-binding motifs may be able to bind to RNA/DNA; on the other hand, the classic DRBPs or DRB domains such as helix-turn-helix, zinc finger or leucine zipper may not always display any RNA/DNA-binding activity (Dvir *et al.*, 2018). Secondly, in the case of RNA/DNA–protein interactions, the binding specificities are rather limited when compared with protein–protein interactions (PPIs), because there are just four main types of nucleotides for DNA or RNA, and theoretically, the number of possible recognition patterns for different nucleotide pairs is fairly limited (Figure 10.1). In other words, for a double-stranded (ds) DNA with

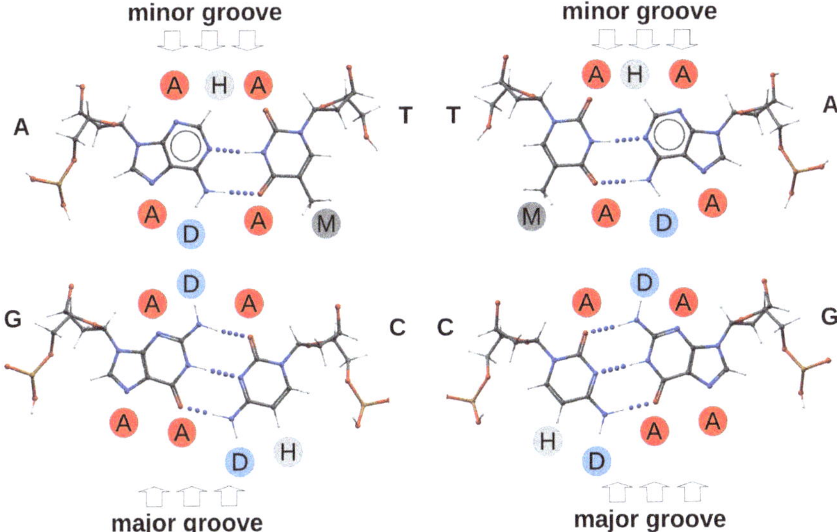

Figure 10.1. Recognition patterns for dsDNA classic base pairs AT, TA, GC and GC. The types of functional groups are assigned with colored circles: A: H-bond acceptor group; D: H-bond donor group; H: hydrophobic group; and M: methyl group.

the standard AT and CG base pairing, there are just two possible patterns on the minor groove and four patterns on the major groove (Figure 10.1). Therefore, recognition typically happens on the major groove, although sequence-specific dsDNA interaction is just one of the possible cases of protein–NA interactions (Figure 10.1).

A different probing question is whether it is possible to distinguish the DNA- and RNA-binding proteins/sites. Indeed, some proteins have been found to interact both with the DNA and the RNA (Hudson and Ortlund, 2014). It was also demonstrated that many algorithms for predicting RNA-binding sites showed similar or even higher accuracies when predicting DNA-binding sites (Miao and Westhof, 2015). The reasons for such overlaps are a lack of structural features that are uniquely characteristic of either RNA or DNA; typically, there are only two such structural features: additional 2′OH groups on the sugars of RNA bases and methyl group in Thymidine in DNA. Regardless, these differences determine the preferred conformations such as the B-form helix for dsDNA and the A-form helix for dsRNA. Based on the comparative analysis of available complex structures, it was assumed that recognition via direct base interaction is typically observed in DNA–protein complexes, whereas in the case of RNA, the main recognition factor is a shape complementarity. There are also known instances of competitive RNA/DNA binding in DRBPs and in such cases, the less specific electrostatic interactions via the phosphate backbone are more likely to occur (Hudson and Ortlund, 2014).

10.3. Docking algorithms and approaches

Protein–RNA/DNA docking problem is more complex than protein–small molecules or even protein–protein docking problems for a number of reasons. First, there are insufficient representative protein–RNA/DNA complexes in PDB, and therefore, it is difficult to evaluate the accuracy of newly developed methods and scoring functions. Moreover, there are very few examples of protein–RNA/DNA complexes that include the individual constituents, i.e., protein and

RNA/DNA in unbound states. Such examples are especially valuable to assess the conformational changes observed upon binding and to assess the algorithms in a real-life scenario. Furthermore, there are few available docking algorithms/programs that are specialized for the docking of oligonucleotides, and therefore, usually the standard tools for protein–small molecules and protein–protein docking are applied to RNA/DNA–protein complexes.

Nevertheless, the empirical potential functions developed for PPIs cannot be directly applied to assess protein–RNA/DNA interaction. Therefore, such a scoring function currently is still to be found, although researchers use a number of different original scoring functions or consensus scoring functions that combine several scoring functions (Pérez-Cano *et al.*, 2017). Next, we will review several original approaches for protein–RNA scoring and docking algorithms (Table 10.1).

10.3.1. *FFT algorithm in GRAMM7, and FTDock*

Docking programs such as GRAMM (Global Range Molecular Matching) (Tovchigrechko and Vakser, 2005) and FTDock (Katchalski-Katzir *et al.*, 1992) are based on the FFT algorithm for interaction surface geometry recognition. Molecular representation in this approach is simplified and reduced to a grid, which allows reduced complexity and also results in a significant speedup in the calculation time. For instance, in GRAMM-X to smooth the inter-molecular energy landscape, the Lennard-Jones potential function was calculated, where the repulsive term for pairwise atom potential (resultant force repelling two atoms from each other) was reduced and the attractive term (resultant force attracting two atoms to each other) was increased (Tovchigrechko and Vakser, 2005). Therefore, the resulting solutions are allowed to have some overlapping of the atomic spheres, which should be corrected during the following steps of the docking protocol, or using additional scoring functions. Overall, the accuracy of GRAMM method depends on the accuracy of the initial structure representation and structural flexibility.

Table 10.1.   List of docking algorithms and scoring programs commonly used for RNA/DNA–protein docking.

Docking program	Docking algorithm	Scoring function	Reference	Availability	Link
GRAMM-X	Fast fourier transformation (FFT)	Knowledge based	(Tovchigrechko and Vakser, 2005)	software (free for academic use) and web-service	http://vakser.compbio.ku.edu/resources/gramm/grammx/
FTDock	FFT	Surface complementarity	(Katchalski-Katzir *et al.*, 1992)	software (free)	http://www.sbg.bio.ic.ac.uk/docking/ftdock.html
DOCK	Distance geometry algorithm	Force-field based	(Kuntz *et al.*, 1982)	software (free for academic use)	http://dock.compbio.ucsf.edu/
DARWIN	Genetic algorithm (GA)	Force-field based	(Taylor and Burnett, 2000)	software (available by request from authors)	—
ITScorePR	—	Knowledge based	(Huang and Zou, 2014)	software (free for academic use)	http://zoulab.dalton.missouri.edu/resources_itscorepr.html
QUASI-RNP and DARS-RNP	—	Knowledge based	(Tuszynska and Bujnicki, 2011)	software (free)	http://iimcb.genesilico.pl/RNP/
RNP-denovo	Monte Carlo fragment assembly	Knowledge based	(Kappel and Das, 2019)	software (free for academic use)	https://www.rosettacommons.org/docs/latest/application_documentation/rna/rna-denovo
RpveScore	—	Weighted combined scoring function	(Zhang *et al.*, 2017)	software (free for academic use)	http://life.bjut.edu.cn/kxyj/kycg/2017116/14845362285362368_1.html
HADDOCK	Docking with interaction restraints	Energy-based weighted function	(van Dijk *et al.*, 2006)	software (free for academic use) and web-server	http://www.bonvinlab.org/software/haddock2.2/

10.3.2. *Distance geometry algorithm in DOCK*

The first historically developed docking program DOCK (Kuntz *et al.*, 1982) has implemented an algorithm based on the representation of the receptor-binding pocket as a bunch of hard spheres and matching of these spheres with the ligand shape. The scoring function for the original method included just two components of interaction potential: spheres overlapping and hydrogen bonds formation. Later implementations of the DOCK program included solutions for ligand flexibility and receptor refinement and also more elaborate scoring functions.

10.3.3. *GA in DARWIN*

DARWIN docking program has implemented a GA for the docking pose search combined with a gradient minimization strategy (Taylor and Burnett, 2000). According to this algorithm from an initial "population" of pose sampling solutions, only those that best fit the ranking function survive and continue to "evolve," or participate in the generation of next level solutions. Each of the solutions may, therefore, be presented by a number of variables which are changed or "mutated" upon the pose sampling. DARWIN program uses CHARMM force fields to calculate the free energy of the resulting complex. The advantage of the GA algorithm is that it provided a means to solve the problems with a large number of variables such as molecular docking, and therefore, this method can be applied for a protein–RNA/DNA docking as well.

10.3.4. *Distance-dependent atomic interaction potential ITScore-PR*

The scoring function in this method is based on the pairwise distance-dependent atomic interaction potentials extracted from the available experimentally obtained crystal structures of protein–RNA complexes. The potentials were obtained using a statistical mechanics-based iterative method, where the potentials for each atom pair were recalculated iteratively until the native structure was determined from the 1000 decoys for each item in the training set (Huang and Zou, 2014).

10.3.5. *Quasi-chemical potential (QUASI-RNP) and the decoys as the reference state potential (DARS-RNP)*

In QUASI-RNP and DARS-RNP methods, both the protein and RNA molecules are transformed in a reduced representation, where each residue or RNA base is substituted by an appropriate type of united atoms, i.e., the groups of several connected atoms specific for a given residue or nucleotide. The initial set of decoys in (Tuszynska and Bujnicki (2011) was obtained by the GRAMM program, which also operates with a reduced molecular representation. Knowledge-based potentials to score the generated poses in this method are calculated using the reverse Boltzmann statistics for the each of the united atoms in a protein–RNA pair within a distance cutoff, as described in (Tuszynska and Bujnicki (2011).

10.3.6. *Simultaneous docking and folding approach for ribonucleoprotein (RNP) -denovo method (Rosetta)*

Rosetta developers recently presented a method for reconstructing the RNA–protein complex that involved combining RNA folding and docking (Kappel and Das, 2019); unlike most of the methods described above, RNP-denovo method can be used in instances when the RNA structure is not available. FARNA (Fragment Assembly of RNA) method (Das and Baker, 2007) is commonly used for RNA structure reconstruction. FARNA is a fragment-based approach, with Monte-Carlo simulation used to obtain a low-energy RNA conformer. The energy is calculated based on a low-resolution knowledge-based function that is a linear combination of the RNA score function, used in FARNA algorithm, with functions, characterizing different types of RNA–protein interactions, as described in Kappel and Das (2019). The FARNA algorithm was modified by adding the rigid-body docking moves (translations and rotations) to the Monte Carlo simulation.

10.3.7. *Combinational scoring functions*

Several studies have been done on the development of knowledge-based scoring functions for protein–RNA interactions. One such scoring function was developed by Zhang *et al.* (, 2017), who developed and tested a combined scoring function RpveScore, which included the protein–RNA pairwise potential for non-ribosomal complexes and six physical interaction-based energy terms (components of the equation). The pairwise potential function was determined as a $60*8$ residue–nucleotide propensity potential and included secondary structure information for both protein and the RNA (Li *et al.*, 2012). First, they defined eight types of secondary structure elements in proteins (3_10-helix, alpha-helix, pi-helix, turn, beta-sheet, beta-bridge, bend and the unclassified) and two in RNA (Watson-Crick and non-Watson-Crick nucleotide pairs). Subsequently, they calculated the probabilities of a given element to occur in the protein–RNA interaction interface based on a non-redundant data set derived from the crystal structures of the protein–RNA complexes.

Pérez-Cano and colleagues performed a scrutinizing analysis of the effect of different energy terms along with the knowledge-based docking scores on the accuracy of protein–RNA docking (Pérez-Cano *et al.*, 2017). Based on their observations, they suggested a combined scoring function that outperformed several other scoring functions in docking tests. They observed that van der Waals potential, electrostatics and shape complementarity-based scores such as FTPdock-Scscore (Katchalski-Katzir *et al.*, 1992) demonstrated a higher predictive ability for protein–RNA docking, compared with the standard protein–protein docking tests. Furthermore, the desolvation energy (energy contribution of the displacement of the solvent molecules from the molecular surface upon binding) was demonstrated to have a higher predictive potency for protein–protein docking and relatively low predictive potency for protein–RNA docking. Their findings, therefore, corroborated the belief that shape complementarity plays a crucial role in protein–RNA interactions (Hudson and Ortlund, 2014).

10.4. Protein and DNA/RNA flexibility problem

A significant issue with protein–RNA/DNA docking is the conformational plasticity of both the protein and the RNA/DNA. Often RNA changes its conformation upon binding to a receptor, and the protein can change its conformation when bound to RNA/DNA as well. Such changes often cannot be found in crystal structures databases or be predicted by modeling. Interestingly, Pérez-Cano *et al.* (2017) in their investigations observed that the electrostatic interactions were especially significant in the instances of highly flexible receptors and displayed increasing importance in protein–DNA interactions depending on the protein flexibility. The authors assumed that in such cases the electrostatics term (energy contribution of the electric potential energy of charged atoms) contributes to the enthalpy changes that can then compensate the conformation entropy changes that occurred upon binding.

10.4.1. *Fragment-based docking approach for a single-stranded RNA-binding problem*

Molecular flexibility often leads to significant complications, especially in the case of single-stranded (ss) RNA/DNA molecules when the interaction happens via an unstructured region (tails or loops), i.e., when the RNA/DNA interacts with the protein via regions that do not have any specific folding or any higher-order structural organization, and therefore, results in too many possible conformations that need to be sampled. To overcome this problem, Chauvot de Beauchene *et al.* (2016) developed a fragment-based docking approach for ssRNA docking. In this method, RNA is spliced into three-nucleotide components, and the conformations of these components are sampled and separately docked on the receptor surface. The groups of triplets, forming the correct sequence, are joined, and the incorrect decoy poses are filtered out. This method does not require any information on the locations of the RNA/DNA-binding sites on the receptor surface but can be applied only to cases where the RNA interacts with receptor via unstructured regions. In this case, it requires full interaction freedom for the RNA fragments and the method cannot consider RNA parts which do not participate in the interaction.

10.4.2. *Information-driven docking approach in HADDOCK (High Ambiguity Driven DOCKing)*

To deal with structural flexibility HADDOCK docking method has implemented a structure relaxation algorithm following rigid docking, the so-called induced fit docking (van Dijk *et al.*, 2006). The docking protocol, therefore, consists of three main steps: rigid, information-driven docking and pose scoring, followed by a refinement, where both the DNA/RNA and protein's active residues undergo relaxation in several stages; and the final complex's refinement in a water shell. HADDOCK program allows the usage of experimental and bioinformatics data to perform data-driven docking. Bioinformatics data, for example, can include sequence conservation and mutational analysis to define RNA/DNA-binding sites on the protein surface. Such information, therefore, can be utilized by the HADDOCK as interaction restraints (Ambiguous Interaction Restraints or AIR), which are affixed on both the active residues of the protein and the DNA/RNA during the first step of the docking procedure (rigid docking).

10.5. Conclusion

A plethora of solutions has been developed to address the challenges associated with protein–RNA/DNA docking over the past years to deal with both the scoring problem and the problem of protein and

DNA/RNA flexibility. Nevertheless, there are no universal solutions, because there is a broad spectrum of protein–DNA/RNA interactions, such as sequence-specific and non-specific, DNA or RNA, double- or single-stranded NA binding or protein–aptamer interaction, and they are governed by different mechanisms.

References

Berman, H.M., Westbrook, J., Feng, Z., *et al.* The protein data bank. *Nucleic Acids Res.* 2000;28(1):235–242.

Das, R., and Baker, D. Automated de novo prediction of native-like RNA tertiary structures. *Proc Natl Acad Sci.* 2007; 104(37):14664–14669.

de Beauchene, I.C., de Vries S.J., and Zacharias, M. Binding site identification and flexible docking of single stranded RNA to proteins using a fragment-based approach. *PLoS Comput Biol.* 2016;12(1):e1004697.

Dvir, S., Argoetti, A. and Mandel-Gutfreund, Y. Ribonucleoprotein particles: Advances and challenges in computational methods. *Curr Opin Struct Biol.* 2018;53:124–130.

Huang, S.Y., and Zou, X. A knowledge-based scoring function for protein-RNA interactions derived from a statistical mechanics-based iterative method. *Nucleic Acids Res.* 2014;42(7):55

Hudson, W.H., and Ortlund, E.A. The structure, function and evolution of proteins that bind DNA and RNA. *Nat Rev Mol Cell Biol.* 2014;15(11):749–760.

Kappel, K., and Das, R. (2019). Sampling native-like structures of RNA-protein complexes through Rosetta folding and docking. *Structure.* 2019;27(1):140–151.

Katchalski-Katzir, E., *et al.* Molecular surface recognition: Determination of geometric fit between proteins and their ligands by correlation techniques. *Proc Natl Acad Sci.* 1992;89(6):2195–2199.

Kumar, M., Gromiha, M.M., and Raghava, G.P.S. Prediction of RNA binding sites in a protein using SVM and PSSM profile. *Proteins.* 2008;71(1):189–194.

Kuntz, I.D., Blaney, J.M., Oatley, S.J., *et al.* A geometric approach to macromolecule-ligand interactions. *J Mol Biol.* 1982;161(2):269–288.

Li, C.H., Cao, L.B., Su, J.G., *et al.* A new residue-nucleotide propensity potential with structural information considered for discriminating protein-RNA docking decoys. *Proteins.* 2012;80(1):14–24.

Miao, Z., and Westhof, E. A large-scale assessment of nucleic acids binding site prediction programs. *PLoS Comput Biol.* 2015:11(12):e1004639.

Pérez-Cano, L., Romero-Durana, M., and Fernández-Recio, J. Structural and energy determinants in protein-RNA docking. *Methods.* 2017:118–119:63–170.

Taylor, J.S., and Burnett, R.M. DARWIN: A program for docking flexible molecules. *Proteins.* 2000;41(2):173–191.

Tovchigrechko, A., and Vakser, I.A. Development and testing of an automated approach to protein docking. *Proteins.* 2005;60(2):296–301.

Tuszynska, I., and Bujnicki, J.M. DARS-RNP and QUASI-RNP: New statistical potentials for protein-RNA docking. *BMC Bioinformatics.* 2011;12(1):348.

van Dijk, M., van Dijk, A.D.J., Hsu, V., *et al.* Information-driven protein-DNA docking using HADDOCK: It is a matter of flexibility. *Nucleic Acids Res.* 2006;34(11):3317–3325.

Zhang, Z., Lu, L., Zhang, Y., *et al.* A combinatorial scoring function for protein–RNA docking. *Proteins.* 2017;85(4):741–752.

Zhao, C., Anklin, C., and Greenbaum, N.L. Chapter Twelve—Use of 19F NMR methods to probe conformational heterogeneity and dynamics of exchange in functional RNA molecules. In Burke-Aguero, D.H., ed. *Methods in Enzymology.* 2014:267–285.

Chapter 11

Recent progress of methodology development for protein–RNA docking

Yun Guo*, Xiaoyong Pan* and Hong-Bin Shen*,†,‡

*Institute of Image Processing and Pattern Recognition,
Shanghai Jiao Tong University, and
Key Laboratory of System Control and Information
Processing, Ministry of Education of China,
Shanghai 200240, China

†Department of Computer Science,
Shanghai Jiao Tong University, Shanghai 200240, China

Protein–RNA interactions are involved in many important biological processes, and knowing the protein–RNA complex structures can provide significant insights into the mechanisms of these interactions. However, experimentally detecting protein–RNA complex structures is time-consuming and costly. Protein–RNA docking is one of the first steps to predict the three-dimensional (3D) protein–RNA complex structures, which aims to model protein–RNA complexes starting from separate structures. In this review, the latest advances in the methodology development for computational analysis of protein–RNA docking have been discussed. In general, the protein–RNA docking mainly comprises two modules of conformation sampling and scoring function design, and their technical details will be described. Evaluation

‡hbshen@sjtu.edu.cn

of the performance of protein–RNA docking requires the benchmark data sets, which have been discussed in this chapter. Finally, the challenges of current protein–RNA docking methods have been briefly discussed, and some future directions for further development in this area have been suggested.

11.1. Introduction

Many functions of intracellular RNAs are achieved via interacting with certain proteins [1–5]. These functions include tRNA transcription, RNA reverse transcription, mRNA translation, posttranscriptional processing of RNA, stabilization of protein complexes, etc. Identifying the interactions between proteins and RNAs is important for understanding the functions of RNAs and proteins. In structural biology, there are many experimental methods for the structural determinations of protein–RNA complexes, such as X-ray crystal diffraction [6], nuclear magnetic resonance (NMR) [7, 8] and cryo-electron microscopy (CryoEM) [9]. However, these experimental methods are usually time-consuming and expensive, and most of the previous efforts are devoted to protein–protein interaction complexes. Till now, the number of protein–RNA complexes in the Protein Data Bank (PDB) is still few. According to our statistics, only 4272 protein–RNA complexes have been recorded in PDB [10] by March 18, 2019; this number is much less than 42,528 protein–protein complexes in the same database, indicating protein–RNA complex studies is still a relatively new area, requiring significant future efforts. The detailed information is summarized in Table 11.1.

According to Nithin *et al.*, there are at least three main factors that may hinder the progress of experimentally structural determination of protein–RNA complexes [11]: (1) Many protein–RNA complexes are short-lived because of the transience of protein–RNA interactions. (2) The chemical character of the RNA component increases the experimental difficulty. The RNA molecules are often structurally heterogeneous and more flexible in conformations than proteins. (3) Protein–RNA complexes have a greater repulsion between molecules due to the negatively charged sugar-phosphate backbone than protein–protein complexes.

Table 11.1.   Statistics on different types of protein-binding complexes in PDB.[a]

Structure determination methods	X-ray	NMR[b]	EM[c]	Others
Protein–protein complexes	38,619	1587	2164	158
Protein–DNA complexes	4912	278	203	7
Protein–RNA complexes	3359	233	669	11

[a]Statistics are collected from PDB released on March 18, 2019.
[b]Statistics of NMR are from solution NMR and solid-state NMR.
[c]Electron microscopy.

Thus, computational analysis, such as protein–RNA docking, is an important alternative way for the understanding of protein–RNA complexes in a much faster way. Compared to the relatively slow and difficult wet-lab experiments, computational docking methods will provide many advantages due to their high speed and low cost. Especially, the computational power has been significantly increased in recent years, which will further help us to search a much bigger interaction space for finding more accurate complex solutions.

Molecular docking is a computational algorithm in the related fields of structural biology. When two molecules combine with each other to form a stable complex, molecular docking predicts the preferred orientations of the two molecules [12]. In general, the larger molecule refers to the receptor, and the other refers to the ligand. Conformation sampling and scoring function design are two main steps for molecular docking. In other words, given two molecular structures, conformation sampling searches for all possible orientations and conformations of the two structures to generate sample models (called poses or decoys), and then these models are ranked according to a designed scoring function to find near-native structures [13].

With the efforts of the last decade, different types of molecular docking methods have been proposed, which can be roughly divided into three groups according to the criteria that either the ligand or the receptor is rigid or flexible in the docking simulations [14, 15]: (1) rigid-body docking: both the receptor and the ligand are rigid molecules; (2) semi-flexible docking: only the receptor is a rigid molecule;

and (3) full-flexible docking: both the receptor and the ligand are flexible molecules [16].

1) For rigid-body docking, the conformations of the receptor and the ligand usually do not change during the docking process. This type of docking is suitable for large molecules and it mainly considers the degree of fitting between the two molecules. Typical algorithms of this type are ZDOCK [17] and HDOCK [18].

2) Semi-flexible docking is often used for the docking of small molecules and macromolecules. During the docking process, the conformations of small molecules can vary within a certain range, but the macromolecules are rigid and cannot change. This type of docking method cannot only examine the influence of flexibility to a certain extent, but also maintain high-computational efficiency. Semi-flexible molecular docking methods are often employed in drug design and virtual screening processes [19]. The AutoDock [20] is one of typical methods in this type.

3) Full-flexible docking is generally used to accurately study the interface change between two docking molecules. As the conformational changes of two molecules are allowed, it will need a longer time.

Although many computational docking algorithms have been developed, most of them belong to rigid-body docking and semi-flexible docking, especially for those methods using proteins as receptors. To date, developing protein–RNA docking methods is still at its infancy stage, although many methods have been designed for the protein–protein docking. Figure 11.1 illustrates the number of published papers on protein–protein docking, protein–DNA docking and protein–RNA docking from 1986 to 2018. As can be seen, the number of protein–RNA docking studies is much fewer than that of protein–protein docking, and also is fewer than protein–DNA docking studies, suggesting that protein–RNA docking is a relatively new area compared to other two types of docking studies. Figure 11.1 also shows us an increasing interest in docking, suggesting that interaction complexes have attracted much more attention than ever.

Most of the existing docking methods are specially designed to model protein–protein interaction complexes, which include GRAMM

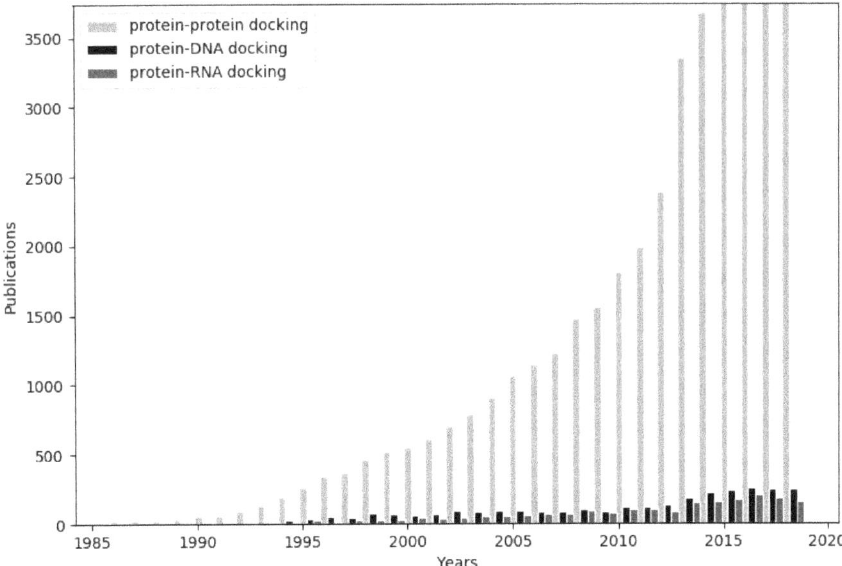

Figure 11.1. The number of published papers about molecule docking from 1986 to 2018. They are extracted from the PubMed-NCBI database by searching keywords "protein–protein docking," "protein–DNA docking," and "protein–RNA docking," respectively.

[21, 22], ZDOCK, RosettaDock [23], FTDock [24], HEX [25], HADDOCK [26], etc. A major computational reason hindering the development of protein–RNA docking is that RNAs usually undergo large conformational changes when binding to proteins [27]. There is an international competition to assess the performance of different methods, known as Critical Assessment of Prediction of Interactions (CAPRI, http://capri.ebi.ac.uk) [28]. In the rounds 13–19 of CAPRI, protein–protein docking methods are also employed for evaluating protein–RNA docking to predict protein–RNA complex structures [29]. Since then, docking algorithms, e.g. HADDOCK, ClusPro [30], ZDOCK, PatchDock [31], RosettaDock, ATTRACT [32], were accordingly adapted for protein–RNA docking. Because of either the incomplete sampling of the conformational space or the inadequacy of the internal scoring functions, these methods often failed to produce native-like poses for protein–RNA docking [33, 34]. This suggests that considering the features of RNAs and their specific

interaction mechanisms with proteins will play an important role for developing an accurate protein–RNA docking algorithm. In recent years, researchers have devoted many efforts in this regard and developed special algorithms for protein–RNA docking, such as the NPDOCK [35], ICM [34], 3dRPC [36].

To objectively evaluate the performance of existing docking methods, benchmark data sets are required, which should be constructed from three-dimensional (3D) atomic coordinates of protein–RNA complexes. Four main benchmark data sets for protein–RNA docking were established in this area. For example, Barik *et al.* developed a non-redundant protein–RNA docking benchmark data set from available unbound and bound structures involving polypeptides and nucleic acid chains in the PDB [37]. In Section 11.2.2.4, we give details about the four benchmark data sets.

11.2. Algorithms of protein–RNA docking

11.2.1. *Protein–RNA docking has its own characteristics*

The characteristics (e.g., shape and docking energy) of docking interfaces between molecules can be used to estimate the reliability of molecule docking methods. From the computational point of view, protein–RNA docking could be more challenging. For instance, when RNAs interact with proteins, their structural conformations would undergo a big change, resulting in a larger conformational space to search. The other difficulty for this problem is that the recognition motif patterns between amino acids and nucleotides are still not clear, which will further increase the challenges for deriving an accurate complex model in an efficient way.

Previous studies have shown that the interfaces between protein–RNA and protein–protein complexes are much different [38–40]. For example, the packing index [41], buried fraction and gap volume index [42, 43] will be different between these two docking types; secondary structure of RNA nucleotides is also found to play an important role in forming interaction with the amino acids [44]. All these unique features of protein–RNA complexes require specific development of corresponding algorithms.

11.2.2. *General pipeline for implementing protein–RNA docking algorithms*

Protein–RNA docking methods aim at finding the most potential interaction conformation between proteins and RNAs. Figure 11.2 briefly depicts the main steps of the protein–RNA docking methods.

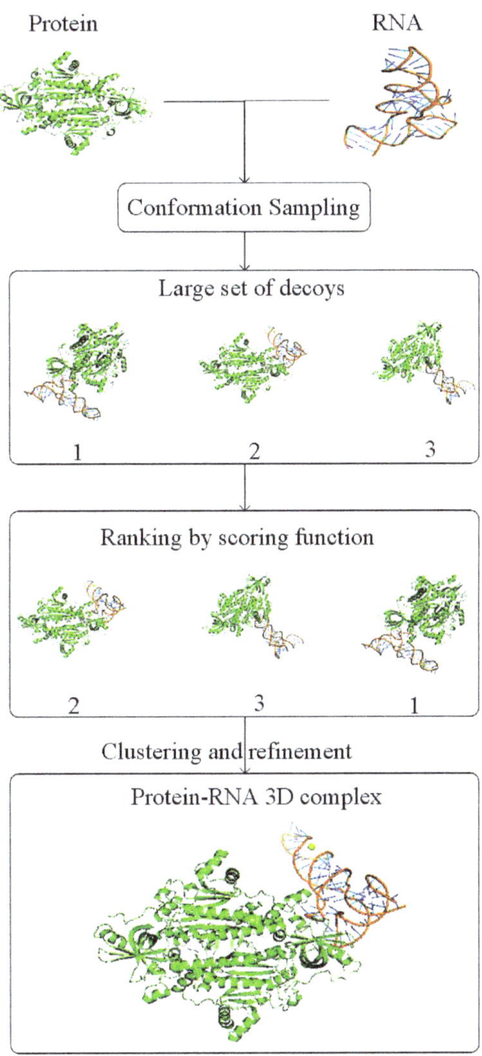

Figure 11.2. A general workflow of protein–RNA docking.

In the first step, protein–RNA docking methods require 3D structures of the proteins and the RNAs as the inputs. If experimentally derived structures do not exist, the predicted structures can be used. After obtaining the real or predicted structures, conformation sampling and scoring function design are two main steps for docking. The conformation sampling is mainly based on searching algorithms, and the scoring function can be designed based on the force field, empirical knowledge or their combinations. Protein–RNA docking methods may focus more on designing proper scoring functions, because similar conformation sampling techniques to protein–protein docking methods can also be applied for protein–RNA docking. Theoretically and ideally, the scoring function will reach a minimum when the docking pose is close to a native interaction complex.

11.2.2.1. *Conformation sampling*

In general, molecular docking begins with a spatial searching for the binding modes of the receptor and the ligand. The ligand and the receptor are treated as two rigid bodies, and then the ligand is translated and rotated to join the centers of mass of two molecules when the receptor is fixed. Lastly, a large set of binding modes are produced using conformational searching, and they are used for generating sample decoys. In protein–RNA docking, the commonly used conformational searching methods mainly include fast Fourier transform (FFT) algorithm, Monte Carlo (MC) algorithm and genetic algorithm (GA).

(1) *FFT:* Katchalski-Katzir *et al.* employed FFT in geometric recognition algorithm [22]. By projecting the receptor and ligand molecules onto a 3D space grid with $N \times N \times N$ (N is the length of each dimension of the 3D grid space, and it is greater than the sum of the lengths of the two molecules) points, and discretizing the two molecules, the algorithm can yield two discrete functions: $f_R(x, y, z)$ denoted as the function of the receptor and $f_L(x, y, z)$ denoted as the function of the ligand. The two functions describe the values on each grid point, respectively (cf. Eq. (1) and Eq. (2)) [22, 24, 45]:

$$f_R(x, y, z) = \begin{cases} 1 & \text{if the point lies on the surface of the receptor} \\ \rho & \text{if the point lies inside the receptor} \\ 0 & \text{if the point lies outside the receptor} \end{cases} \quad (1)$$

$$f_L(x, y, z) = \begin{cases} 1 & \text{if the point lies on the surface of the ligand} \\ \delta & \text{if the point lies inside the ligand} \\ 0 & \text{if the point lies outside the ligand} \end{cases} \quad (2)$$

where ρ and δ are constant values. x, y, z are the points in the 3D grid with *x*-, *y*- and *z*-axes (x, y, z $\in \{1.2, ..., N\}$).

The number of shifting steps for the ligand related to the receptor can be described via a shift vector $\{\alpha, \beta, \gamma\}$. The correlation function of each translation can be defined as follows:

$$C(\alpha, \beta, \gamma) = \sum_{x=1}^{N} \sum_{y=1}^{N} \sum_{z=1}^{N} f_R(x, y, z) f_L(x + \alpha, y + \beta, z + \gamma) \quad (3)$$

Due to the high cost of Eq. (3), FFT is used to speed up the process of sampling. Taking a 2D grid as an example, Figure 11.3 illustrates the process of using correlation function to find the matched geometry. When the geometry of the receptor and the ligand are perfectly complementary, the correlation function reaches its maximum value. Similarly, after each translational move, the ligand is rotated by searching all the rotational space.

(2) *MC*: The Monte Carlo algorithm is a statistical simulation method, which can proceed the following three steps for protein–RNA docking. First, it acquires the rotational and translational directions randomly. Second, the ligand molecule is rotated and translated related to the fixed receptor molecule to generate some binding modes. Third, the rotational and translational directions are continuously modified accordingly, and a probability model, i.e., Bayesian probability model, can be used to determine whether the modification is accepted or not. In this field, the RosettaDock software is a widely used MC-based docking algorithm [46, 47]. However, as the MC algorithm is a heuristic algorithm, it easily suffers to local optimal

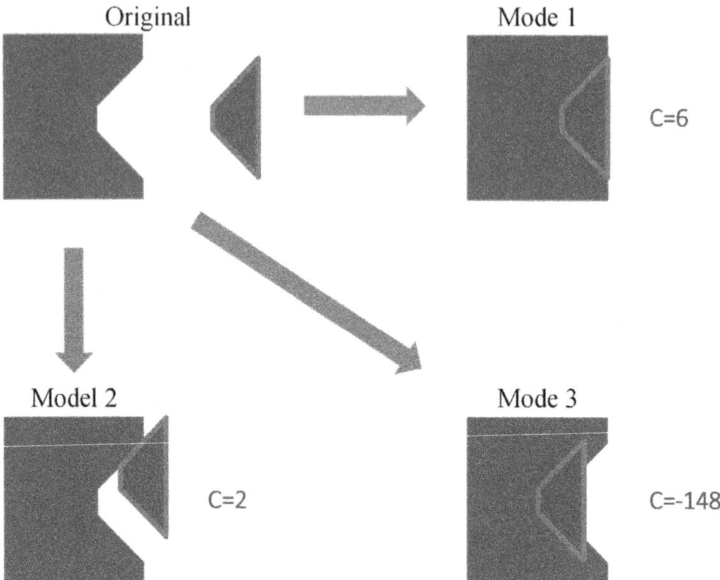

Figure 11.3. Three potential modes in the geometric recognition process. Model 1: the ligand molecule fits well with the receptor molecule; Model 2: the ligand and receptor molecules match partly; Model 3: the ligand and receptor molecules intersect. The parameters ($\rho = -15$, $\delta = 1$) are from Katchalski-Katzir *et al.*

solutions. To avoid this issue, simulated annealing could be combined with the MC algorithm [48].

(3) *GA*: The genetic algorithm randomly generates some conformations and treats them as individuals in the process of biological evolution, which is based on analog evolution. It will further assess current individuals according to a given fitness criteria and use the probabilistic methods to select the individuals with the fitness value over a predefined threshold. Then these individuals are used as seeds for the next generation via crossover and mutation. For example, Morris *et al.* applied Lamarckian GA during conformational searching and achieved reliable results [49]. The conformational parameters (translation and rotation) are analogous to genes in the evolution. For example, AutoDock has implemented the conformation sampling module using a GA algorithm [20, 50].

11.2.2.2. *Scoring function design*

The scoring function of molecular docking is used to evaluate the large number of binding patterns generated during the conformational searching phase and further select the near-native structures. The scoring function can be generally divided into four types: force-field-based terms, empirical terms, knowledge-based terms and their combinations.

Force-field terms: A force-field term generally consists of three parts: an atom type, a potential function and force field parameters. The force-field-based terms are used to measure the potential energy of a coarse-grained particle or a system of atoms, including van der Waals and electrostatic. Two basic functional forms exist: bonded terms (interactions between covalent bonds and atoms) and non-bonded terms in molecular mechanics. The van der Waals and electrostatic terms belong to the latter. Up to now, many force-field functions for different systems have been designed, e.g., AMBER (Assisted Model Building and Energy Refinement) [51], CHARMM (Chemistry at HARvard Molecular Mechanics) [52].

Empirical terms: The empirical terms are mainly based on the number of various interactions between the protein and RNA interfaces [53]. Experience-based scoring functions commonly use multiple regression methods to fit the contribution of various physical properties to binding energy. These properties include hydrophobic terms, the number of hydrogen bonds and rotatable bonds immobilized in the formation of complexes.

Knowledge-based terms: Knowledge-based terms are based on statistical potentials of interactions at the interface of protein–RNA complexes. Due to more and more solved complex structures, this type of data-driven knowledge attracts more and more attention. In recent years, researchers have proposed several knowledge-based statistical potential energy functions, and part of them are illustrated in Figure 11.4.

The Varani group developed two types of scoring functions (H-bonding potential and all-atom potential) for protein–RNA

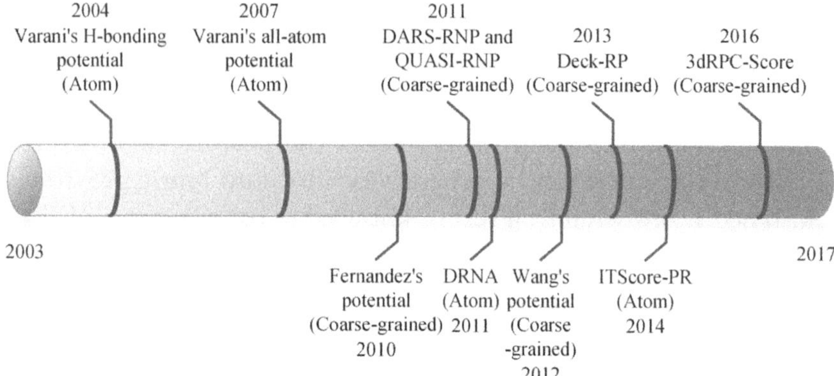

Figure 11.4. The development of the statistical potential energy function. "Atom" and "coarse-grained" are the representations of the molecules.

interactions in 2004 and 2007, respectively. The H-bonding potential depends on the orientation and distance of hydrogen bonding, and it only shows a part of interactions at the interface [54, 55]. The all-atom potential is able to distinguish native poses of complexes from other inaccurate docking poses [56].

In 2010, Fernandez group obtained a 20×4 residue–nucleotide pairwise potential based on 282 non-redundant protein–RNA complexes [57]. They generated decoys based on FTDock and ranked these decoys with the pairwise potential. The result showed that the potential facilitates the separation of near-native structures from the decoys generated by FTDock.

In 2011, two knowledge-based potentials, Quasi-chemical potential (QUASI-RNP) and decoys as the reference-state potential (DARS-RNP), were developed to score protein–RNA models after docking by Tuszynska *et al.* Both were coarse-grained representations for the docking molecules based on the distance-dependent term, angular-dependent term, edge-dependent term and penalty for steric clashes [58]. The difference is that QUASI-RNP takes the mole fractions of residues as the reference state and DARS-RNP uses the decoys.

Other energy functions include DFIRE-based energy function [59], a 60×8 residue–nucleotide pairwise potential by combining secondary structure states [60], DECK-RP built on a protein–protein potential function DECK (Distance- and Environment-dependent, Coarse-grained, Knowledge-based) [61], ITScore-PR, whose pairwise atomic interaction potentials depend on the distance in known protein–RNA complex structures [62] and 3dRPC-Score based on the conformations of residue–base pairs [63, 64]. Among the above-mentioned methods, the H-bonding potential and all-atom potential will have the lowest thresholds for discriminating decoys, which can distinguish the reference structures with a RMSD<3Å and <5Å, respectively. It is also not escaped our notice that DARS-RNP, QUASI-RNP, Deck-RP, ITScore-PR and 3dRPC have provided the standalone software to download.

Combinations: Combining force-field, empirical and knowledge-based terms can generally achieve better results when ranking the decoys. For instance, Zhang *et al.* developed a weighted consensus scoring function RpveScore that combines van der Waals, electrostatic and Wang's potential [65]. They optimize the combined weights using a multiple linear regression method, which fits the scoring function to the ligand's root mean square deviation (L_rmsd) of the bound docking poses from the Benchmark II (http://life.bsc.es/pid/protein-rna-benchmark). The results show that RpveScore yields better performance.

11.2.2.3. *Clustering and refinement*

Many docking algorithms not only output a single structure with the best score, but also give a set of representative structures of several largest clusters. The reason is that combining multiple structures with good scores will be more accurate than the single structure with the best score. For example, NPDOCK clustering the structures based on their RMSDs and selects the representative structures of three largest clusters; The ICM applies binary contact fingerprints (CFPs) and Tanimoto distances [66] for clustering structures. In addition, refinement is an effective way to further optimize protein–RNA interaction complexes, which will lead to a more accurate result. Commonly used refinement algorithms are MC and molecular dynamics (MD) simulation, e.g., RosettaDock and NPDock softwares.

11.2.2.4. *Benchmark data sets for evaluating protein–RNA docking*

According to the states of the molecular structure before docking, there are three types of docking: (1) both the protein and RNA structures are unbound states, denoted as unbound–unbound (UU) docking; (2) only the protein structures are unbound states, denoted as unbound–bound (UB) docking; (3) only the RNA structures are unbound states, denoted as bound–unbound (BU) states. In order to promote the research of protein–RNA docking, some benchmark data sets have been constructed to evaluate protein–RNA docking methods. In this section, we introduce four benchmark sets developed in this area, which use different rules to select structures from existing protein–RNA complexes in PDB. Table 11.2 lists the information about the four benchmark sets, denoted as PRD45, PRD106, PRD72 and PRD126.

In 2012, a benchmark set of PRD45 was constructed for protein–RNA docking, which is a non-redundant benchmark data set derived from available unbound and bound structures involving polypeptides and nucleotide chains in the PDB [37]. This data set is composed by 45 molecular complex structures (9 UB cases and 36 UU cases) superior to 3.0 Å from X-ray validation or the average structure of NMR ensembles, where the sequences are at least 30 amino acids and have been removed to 35% sequence identity. These 45 protein–RNA complexes can be further divided as 34 rigid-body, 8 semi-flexible and 3 full flexible classes. In 2017, the PRD45 data set was updated [67],

Table 11.2. The information of four protein–RNA docking benchmark data sets.[a]

Benchmark	Total cases	UU cases	UB cases	BU cases	Other cases	Year	Reference
PRD45	45	36	9	0	0	2012	[37]
PRD106	106	62	9	0	35	2012	[27]
PRD72	72	52	20	0	0	2013	[68]
PRD126	126	21	95	10	0	2017	[67]

[a]The table lists the number of total test cases, unbound–unbound, unbound–bound, bound–unbound, other cases and the year of publication in the benchmark set.

and a new benchmark set PRD126 was generated, which consists of 21 UU cases, 95 UB cases and 10 BU cases. Compared to PRD45, there are 12 new UU cases, 59 new UB cases and 10 BU cases in this benchmark.

In the same year, Perez-Cano *et al.* extended the benchmark set using homology modeling and experimental data to construct a new protein–RNA docking benchmark set PRD106, which contains 106 protein–RNA complexes [27]. Of them, 35 cases have at least one molecular structure constructed from homology-based modeling. They consist of 3 model–model (both the protein and RNA structures are from homology-based modeling), 5 unbound–model (the protein structures are unbound states, whereas the RNA structures are from homology-based modeling), 8 model–unbound (the protein structures are from homology-based modeling, whereas RNA structures are unbound states) and 19 model–bound (the protein structures are from homology-based modeling, whereas RNA structures are bound states). The remaining 71 protein–RNA complexes (9 UB cases and 62 UU cases) are collected from crystallographic or NMR structures. In this benchmark set, according to the flexibility of proteins and RNAs, 106 protein–RNA complexes are further divided into three groups: easy (64 cases), intermediate (24 cases) and difficult (18 cases).

A non-redundant benchmark data set called PRD72 has been constructed by Huang and Zou in 2013 [68], which are collected from PDB with X-ray crystal structures of protein–RNA complexes with resolutions higher than 4.0 Å. Sequence identity has been removed in this data set and further clustering process has been performed on this data set to pick a more representative complex sample. Of the final 72 protein–RNA complexes, there are 52 UU cases and 20 UB cases. Similarly, the benchmark set is also divided into easy, intermediate and difficult groups with 49, 16 and 7 complexes, respectively.

Benchmark data sets are critically important for evaluating the docking algorithms. With more protein–RNA complexes deposited in the PDB database, it is expected that larger benchmark data sets will be constructed and used in this field.

11.3. On existing protein–RNA docking software

As shown in Figure 11.1, protein–protein dockings are studied much earlier than protein–RNA dockings. Most existing protein–RNA docking methods are modified from other protein–protein docking algorithms. A few of these methods provide easy-of-use Web servers for protein–RNA docking, and Figure 11.5 shows a comparison for them. Apart from NPDOCK [35], HDOCK [18] and 3dRPC [36], other methods are modified from corresponding protein–protein docking algorithms.

The NPDOCK is a computational pipeline consisting of rigid-body docking, scoring of decoys, clustering of the best-scored binding modes and refinement of the most promising predictions. NPDOCK uses GRAMM to perform a global searching and generate multiple decoys. After the conformation sampling, it filters out some decoys based on the predefined criteria, e.g., the distance between the set of residues. Then statistical potential in DARS-RNP is used to rank these decoys. The poses with the good scores are clustered into

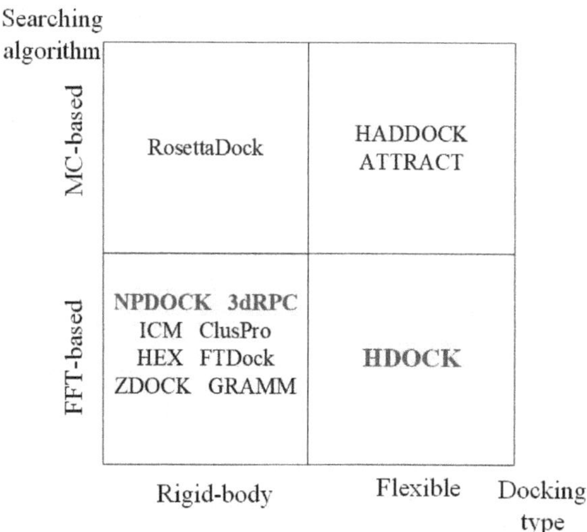

Figure 11.5. Comparison of existing protein–RNA docking methods. The bold ones are developed specifically for protein–RNA docking and others are modified from protein–protein docking algorithms.

three groups and the representatives of each group are selected. Lastly, these representatives are refined by an MC method to obtain the final 3D protein–RNA complex structures.

The HDOCK performs the global searching using a hierarchical FFT-based method [69], which uses an improved shape-based pair-wise term as the scoring function [70]. Moreover, HDOCK supports protein sequence inputs when the protein structures are unknown. For sequence inputs without structures, HDOCK will search the homologous structures in the PDB as the templates using the HHSuite package [71] and then use these templates for subsequent modeling based on MODELLER [72].

The 3dRPC uses the rigid-body docking program RPDOCK for sampling the conformational space of protein–RNA complexes [38]. During the RPDOCK process, proteins and RNAs are considered as receptors and ligands, respectively. The RNA is translated and rotated relative to the protein, and their geometric score and electrostatic score are calculated for each binding mode, where the electrostatic score is derived from the AMBER force field. The entire searching process is accelerated by the FFT algorithm. The 3dRPC provides two types of scoring functions: Deck-RP and 3dRPC-Score, which can work well in real applications [64].

The ICM is adapted for protein–RNA docking by adding an electrostatic potential energy to the original docking protocol for protein–protein docking and then optimizes the scoring function with van der Waals terms, electrostatic terms and solvent terms. Similarly, original protein–protein docking version of RosettaDock was also extended to protein–RNA docking [73]. First, the RosettaDock with a low-resolution model was extended to RNAs for searching and scoring, and the Protein–RNA Interface Database (PRIDB) [74] was used as a reference data set to generate docking conformations. Then, a supervised machine learning model for high-resolution scoring function optimization was applied in the RosettaDock. The Receiver Operating Characteristic(ROC)-based Genetic LearneR (ROGER) algorithm [75] is applied to optimize the area of ROC curve. Finally, the docking conformations can be predicted to be three types of structures (near-native structures, decoys and test decoys) by the trained model.

Software of GRAMM, FTDock, HAADOCK, ClusPro and HEX accept RNA and protein inputs in their molecular docking programs. The Zacharias group designs a coarse-grained force field for protein–RNA docking based on their protein–protein docking model ATTRACT [76–79]. There are also other improvements on protein–RNA dockings. For example, Iwakiri *et al.* combines AMBER and CHARMM to integrate the physicochemical information of RNAs into ZDOCK by MD simulation of nucleotides, thereby improving the accuracy of rigid-body docking prediction [33].

11.4. Conclusion and outlook

Protein–RNA docking has attracted increasing attentions in recent years (Figure 11.1), and several methods designed for protein–RNA docking have been proposed and yielded much progress. However, some challenges remain for further methodology development of protein–RNA docking, for instance:

(1) The benchmark set for evaluating protein–RNA docking methods is still small, which may hinder the large-scale simulations of protein–RNA docking.
(2) Since the molecular conformation may change in the case of proteins and RNAs binding, how to effectively taking into account this dynamic change is a challenge for docking algorithms.
(3) For those proteins or RNAs without 3D structures in PDB, the accuracy for simulating them is still low, A homology-based model can be used to infer the structures, but this will introduce some uncertainty and noise into subsequent protein–RNA docking.

Protein–RNA docking is not yet well investigated compared to the protein–protein docking. To date, the research on protein–RNA docking is still actively exploring. Deep learning has been widely used in computational biology [80–83] and achieved remarkable results. With increasing protein–RNA complexes in PDB, we expect the data-driven deep learning to be more applied for protein–RNA docking in the near future. For example, the knowledge-based terms in the scoring function using machine learning model can be more accurate, and we can thus consider updating the conventional machine learning models using the deep learning models. At least four future directions can be further explored in this area: (1) the design of more accurate scoring functions for different types protein–RNA docking; (2) the exploration of multi-objective energy function using parallel optimization protocol for more accurate conformation search; (3) the study of various conformational changes of RNAs; and (4) with more advanced sequence-based prediction tools developed for predicting the interaction motifs and residues between RNAs and proteins, this prior knowledge is expected to be helpful for speeding up the protein–RNA docking conformation search.

Acknowledgments

This work was supported by the Science and Technology Commission of Shanghai Municipality (No. 17JC1403500).

References

1. Re, A., Joshi, T., Kulberkyte, E., *et al.* RNA-protein interactions: An overview. *Methods Mol Biol.* 2014;1097:491–521.
2. Chen, Y., and Varani, G. Protein families and RNA recognition. *FEBS J.* 2005;272(9):2088–2097.
3. Glisovic, T., Bachorik, J.L., Yong, J., *et al.* RNA-binding proteins and post-transcriptional gene regulation. *FEBS Lett.* 2008;582(14): 1977–1986.
4. Gromiha, M.M., Yokota, K., and Fukui, K. Understanding the recognition mechanism of protein-RNA complexes using energy based approach. *Curr Protein Pept Sci.* 2010;11(7):629–638.
5. Gromiha, M.M., and Fukui, K. Scoring function based approach for locating binding sites and understanding recognition mechanism of protein-DNA complexes. *J Chem Inf Model.* 2011;51(3):721–729.
6. Ke, A., and Doudna, J.A. Crystallization of RNA and RNA-protein complexes. *Methods.* 2004;34(3):408–414.
7. Theimer, C.A., Smith, N.L., and Khanna, M. NMR studies of protein-RNA interactions. *Methods Mol Biol.* 2012;831:197–218.

8. Scott, L.G., and Hennig, M. RNA structure determination by NMR. *Methods Mol Biol.* 2008;452:29–61.

9. Carroni, M., and Saibil, H.R. Cryo electron microscopy to determine the structure of macromolecular complexes. *Methods.* 2016;95:78–85.

10. Burley, S.K., Berman, H.M., Kleywegt, G.J., *et al.* Protein data bank (PDB): The single global macromolecular structure archive. *Methods Mol Biol.* 2017;1607:627–641.

11. Nithin, C., Ghosh, P., and Bujnicki, J.M. Bioinformatics tools and benchmarks for computational docking and 3D structure prediction of RNA-protein complexes. *Genes (Basel).* 2018;9(9).

12. Lengauer, T., and Rarey, M. Computational methods for biomolecular docking. *Curr Opin Struct Biol.* 1996;6(3):402–406.

13. Vajda, S., Hall, D.R., and Kozakov, D. Sampling and scoring: A marriage made in heaven. *Proteins.* 2013;81(11):1874–1884.

14. Gschwend, D.A., Good, A.C., and Kuntz, I.D. Molecular docking towards drug discovery. *J Mol Recognit.* 1996;9(2):175–186.

15. Lamb, M.L., and W.L. Jorgensen. Computational approaches to molecular recognition. *Curr Opin Chem Biol.* 1997;1(4):449–457.

16. Dar, A.M., and Mir, S. Molecular docking: Approaches, types, applications and basic challenges. *J Anal Bioanal Tech.* 2017;8:2.

17. Chen, R., Li, L., and Weng, Z. ZDOCK: An initial-stage protein-docking algorithm. *Proteins.* 2003;52(1):80–87.

18. Yan, Y., Zhang, D., Zhou, P., *et al.* HDOCK: A web server for protein–protein and protein-DNA/RNA docking based on a hybrid strategy. *Nucleic Acids Res.* 2017;45(W1):W365–W373.

19. Trott, O., and Olson, A.J. AutoDock Vina: Improving the speed and accuracy of docking with a new scoring function, efficient optimization, and multithreading. *J Comput Chem.* 2010;31(2):455–461.

20. Rizvi, S.M.D., Shakil, S., and Haneef, M. A simple click by click protocol to perform docking: Autodock 4.2 made easy for non-bioinformaticians. *EXCLI J.* 2013;12:831–857.

21. Tovchigrechko, A., and Vakser, I.A. GRAMM-X public web server for protein–protein docking. *Nucleic Acids Res.* 2006;34:W310–W314.

22. Katchalski-Katzir, E., Shariv, I., Eisenstein, M., *et al.* Molecular surface recognition: Determination of geometric fit between proteins and their ligands by correlation techniques. *P Natl Acad Sci USA.* 1992;89(6):2195–2199.

23. Schueler-Furman, O., Wang, C., and Baker, D. Progress in protein–protein docking: Atomic resolution predictions in the CAPRI experiment using RosettaDock with an improved treatment of side-chain flexibility. *Proteins.* 2005;60(2):187–194.

24. Gabb, H.A., Jackson, R.M., and Sternberg, M.J.E. Modelling protein docking using shape complementarity, electrostatics and biochemical information. *J Mol Biol.* 1997;272(1):106–120.

25. Ritchie, D.W., and Kemp, G.J.L. Protein docking using spherical polar Fourier correlations. *Proteins.* 2000;39(2):178–194.

26. Dominguez, C., Boelens, R., and Bonvin, A.M.J.J. HADDOCK: A protein–protein docking approach based on biochemical or biophysical information. *J Am Chem Soc.* 2003;125(7):1731–1737.

27. Perez-Cano, L., Jimenez-Garcia, B., and Fernandez-Recio, J. A protein-RNA docking benchmark (II): Extended set from experimental and homology modeling data. *Proteins.* 2012;80(7):1872–1882.

28. Janin, J., Henrick, K., Moult, J., *et al.* CAPRI: A critical assessment of predicted interactions. *Proteins.* 2003;52(1):2–9.

29. Janin, J. The targets of CAPRI rounds 13–19. *Proteins.* 2010;78(15): 3067–3072.

30. Kozakov, D., Hall, D.R., Xia, B., *et al.* The ClusPro web server for protein–protein docking. *Nat Protoc.* 2017;12(2):255–278.

31. Schneidman-Duhovny, D., Inbar, Y., Nussinov, R., *et al.* PatchDock and SymmDock: Servers for rigid and symmetric docking. *Nucleic Acids Res.* 2005;33:W363–W367.

32. Zacharias, M. ATTRAACT: Protein–protein docking in CAPRI using a reduced protein model. *Proteins.* 2005;60(2):252–256.

33. Iwakiri, J., Hamada, M., Asai, K., *et al.* Improved accuracy in RNA-protein rigid body docking by incorporating force field for molecular dynamics simulation into the scoring function. *J Chem Theory Comput.* 2016;12(9):4688–4697.

34. Arnautova, Y.A., Abagyan, R., and Totrov, M. Protein-RNA docking using ICM. *J Chem Theory Comput.* 2018;14(9):4971–4984.

35. Tuszynska, I., Magnus, M., Jonak, K., *et al.* NPDock: A web server for protein-nucleic acid docking. *Nucleic Acids Res.* 2015;43(W1): W425–W430.

36. Huang, Y., Li, H., and Xiao, Y. 3dRPC: A web server for 3D RNA-protein structure prediction. *Bioinformatics.* 2018;34(7):1238–1240.

37. Barik, A., C.N., P.M., *et al.* A protein-RNA docking benchmark (I): Nonredundant cases. *Proteins.* 2012;80(7):1866–1871.
38. Huang, Y.Y., Liu, S., Guo, D., *et al.* A novel protocol for three-dimensional structure prediction of RNA-protein complexes. *Sci Rep.* 2013;3:1887.
39. Allers, J., and Shamoo, Y. Structure-based analysis of protein-RNA interactions using the program ENTANGLE. *J Mol Biol.* 2001;311(1): 75–86.
40. Perez-Cano, L., Solernou, A., Pons, C., *et al.* Structural prediction of protein-RNA interaction by computational docking with propensity-based statistical potentials. *Pac Symp Biocomput.* 2010;2010:293–301.
41. Bahadur, R.P., Zacharias, M., and Janin, J. Dissecting protein-RNA recognition sites. *Nucleic Acids Res.* 2008;36(8):2705–2716.
42. Jones, S., Daley, D.T., Luscombe, N.M., *et al.* Protein-RNA interactions: A structural analysis. *Nucleic Acids Res.* 2001;29(4):943–954.
43. Jones, S., van Heyningen, P., Berman, H.M., *et al.* Protein-DNA interactions: A structural analysis. *J Mol Biol.* 1999;287(5):877–896.
44. Iwakiri, J., Tateishi, H., Chakraborty, A., *et al.* Dissecting the protein-RNA interface: The role of protein surface shapes and RNA secondary structures in protein-RNA recognition. *Nucleic Acids Res.* 2012;40(8): 3299–3306.
45. Zhang, Q., Feng, T., Xu, L., *et al.* Recent advances in protein–protein docking. *Curr Drug Targets.* 2016;17(14):1586–1594.
46. Gray, J.J., Moughon, S., Wang, C., *et al.* Protein–protein docking with simultaneous optimization of rigid-body displacement and side-chain conformations. *J Mol Biol.* 2003;331(1):281–299.
47. Chaudhury, S., Berrondo, M., Weitzner, B.D., *et al.* Benchmarking and analysis of protein docking performance in Rosetta v3.2. *PLOS ONE.* 2011;6(8):e22477.
48. Vanderbilt, D., and Louie, S.G. A Monte carlo simulated annealing approach to optimization over continuous variables. *J Comput Phys.* 1984;56(2):259–271.
49. Morris, G.M., Goodsell, D.S., Halliday, R.S., *et al.* Automated docking using a Lamarckian genetic algorithm and an empirical binding free energy function. *J Comput Chem.* 1998;19(14):1639–1662.
50. Trott, O., and Olson, A.J. Software news and update AutoDock Vina: Improving the speed and accuracy of docking with a new scoring function, efficient optimization, and multithreading. *J Comput Chem.* 2010;31(2):455–461.

51. Case, D.A., Cheatham, T.E. 3rd, Darden, T., *et al.* The Amber biomolecular simulation programs. *J Comput Chem.* 2005;26(16): 1668–1688.

52. Brooks, B.R., Brooks, C.L. 3rd, Mackerell, A.D. Jr., *et al.* CHARMM: The biomolecular simulation program. *J Comput Chem.* 2009;30(10): 1545–1614.

53. Bohm, H.J. Prediction of binding constants of protein ligands: A fast method for the prioritization of hits obtained from de novo design or 3D database search programs. *J Comput Aided Mol Des.* 1998; 12(4):309–323.

54. Chen, Y., Kortemme, T., Robertson, T., *et al.* A new hydrogen-bonding potential for the design of protein-RNA interactions predicts specific contacts and discriminates decoys. *Nucleic Acids Res.* 2004;32(17): 5147–5162.

55. Treger, M., and Westhof, E. Statistical analysis of atomic contacts at RNA-protein interfaces. *J Mol Recognit.* 2001;14(4):199–214.

56. Zheng, S., Robertson, T.A., and Varani, G. A knowledge-based potential function predicts the specificity and relative binding energy of RNA-binding proteins. *FEBS J.* 2007;274(24):6378–6391.

57. Perez-Cano, L., Solernou, A., Pons, C., *et al.* Structural prediction of protein-RNA interaction by computational docking with propensity-based statistical potentials. *Pac Symp Biocomput.* 2010;293–301.

58. Tuszynska, I., and Bujnicki, J.M. DARS-RNP and QUASI-RNP: New statistical potentials for protein-RNA docking. *BMC Bioinformatics.* 2011;12:348.

59. Zhao, H.Y., Yang, Y.D., and Zhou, Y.Q. Structure-based prediction of RNA-binding domains and RNA-binding sites and application to structural genomics targets. *Nucleic Acids Res.* 2011;39(8):3017–3025.

60. Li, C.H., Cao, L.B., Su, J.G., *et al.* A new residue-nucleotide propensity potential with structural information considered for discriminating protein-RNA docking decoys. *Proteins.* 2012;80(1):14–24.

61. Liu, S.Y., and Vakser, I.A. DECK: Distance and environment-dependent, coarse-grained, knowledge-based potentials for protein–protein docking. *BMC Bioinformatics.* 2011;12:280.

62. Huang, S.Y., and Zou, X.Q. A knowledge-based scoring function for protein-RNA interactions derived from a statistical mechanics-based iterative method. *Nucleic Acids Res.* 2014;42(7):e55.

63. Huang, Y., Li, H., and Xiao, Y. Using 3dRPC for RNA-protein complex structure prediction. *Biophys Rep.* 2016;2(5):95–99.

64. Li, H.T., Huang, Y.Y., and Xiao, Y. A pair-conformation-dependent scoring function for evaluating 3D RNA-protein complex structures. *PLOS ONE.* 2017;12(3):e0174662.

65. Zhang, Z., Lu, L., Zhang, Y., *et al.* A combinatorial scoring function for protein-RNA docking. *Proteins.* 2017;85(4):741–752.

66. Rogers, D.J., and Tanimoto, T.T. A computer program for classifying plants. *Science.* 1960;132(3434):1115–1118.

67. Nithin, C., Mukherjee, S., and Bahadur, R.P. A non-redundant protein-RNA docking benchmark version 2.0. *Proteins.* 2017;85(2):256–267.

68. Huang, S.Y., and Zou, X.Q. A nonredundant structure dataset for benchmarking protein-RNA computational docking. *J Comput Chem.* 2013;34(4):311–318.

69. Huang, S.Y., and Zou, X.Q. MDockPP: A hierarchical approach for protein–protein docking and its application to CAPRI rounds 15–19. *Proteins.* 2010;78(15):3096–3103.

70. Yan, Y.M., and Huang, S.Y. A new pairwise shape-based scoring function to consider long-range interactions for protein–protein docking. *Biophys J.* 2017;112(3):470a–470a.

71. Remmert, M., Biegert, A., Hauser, A., *et al.* HHblits: Lightning-fast iterative protein sequence searching by HMM-HMM alignment. *Nat Methods.* 2012;9(2):173–175.

72. Sali, A., and Blundell, T.L. Comparative protein modelling by satisfaction of spatial restraints. *J Mol Biol.* 1993;234(3):779–815.

73. Guilhot-Gaudeffroy, A., Froidevaux, C., Azé, J., *et al.* Protein-RNA complexes and efficient automatic docking: Expanding RosettaDock possibilities. *PLOS ONE.* 2014;9(9).

74. Lewis, B.A., Walia, R.R., Terribilini, M., *et al.* PRIDB: A protein-RNA interface database. *Nucleic Acids Res.* 2011;39:D277–D282.

75. Bernauer, J., Azé, J., Janin, J., *et al.* A new protein–protein docking scoring function based on interface residue properties. *Bioinformatics,* 2007;23(5):555–562.

76. Setny, P., and Zacharias, M. A coarse-grained force field for protein-RNA docking. *Nucleic Acids Res.* 2011;39(21):9118–9129.

77. Zacharias, M. Protein–protein docking with a reduced protein model accounting for side-chain flexibility. *Protein Sci.* 2003;12(6):1271–1282.

78. Zacharias, M. Protein–protein docking with a reduced protein model accounting for side chain and loop flexibility. *Biophys J.* 2002;82(1):9A–9A.

79. Fiorucci, S., and Zacharias, M. Binding site prediction and improved scoring during flexible protein–protein docking with ATTRACT. *Proteins.* 2010;78(15):3131–3139.

80. Pan, X., and Shen, H.-B. Predicting RNA–protein binding sites and motifs through combining local and global deep convolutional neural networks. *Bioinformatics.* 2018;34(20):3427–3436.

81. Pan, X., and Shen, H.B. RNA-protein binding motifs mining with a new hybrid deep learning based cross-domain knowledge integration approach. *BMC Bioinformatics.* 2017;18(1):136.

82. Pan, X.Y., Rijnbeek, P., Yan, J., *et al.* Prediction of RNA-protein sequence and structure binding preferences using deep convolutional and recurrent neural networks. *BMC Genomics.* 2018;19:511.

83. Ching, T., Himmelstein, D.S., Beaulieu-Jones, B.K., *et al.* Opportunities and obstacles for deep learning in biology and medicine. *J R Soc Interface.* 2018;15(141).

Part III
Protein–Ligand Interactions

Chapter 12

Protein–carbohydrate complexes: Binding site analysis, prediction, binding affinity and molecular dynamics simulations

K. Veluraja[*,‡], N. R. Siva Shanmugam[†], J. Jino Blessy[†], R. A. Jeyaram[*], B. Lalithamaheswari[*] and M. Michael Gromiha[†,‡]

Research Laboratory of Molecular Biophysics, School of Advanced Sciences, Vellore Institute of Technology, Vellore 632014, Tamil Nadu, India

†*Department of Biotechnology, Bhupat and Jyoti Mehta School of Biosciences, Indian Institute of Technology Madras, Chennai 600036, Tamil Nadu, India*

Carbohydrates are considered the molecules of molecular recognition because of their influential roles in interacting with viruses, toxins and cell–cell adhesion phenomena. The recognition phenomena are due to the diversified sequences and three-dimensional structures of carbohydrates and their conformational dynamics, binding site residues of proteins and noncovalent interactions between them. This review provides (1) different databases of carbohydrates, which have vital roles in studying protein–carbohydrate interactions, (2) the role of molecular dynamics (MD) simulations for deducing

‡kvrajamsu@gmail.com, veluraja.k@vit.ac.in; gromiha@iitm.ac.in

conformational models of carbohydrates as well as studying the interactions between proteins and carbohydrates and (3) design of inhibitors to avoid pathogenic interactions.

12.1. Introduction

The biological world is dominated by proteins, carbohydrates, nucleic acids and lipid molecules (Kanie and Kanie, 2017). Among them, carbohydrate is a wonder molecule, which can occur on its own as monosaccharides, oligosaccharides and polysaccharides and along with other biological molecules as an integral structure. It interacts with proteins as glycoproteins, lipids as glycolipids and nucleic acids as ribose or deoxyribose sugars. Protein–carbohydrate interactions are mediated with noncovalent interactions such as electrostatic, van der Waals, hydrogen bonding and hydrophobic interactions. Functionally, carbohydrates act as receptors for various toxins and viruses, involved in cell-adhesion, cell trafficking, egg–sperm interaction, bacterial, fungi and parasite adhesion, cell–matrix interaction, clearance of damaged glycoconjugates and cells, signal transduction, host–pathogen recognition and inflammation (Fernández-Alonso *et al.*, 2012; Varki, 2017). Nature has created sugar molecules for carrying out these numerous biological recognition processes due to the fact that with minimal number of sugar molecules, diversified structures can be produced by altering the anomers, glycosidic linkages and branching.

Carbohydrates are ubiquitous and perform an intriguing role inside the cell as well as on the cell surface. Carbohydrates bound to proteins present on the cell surface can easily interact with toxins, viruses and the surface of other cells. Glycoproteins hold a major role in biological recognition events through protein–carbohydrate interactions, which include cell–cell recognition and cell development (Murrey and Hsieh-Wilson, 2008). Over the last decade, studies have shown that specific proteins, which recognize carbohydrates, also participate in biological processes including pathological and physiological functions in the cell. In many cases, sugars provide stability to proteins, especially to plant lectins (Kumar *et al.*, 2012).

Interactions between proteins and carbohydrates are usually the first step in infection, via the attachment of parasites, fungi, bacteria and viruses to the carbohydrate segments of the host cells, an example for cell adhesion. In many cases, protein–carbohydrate interactions are not simple events but only the first step in a series of events and often leading to a complex signaling cascade. Different types of carbohydrate-binding proteins such as enzymes, anticarbohydrate antibodies, sugar transporters and lectins are structurally diverse, differing markedly in size, with distinct tertiary and quaternary structure. Consequently, the structures of their binding pockets are also different (Sharon, 2006). Understanding of the key structural details at the atomic and molecular levels is of paramount importance to effectively design molecules for therapeutic purposes.

Interactions between proteins and carbohydrates as well as their binding affinity are studied through various experimental techniques such as NMR spectroscopy, surface plasmon resonance (SPR), isothermal titration calorimetry (ITC), fluorescence spectroscopy, frontal affinity chromatography and so on (Linman *et al.*, 2008; Fernández-Alonso *et al.*, 2012; von Schantz *et al.*, 2012; Sly and Conboy, 2014; Kasai, 2014). Computational methods that include molecular modeling and molecular dynamics (MD) simulations have been widely used to investigate the atomistic level interactions between proteins and carbohydrates.

This review aims to provide an overview of current knowledge and recent updates in the field of protein–carbohydrate interactions. Specifically, we survey the databases available for carbohydrates, carbohydrate-binding proteins and protein–carbohydrate complexes. In addition, specific interactions dominating the interface and factors influencing the binding affinity of protein–carbohydrate complexes will be outlined along with prediction methods for identifying the binding sites and free energy of binding. Further, applications of MD simulations for understanding the recognition of protein–carbohydrate complexes will be described.

12.2. Databases for carbohydrates and protein–carbohydrate complexes

Databases specifically designed for protein–carbohydrate complexes are limited. However, several databases have been developed for carbohydrate-binding proteins, carbohydrates and integrated resources with other complexes such as protein–protein, protein–nucleic acid and protein–ligand. Table 12.1 lists a set of databases for structures and sequences of carbohydrate-binding proteins, binding affinity, specific carbohydrates and integrated resources.

Table 12.1.   List of databases related to protein–carbohydrate interactions.

Name	URL	Description
Structural Databases		
PDB	https://www.rcsb.org/	3D structures of proteins, nucleic acids and biomolecular complexes
PROCARB	http://www.procarb.org/procarbdb/	Carbohydrate-binding protein structures
LectinDB	http://proline.physics.iisc.ernet.in/lectindb/	Structure and sequence information on plant lectins
UniLectin3D	https://www.unilectin.eu/unilectin3D/	3D structures of lectins
Sequence Databases		
LectinDB	http://proline.physics.iisc.ernet.in/lectindb/	Structure/sequence on plant lectins
CAZy	http://www.cazy.org/	Carbohydrate active enzymes
Mucin Database	http://www.medkem.gu.se/mucinbiology/databases/	Databases of mucin genes, transcripts, protein sequences and functional domains
O-GlycBase	http://www.cbs.dtu.dk/databases/OGLYCBASE/	O- and C-glycosylated proteins
GBP	http://www.functionalglycomics.org/glycomics/molecule/jsp/gbpMolecule-home.jsp	GBPs; lectin database
Binding Affinity Databases		
PDBBind	http://www.pdbbind.org.cn/	Binding affinity data biomolecular complexes
BindingDB	https://www.bindingdb.org/bind/index.jsp	Binding affinities of protein-ligands

Table 12.1. (*Continued*)

Name	URL	Description
Carbohydrates Databases		
GlycoSuiteDB	http://www.functionalglycomics. org/glycomics/molecule/ jsp/carbohydrate/ carbMoleculeHome.jsp	Glycan structures database
Glycosyltransferases	http://www.functionalglycomics. org/glycomics/molecule/ jsp/glycoEnzyme/ geMolecule.jsp	Glycosyltransferases with glycan linkages
SugarBindDB	https://sugarbind.expasy.org/	Sugar-binding database.
CSDB	http://csdb.glycoscience.ru/ database/index.html	Carbohydrate structures, taxonomy, bibliography and NMR data
MonosaccharideDB	http://www.monosaccharidedb. org/	Monosaccharide database
GlycoPattern	https://glycopattern.emory. edu/	Glycan array data under the Consortium for functional glycomics.
SuperSweet	http://bioinformatics.charite. de/sweet/	Collection of sweetening agents
3DSDSCAR	http://www.3dsdscar.org/	Structural database for SIA-containing carbohydrates
Web Portals and Integrated Resources		
GlyGen	https://www.glygen.org/	Computational and informatics resources for glycoscience
GlyTouCan	https://glytoucan.org/	Glycan structure repository
Glycosciences.de	http://www.glycosciences.de/	Databases and tools to support glycobiology and glycomics research
UniCarb-DB	https://unicarb-db.expasy.org/	The glycomic database and repository
Glyco3D	http://glyco3d.cermav.cnrs.fr/ home.php	Portal for structural glycosciences

Accessed on July 19, 2019.

12.2.1. *Structural databases of protein–carbohydrate complexes and carbohydrate-binding proteins*

Protein Data Bank (PDB) contains three-dimensional (3D) structures of proteins, nucleic acids and biomolecular complexes, which includes protein–carbohydrate complexes (Burley *et al.*, 2018). Currently, PDB has 13,643 protein–carbohydrate complex structures (as of July 19, 2019), which are used for analyzing the preference of amino acid residues and important interactions at the binding interface. Further, structural databases for specific types of carbohydrate-binding proteins (e.g., lectins) or organism-based databases are reported in the literature (Chandra *et al.*, 2006; Malik *et al.*, 2010; Bonnardel *et al.*, 2018). The PROCARB contains experimentally determined 3D structures derived from PDB and manual verification of existence and identification of carbohydrate ligands (Malik *et al.*, 2010). LectinDB contains structural, sequence and functional information on animal, fungal, plant, viral and bacterial lectins (Chandra *et al.*, 2006). The UniLectin3D provides curated information on 3D structures that are grouped into families based on the carbohydrate-binding domains (Bonnardel *et al.*, 2018).

12.2.2. *Sequence databases*

Specific databases have been developed on different classes of carbohydrate-binding proteins. CAZy database describes the families of structurally related catalytic and carbohydrate-binding modules (or functional domains) of enzymes that degrade, modify or create glycosidic bonds (Cantarel *et al.*, 2008). It has five major families such as glycoside hydrolase, glycosyltransferase, polysaccharide lyase, carbohydrate esterase and carbohydrate-binding module. Mucin database has a compilation of mucin sequences (glycosylated proteins) (glycoconjugates) with heavy molecular weight in different organisms such as human, mouse, chicken, horse, pig, ferret, frog and zebrafish. It has sequences, domain information, mRNA and cross-references (www.medkem.gu.se/mucinbiology/databases/). O-GlycBase is a database of O- and C-glycosylated proteins with experimentally verified glycosylation sites (Gupta *et al.*, 1999). It has sequence information and amino acid residue numbers, which are involved in O- (Ser and Thr), N- (Asn) and C- (Trp) glycosylation. Glycan-binding protein (GBP) is an integrated database of lectins, which covers three classes such as C-type lectins, galectins and siglecs, and provides information on genome, proteome and glycome levels. LectinDB has sequence information on different kingdoms such as animal, plant, bacteria, fungal and virus lectins (Chandra *et al.*, 2006).

12.2.3. *Binding affinity databases*

Protein–carbohydrate-binding affinity data are available in generic databases on binding affinity, which have the compilation of protein–protein, protein–nucleic acid, protein–carbohydrate and protein–ligand complexes. PDBbind is a comprehensive collection of experimentally measured binding affinity data along with structural information for all types of biomolecular complexes (Wang *et al.*, 2004). PDBbind provides binding data of about 20,000 biomolecular complexes, including >500 data for protein–carbohydrate complexes. BindingDB is a public, Web-accessible database of measured binding affinities, focusing mainly on the interactions of proteins considered to be drug targets with small, drug-like molecules (Liu *et al.*, 2006). It has about 1.7 million binding data from 7000 protein targets and 750,000 small molecules including more than 2200 data on protein–carbohydrate complexes. BindingDB has options for browsing, searching and downloading the data, and it can be searched with the name of a protein, PMID, chemical similarity and substructure.

12.2.4. *Carbohydrates databases*

Carbohydrates vary from simple to complex structures with multiple conformations. Due to their branching in different positions, they have different specificity and selectivity toward proteins or other molecules. In protein–carbohydrate interactions, carbohydrates are found as either glycoproteins or glycoconjugates with O-, N- or C-linked. Accordingly, GlycoSuiteDB and Glycosyltransferases databases have been developed for glycans and glycosyltransferases, respectively (Cooper *et al.*, 2001). These databases provide information on molecular substructures, molecular weight, composition and glycan linkages. GlycoPattern is a computational resource to search and analyze glycan data (Agravat *et al.*, 2014). SugarBindDB provides information on known carbohydrate sequences to which pathogenic organisms (bacteria, toxins and viruses) specifically adhere (Mariethoz *et al.*, 2015). CSDB contains manually curated natural carbohydrate structures, taxonomy, bibliography and NMR (Toukach and Egorova, 2015). SuperSweet is a comprehensive knowledgebase for existing carbohydrates and artificial sweeteners, which includes a modeled sweet taste receptor and docked binders (Ahmed *et al.*, 2010). 3D structural database for sialic acid (SIA)-containing carbohydrates (3DSDSCAR) contain conformational models for a given glycan derived through MD simulations (Veluraja *et al.*, 2010). It has structures for carbohydrates belonging to different conformers with varying relative energy. MonosaccharideDB provides monosaccharides with its chemical information, notation across various databases and its atoms with connection data (www.monosaccharidedb.org).

12.2.5. *Web portals and integrated resources*

There are few Web portal and repositories for glycan structures with integrated resources. GlyGen is a data integration and dissemination project for carbohydrates and glycoconjugates, which retrieves information from multiple international data sources and integrates related data (www.glygen.org/). GlyTouCan is a freely available glycan structure repository that assigns globally unique accession numbers to any glycan, independent of the level of information provided by the experimental method used to identify the structure(s). It has options to search for glycan structures and motifs (Tiemeyer *et al.*, 2017). The Glycosciences.de Web portal provides a list of databases and tools to support glycobiology and glycomics research, focusing on 3D structures and references to PDB entries that feature carbohydrates (Böhm *et al.*, 2018). These PDB references enable a targeted search for PDB structures with specific glycans, which cannot be performed easily at PDB site. The UniCarb-DB has structural and experimental mass spectrometry (MS)-glycomic data such as structures, MS/MS fragmentation with complete peak lists, biological contexts and experimental metadata (Campbell *et al.*, 2014). It also provides LC-MS/MS spectra for N- and O-linked glycans released from glycoproteins that were annotated manually. Glyco3D is a repository, covering 3D features of simple and complex saccharides, glycosyltransferases, lectins, monoclonal antibodies against carbohydrates and glycosaminoglycan-binding proteins (Pérez *et al.*, 2015). It has a user-friendly graphical user interface, which offers several search options, visualization and downloading the data.

12.3. Identification of binding site residues in protein–carbohydrate complexes

The binding site residues in biomolecular complexes including protein–carbohydrate complexes are identified using three general methods such as (i) distance between any heavy atoms, (ii) accessible surface area upon binding and (iii) energy-based approach.

12.3.1. *Distance-based criterion*

In distance-based criterion, an amino acid residue in a protein–carbohydrate complex is binding if any of its heavy atoms is within a specific cutoff distance from any heavy atom in the sugar (Malik and Ahmed, 2007). A distance of 3.5 Å is widely used to identify the binding site residues (Malik *et al.*, 2010; Shanmugam *et al.* 2018).

12.3.2. *ASA-based method*

In this method, an amino acid residue is identified as binding if the difference between the accessible surface area of the residue in a protein–carbohydrate complex and that in unbound form is more than a specific cutoff ASA (the widely used cutoff ASAs in the literature are 0.1 Å^2 and 1 Å^2).

12.3.3. *Energy-based approach*

In this method, if the interaction energy (sum of electrostatic and van der Waals interactions) between an amino acid residue in a protein and all heavy atoms in the partner carbohydrate is less than −1 kcal/mol, then the residue is identified as binding (Gromiha *et al.*, 2014).

12.4. Analysis of binding sites

Carbohydrates interact with proteins using several types of intermolecular interactions. The interaction of apolar groups of carbohydrates with aromatic amino acid residues, known as dispersion interaction or CH/π interaction, is reported to be important in protein–carbohydrate complexes (Wimmerová *et al.*, 2012). Binding propensity analysis of a set of nonredundant protein–carbohydrate complexes indicates the dominance of Trp to interact with carbohydrates through aromatic–aromatic interactions and charged residues to form electrostatic interactions (Gromiha *et al.*, 2014). In carbohydrate-binding pockets of proteins, aliphatic and hydrophobic residues are disfavored, whereas aromatic side chains are enriched. Specifically, Trp is shown to have an increased prevalence of ninefold (Hudson *et al.*, 2015). Further, carbohydrate CH groups interact preferentially with aromatic residues. Although aromatic residues play a major role in protein–carbohydrate interactions, other residues are also reported to be important in binding (Hancock *et al.*, 2002).

12.4.1. *Residues involved in binding and stabilizing in protein–carbohydrate complexes*

Shanmugam *et al.* (2018) analyzed a set of protein–carbohydrate complexes to explore the residues, which are involved in both binding and stabilizing the complex. Stabilizing residues (SRs) are identified using structural properties such as surrounding hydrophobicity (>16 kcal/mol), long-range order (>0.01), conservation score (>6) and belong to a stabilization center (Gromiha *et al.*, 2004), and are identified using the server, SRide (http://sride.enzim.hu/).

A residue is said to be binding if it has at least one contact with any of the heavy atoms in the carbohydrate within a cutoff distance of ≤3.5 Å. Residues that are common in both stabilizing and binding are known as key residues (KRs).

12.4.2. *Frequency of occurrence of stabilizing, binding and KRs*

The frequency of occurrence of SRs, BRs, KRs in protein–carbohydrate complexes and the results are shown in Figure 12.1. SRs are dominated by hydrophobic residues as they are buried inside the protein complex, whereas binding residues (BRs) are dominated by polar and charged residues. Further, acidic amino acids are more preferred than positively charged amino acids in the binding sites of protein–carbohydrate complexes. Specifically, Glu is preferred in protein–carbohydrate complexes but not in the protein–protein complexes. This result is supported by the study of Brosnan (2000), which showed the importance of Glu at the interface of carbohydrate metabolism. The normalized occurrence of each amino acid residue at the binding sites with their respective occurrence in all protein–carbohydrate complexes showed that the residues, Trp, Tyr and His are preferred at the binding sites, emphasizing the importance of aromatic interactions between aromatic residues and sugar rings of carbohydrates as reported in the literature (Chen *et al.*, 2013, Hudson

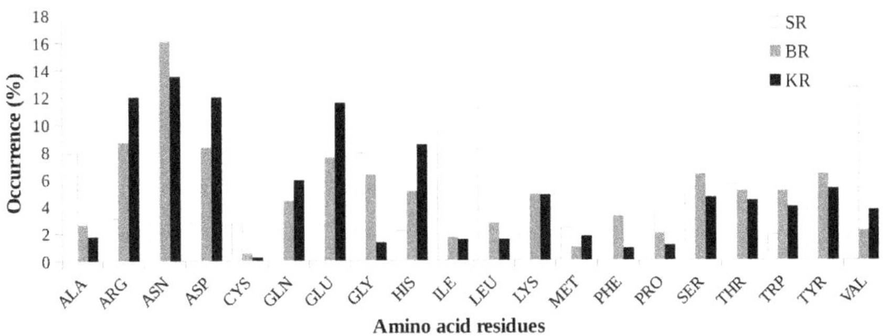

Figure 12.1. Frequency of occurrence of 20 amino acid residues in stabilizing, binding and KRs (Shanmugam *et al.* 2018).

et al., 2015). KRs are enriched with polar and charged residues, which include Asn, Arg, Asp, Glu, His, Lys and Gln.

12.5. Binding affinity

Binding affinity is an important factor for drug design and application-oriented research (Kairys *et al.*, 2019; Parenti and Rastelli, 2012). The binding affinity of protein–carbohydrate complexes are formed by noncovalent interactions between a protein and a ligand at the interface and are mainly influenced with hydrogen bonds, van der Waals, aromatic and electrostatic interactions. Molecular interactions are quantified with dissociation constant (K_d), association constant (K_a) and binding free energy (ΔG_{bind}). Typically, protein–carbohydrate interactions have low affinity, with dissociation constants in the range of mM to μM (Kapoor *et al.*, 2003; Nakamura-Tsuruta *et al.*, 2007) and are due to the size of carbohydrate-binding site of the protein. Multivalent binding is typically involved to compensate for the intrinsically weak nature of protein–carbohydrate interactions, where proteins use multiple binding sites to hold many carbohydrates that are linked together. The spatial orientation of carbohydrate recognition domains (CRDs) in multimeric lectins plays a central role in governing their binding affinity and specificity. Such information is also important for designing potent multivalent inhibitors and it requires high-resolution lectin crystal structures. Further, intrinsic conformational dynamics of proteins has been suggested to play crucial roles in ligand binding and dissociation. Seo *et al.* (2014) revealed the interplay between the conformational change and binding affinity in maltose-binding protein (MBP)—maltose complex.

12.6. Prediction of binding site residues

Several computational methods have been developed to predict carbohydrate-binding proteins and their binding sites, and are classified into two categories based on sequence and structure (Zhao *et al.*, 2018). The main strength of the sequence-based approaches for binding site prediction is the ability of determining carbohydrate-binding motifs in proteins, which may not have the same fold. Sequence-based methods utilize physiochemical properties of amino acids, preferred residues in binding, conservation score through multiple sequence alignment, PSSM and evolutionary information for prediction (Panchenko *et al.*, 2004; Malik *et al.*, 2010; Huang *et al.*, 2012).

Malik and Ahmad (2007) developed a neural network-based method, CBS-Pred, using amino acid features for predicting the binding sites. It predicts carbohydrate-binding sites in proteins, both from protein single sequences directly and using position-specific scoring matrices. PreMieR is a similarity-based approach for predicting mannose-interacting residues in proteins using composition and PSSM profiles of patterns (Agarwal *et al.*, 2011). GlycoPP and GlycoEP utilize PSSM along with average surface accessibility for prediction of N- and O-glycosylation sites (Chauhan *et al.*, 2012; Chauhan *et al.*, 2013). MOWGLI is an ensemble-based approach for predicting mannose-interacting residues (Pai and Mondal, 2016). Sugar-Binding Residue Predictor is a tool using support vector machines for predicting acid sugar binding, nonacidic sugar binding and their combinations (Banno *et al.*, 2017). SPRINT-Gly predicts *N*- and mucin-type *O*-linked glycosylation sites in mammalian glycoproteins using deep learning neural networks and SVM classifiers (Taherzadeh *et al.*, 2019). SPRINT-CBH is a *de novo* approach that predicts BRs directly from protein sequences by using sequence information and predicted structural properties (Taherzadeh *et al.*, 2016).

Structure-based methods are exploited through interactions type, contact potentials, accessible surface area, atomic density and so on (Shionyu-Mitsuyama *et al.*, 2003; Nassif *et al.*, 2009). These are classified into geometry- and energetic-based approaches in which geometry-based approaches identify BRs by searching for pockets/cavities in a protein structure, whereas energy-based approaches identify BRs by using various interaction energies (Dukka, 2013). Further, models have been proposed to predict the binding sites by using both sequence and structural properties (Taroni *et al.*, 2000). SPOT-CBP (http://sparks-lab.org/yueyang/server/SPOT-CBP/) identifies carbohydrate-recognizing proteins and their binding amino acid residues using structural alignment program SPalign and binding affinity scoring, obtained with knowledge-based statistical potentials based on distance-scaled finite-ideal gas reference state (DFIRE) (Zhao *et al.*, 2014). InCa-SiteFinder uses the van der Waals energy of a protein–probe interaction and amino acid propensities to locate carbohydrate-binding sites (Kulharia *et al.*, 2009). IMCBLab_PCA predicts noncovalent carbohydrate-binding sites based on 3D probability density maps describing the distributions of 36 noncovalent interacting atom types around protein surfaces (Tsai *et al.*, 2012). Taroni *et al.* (2000) used sequence and structural properties such as solvation potential, residue propensity, hydrophobicity, planarity, protrusion and relative accessible surface area to predict carbohydrate-binding sites. The available carbohydrate-binding site prediction servers are listed in Table 12.2.

12.7. Molecular dynamics simulations and recognition mechanism

MD simulations is a computational technique to investigate the physical movement of atoms and molecules based on the classical and quantum mechanics (Rapaport, 2004). The conformation of carbohydrates, protein–carbohydrate interactions, protein folding and protein stability are investigated through MD simulations. It provides vital information on dynamics of protein–carbohydrate interactions at atomistic level (Liu *et al.*, 2003). The forces and potential energies between the protein and carbohydrate are calculated using molecular mechanics force fields.

In MD simulations, protein structures are optimized using the force fields CHARMM36 (Best *et al.*, 2012) and OPLS (Harder *et al.*, 2016). In addition, force fields for lipids (Lyubartsev and Rabinovich, 2016) and nucleic acids (Hart *et al.*, 2012; Denning *et al.*, 2011; Soares *et al.*, 2005) have been developed. Further, carbohydrate parametrization in OPLS–AA–SEI force field for α-D-glucose has been carried out and the conformations of α-D-glucose is shown to be accurate compared with experimental results (Kony *et al.*, 2002). A significant upgrade has taken place for a tremendous

Table 12.2.   List of carbohydrate-binding site prediction servers.

Server/ Software	Type of binding sites	Link	References
Sequence Based			
CBS-Pred	Carbohydrate	http://sciwhylab.jnu.ac.in/shandar/servers/cbs-pred/index.html	Malik and Ahmad, (2007)
PreMieR	Mannose	http://crdd.osdd.net/raghava/premier/index.php	Agarwal *et al.* (2011)
GlycoPP	Glycosites	http://crdd.osdd.net/raghava/glycopp/	Chauhan *et al.* (2012)
GlycoEP	Glycosites	https://webs.iiitd.edu.in/raghava/glycoep/submit.html	Chauhan *et al.* (2013)
MOWGLI	Mannose	https://sites.google.com/site/sukantamondal/software	Pai and Mondal. (2016)
SBRP	Acidic and nonacidic sugars	https://zenodo.org/record/61513#.XRnJ_SYek5k	Banno *et al.* (2017)
SPRINT-Gly	N- and O-linked glycosylation	http://sparks-lab.org/server/SPRINT-Gly/	Taherzadeh *et al.* (2019)
SPRINT-CBH	Carbohydrate	http://sparks-lab.org/server/SPRINT-CBH/	Taherzadeh *et al.* (2016)
Structure Based			
InCa-SiteFinder	Inositol and carbohydrate	http://www.modelling.leeds.ac.uk/InCaSiteFinder/	Kulharia *et al.* (2009)
ISMBLab_PCA	Carbohydrate	http://ismblab.genomics.sinica.edu.tw/predict.php?pred=PCA	Tsai *et al.* (2012)

set of carbohydrate parameter types through the current development of GLYCAM (Kirschner *et al.*, 2008) force field in AMBER. Generalized AMBER Force Field (GAFF) is also having an updated parameter data set for carbohydrates. In all the MD simulations packages, there is an automatic tool for creating topologies of the molecules. The molecular topology files for carbohydrates and glycoproteins are created by GLYCAM Web http://glycam.org) and CHARMM-GUI for PDB structure files with glycans through Glycan Reader (Jo *et al.*, 2008; 2011). The tool, doGlycans (Danne *et al.*, 2017), generates topology files for glycosylated proteins, glycolipids and carbohydrate polymers with GROMACS package. However, accurate force field for carbohydrates is still lacking.

12.7.1. *Conformational analysis of monosaccharides*

Qian *et al.* (2005) proposed *ab initio* MD simulations based on Car–Parrinello (CPMD) approach (Car and Parrinello, 1985). They kept β-D-glucose or β-D-xylose in a unit cell, surrounded by 32 water molecules to understand the degradation mechanisms, which strongly depend on water molecules in acidic media.

Chen *et al.* (2012) analyzed the dynamics of β-D-glucopyranose with imidazole in an aqueous environment using CHARMM force field (Guvench *et al.*, 2008) for glucose, and CHARMM27 (Vanommeslaeghe *et al.*, 2010) for imidazole. They observed that glucose has high tendency to associate with imidazole through face-to-face stacking and intermolecular hydrogen bonding between the hydroxyl groups of glucose and the N3 atom of imidazole stabilize the structure.

Blessy and Sharmila (2015) carried out MD simulations for cholera toxin (CT)-Neu5Gc complex and reported that the residues His13, Trp88, Asn90, Lys91, Gln61 and Gln56 show stable intermolecular hydrogen bond interactions with Neu5Gc. One additional water bridge was formed between Gln56- water and oxygen atom (O) of Neu5Gc in the complex during the simulation. The binding orientation of Neu5Gc favorably fits in the binding pocket of CT.

Kim and Cho (2016) demonstrated the binding specificity of GGBP (glucose/galactose-binding protein) with β-D-glucose and β-D-galactose using MD simulations. They reported that water-mediated hydrogen bond plays a vital role in GGBP-glucose/galactose complexes. Glucose showed more favorable binding energies with GGBP than galactose, which is attributed to the conformational change of Asp14 in galactose–GGBP complex.

Romero *et al.* (2016) analyzed MD simulations for dimer Galectin-1 (Gal-1)-lactose in bound and free states using AMBER14 package (Case *et al.*, 2014) with ff99SB parameters for proteins (Hornak *et al.*, 2006) and glycam_99d parameters for carbohydrates (Kirschner *et al.*, 2008). From the dynamics they concluded that lactose-binding affects the dimer formation. Lactose bound Gal-1 subunits acquire larger conformational freedom upon dimer dissociation than free subunits, and this conformational change was centered on the opposite face (F-face) to the ligand (Sugar-face). They suggested new insights into the detailed understanding of galectin sugar binding and dimerization processes.

Veluraja and Margulis (2005) investigated the MD simulations for sialyl Lewis^x (SLe^x) and selectinE-SLe^x complex. From the selectinE-SLe^x complex, they predicted two possible binding modes in an aqueous environment. The direct and water-mediated hydrogen bonding interactions stabilized the two binding modes of selectinE-SLe^x complex. They observed large configurational flexibility in Lys99 and Lys111, which may play a significant role in the binding of SLe^x by selectinE. The binding site residue Lys111 showed a large conformational fluctuation and due to the flexibility, it forms a direct hydrogen bond with N-acetamido group of GlcNAc and O2 hydroxyl of Fuc. This finding may be helpful to understand the binding specificity of selectinE-SLe^x complex in an aqueous environment.

Jeyaram *et al.* (2019) studied the conformation of H5 of H5N1 influenza A complexed with SIA and fluorinated SIA. The binding free energy (MM-PBSA) and normalized pair interaction energy (NAMD pair interaction) between the SIA–N-acetylneuraminic acid and fluorinated SIA complexed with H5 showed better binding efficacy for fluorinated SIA. The conformational results showed that the SIA and fluorinated SIA complexes are stabilized by intermolecular hydrogen bonding and water-mediated hydrogen bonding. They concluded that the fluorinated SIA may act as an inhibitor for antiviral drugs against H5 of H5N1 influenza A virus.

12.7.2. *Conformational structures for disaccharides and oligosaccharides*

The carbohydrates are the combination of the monomers such as glucose (Glc), galactose (Gal) and N-acetyl-glucosamine (GlcNAc). Lactose residues present at the reducing end of the monomers and SIA and fucose exist at the nonreducing end (Hester *et al.*, 2013; Morrow *et al.*, 2005; Sosa *et al.*, 2002). SIA is the most significant cell surface carbohydrate found at the terminal position of N-glycans, O-glycans in glycoproteins and glycolipids. SIA exists in diverse forms that arise from the modifications of C-5 position in N-acetylneuraminic acid and additional substitutions at the hydroxyl group on the C-4, C-7, C-8 and C-9, and the other diversified forms are generated by the glycosidic linkages at the C-2 position to the penultimate sugar. In oligosaccharides, the most commonly found terminal sialyldisaccharide forms are Neu5Acα(2–3)Gal (SIA23GAL) and Neu5Acα(2–6)Gal (SIA26GAL) (Selvin *et al.*, 2012). The typical structures of disaccharides are shown in Figure 12.2. The diversified conformational structure arises due to the freedom of rotation associated with the glycosidic tortional angles.

The glycosidic torsional angle propensity map was computed from the MD simulations trajectories. The preferred glycosidic torsional angles ($\phi_{SA23GAL}$, $\psi_{SA23GAL}$) of SIA23GAL are (–100°, –50°), (–70°, 0°) and (–150°, –30°), and the respective conformational structures are shown in Figure 12.3. These structures are stabilized by direct and water-mediated hydrogen bonding interactions between the sugar molecules (Selvin *et al.*, 2012).

Oligosaccharides are indigestible by human gut (Bode, 2012; Turin and Ochoa, 2014) and are not used as a nutrient: it assists non-nutritional factors such as development of immune system and inhibition of pathogens in infants (Newburg and Neubauer, 1995). Hence, the investigations are focused on the structural aspects of human milk

(A)

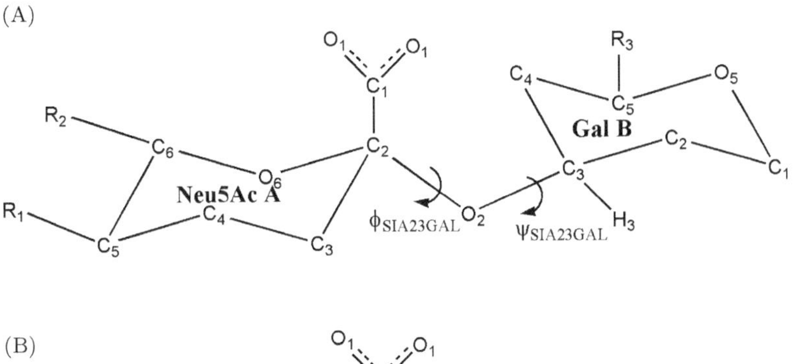

(B)

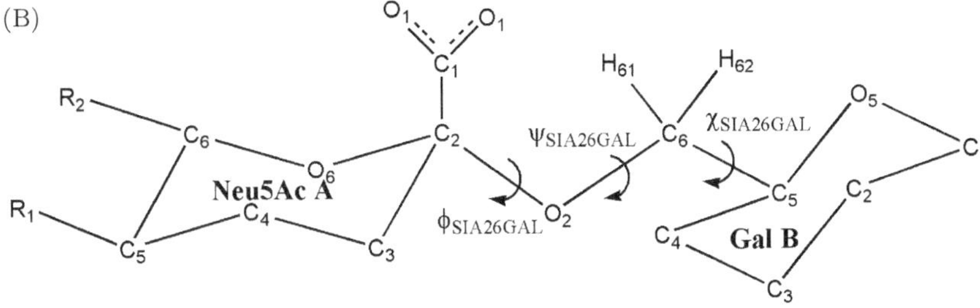

Figure 12.2. Schematic representation of typical disaccharide structure (Selvin *et al.* 2012). (A) Neu5Acα(2–3)Gal ($\Phi_{SIA23GAL}$: C1-C2-O2-C3; $\Psi_{SIA23GAL}$:C2-O2-C3-H3) and (B) Neu5Acα(2–6)Gal ($\Phi_{SIA26GAL}$: C1-C2-O2-C6; $\Psi_{SIA26GAL}$: C2-O2-C6-H61; $\omega_{SIA26GAL}$: O2-C6-C5-O5). H61 hydrogen makes an angle of 120° (H61-C6-C5-O5) when $\omega_{SIA26GAL}$ = 0°. R_1 = Acetamido group, R_2 = Glycerol side chain and R_3 = CH$_2$OH.

oligosaccharides (HMOs) and their interactions with toxins. The protocol for deducing 3D structures of oligosaccharides has been explained by taking one of the HMO-containing SIAs (Sialylα(2–6) lactose (SLac) or Neu5Acα(2–6)Galβ(1–4)Glc).

The structural conformation of SLac is studied through MD simulations in aqueous environment using AMBER9 software through GAFF. This HMO has two major conformational models, $(\phi_n\psi_n)_{n=1,2}$: conformer 1 [(−70°,70°) and (60°, 0°)] and conformer 2 [(−150°,70°) and (60°, 0°)] and the 3D models are shown in Figures 12.4(A) and 12.4(B), respectively. This structure was stabilized by the direct hydrogen bonding between galactose and glucose molecules.

In the field of glycobiology, 3DSDSCAR (3D structural databases for a SIA-containing carbohydrates) is a unique conformational database for carbohydrates and the proposed 3D structures are obtained

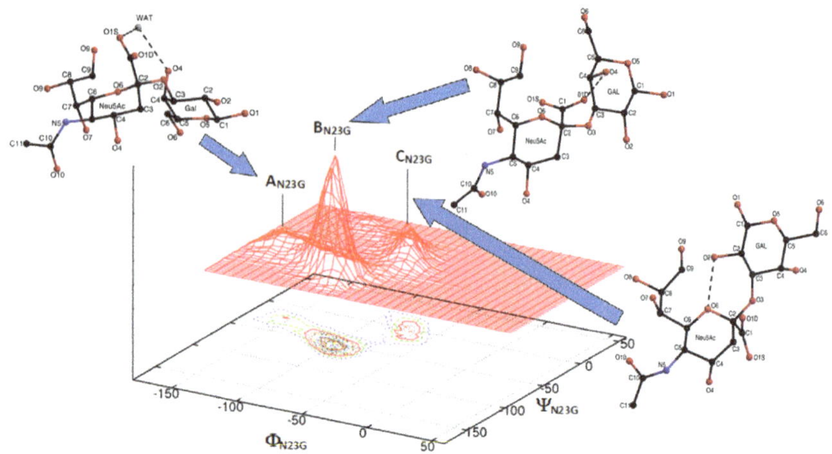

Figure 12.3.   The glycosidic torsional propensity map of Neu5Ac(2–3)Gal and its corresponding conformational models.

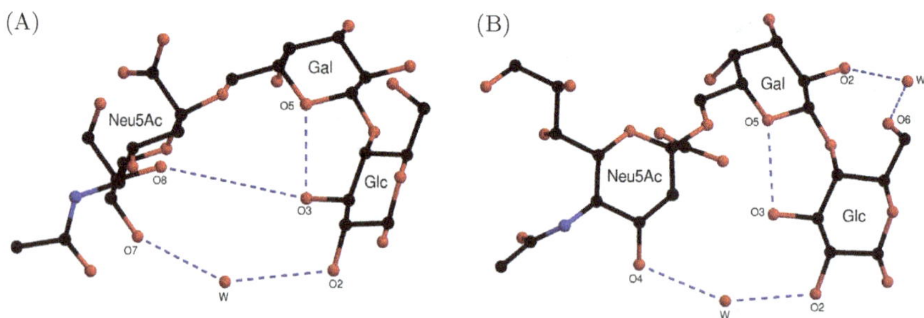

Figure 12.4.   The conformational model of Sialylα(2–6) lactose. (A) Conformer 1 and (B) Conformer 2.

through MD simulations. The structural coordinates are in the downloadable PDB format (www.3dsdscar.org) (Veluraja *et al.*, 2010). At present, this database contains a total of **60** conformational models.

12.7.3. *Carbohydrate binding with toxins and viruses*

The major receptor sites for toxins (CT, heat labile enterotoxin, botulinum neurotoxin, pertussis toxin) and viruses (influenza virus, norovirus, rotavirus and simian virus) are cell surface carbohydrates.

Influenza viruses are classified into influenza A, B, C and D. Influenza A virus is more virulent than other types and has the ability to infect various animals and birds such as pig, horse, duck, whale and humans (Ferguson *et al.*, 2016; Foni *et al.*, 2017; Jeyaram *et al.*, 2019; Blessy *et al.*, 2019; Asha and Kumar, 2019). The viral infection initiated through hemagglutinin (HA) and neuraminidases (NA) of viral protein by recognizing the sialyl disaccharide as receptor cell (Durrant *et al.*, 2016; Sharma *et al.*, 2016; Phanich *et al.*, 2018). Influenza A virus caused few remarkable pandemics in past and present centuries; currently H1N1 and H5N1 influenza viruses are emergent threats to public health issues (Singh and Soliman, 2015). Worldwide only two NA inhibitors (oseltamivir and zanamivir) and M2 ion channel protein inhibitors (amantadine and rimantadine) are approved to treat influenza infection, but the mutations of influenza create resistant to available vaccines (Han and Mu, 2013; Guan *et al.*, 2017; Breschkin, 2013). Hence, the development of therapeutics is necessary for prevention and vaccination against influenza.

One of the possibilities of controlling influenza viral infection is blocking the SIA-binding site of HA or NA by SIA analog compounds. HA is responsible for viral entry and membrane fusion. If HA is blocked at its SIA binding site by small molecules, the viral entry process will be stopped and the penetration of virus into the host cell can be prevented. Hence, HA is an important target to design influenza viral drugs to prevent its infection (Alberto *et al.*, 2013). In addition, targeting the NA protein will block the release of progeny virus on the host cell and prevents the new host cell from the infection. The interactions between protein and carbohydrate are a prior condition to initiate the biochemical reactions or biological processes and the carbohydrate-binding protein is an important target for designing antiviral drugs or inhibitors. The cost-effective computational methods are the good alternatives to the experimental methods and there is a possibility that it may overcome the difficulties associated with experiments. Computational methods are intensively used to study the conformational aspects of carbohydrates and protein—carbohydrate interactions (Yokoyama *et al.*, 2017). Molecular modeling, MD simulations and quantum mechanical calculations are some of the important computational methods that can be used to design inhibitors for protein molecules (Murugan *et al.*, 2015; Shu *et al.*, 2011).

Priyadarzini *et al.* (2012) studied the binding modes of sialyl disaccharides at the binding pockets of HAs (H1, H3, H5 and H9) through MD simulations. Based on the pair interaction energy, binding free energy and hydrogen bonding analyses they have concluded that the order of binding specificity of Neu5Acα(2–3)Gal is H3 > H5 > H9 > H1 and Neu5Acα(2–6)Gal is H1 > H3 > H5 > H9. Structural stabilization of H3—Neu5Acα(2–3)Gal and H1—Neu5Acα(2–6)Gal is shown in Figure 12.5.

Murugan *et al.* (2017) studied the binding specificity of fluorine-substituted Neu5Acα(2–3)Gal and Neu5Acα(2–6)Gal with H1 of H1N1 influenza A virus through MD simulations and reported that the fluorinated Neu5Acα(2–6)Gal has better binding affinity than fluorinated Neu5Acα(2–3)Gal. The MD simulations revealed that the modification of sialyldisaccharides (N26G and FN26G) could be a promising inhibitor for HA influenza (Priyadarzini *et al.*, 2012; Murugan *et al.*, 2017). Recently, Jeyaram *et al.* (2019) studied the effect of fluorine substitution on different hydroxyls (O2, O4, O7, O8 and O9) of SIA complexed with the H5 HA of H5N1 influenza A virus and showed that the order of binding specificity against H5 HA is SIA-F9 >> SIA-F2 > SIA-F7 > SIA-F4 ≈ SIA-F8 ≈ SIA.

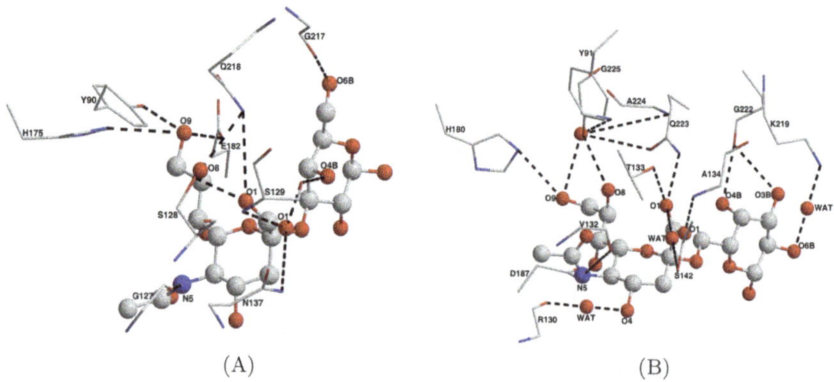

Figure 12.5. Structural stabilization of (A) H3—Neu5Acα(2–3)Gal and (B) H1—Neu5Acα(2–6)Gal (Priyadarzini *et al.* 2012).

Behera *et al.* (2015) studied the binding mode and stability of Neu5Acα(2–3)Gal and Neu5Acα(2–6)Gal complexed with H5 HA1 of H5N1 influenza A virus through MD simulations and they analyzed differential interactions with both wild-type and mutated HA of influenza virus. In comparison with the wild-type influenza, a single mutated HA (S221P) complexed with Neu5Acα(2–3)Gal has less binding affinity and Neu5Acα(2–6)Gal has comparable binding affinity to wild-type. The double mutated structure (S221P, K216E) has enhanced binding affinity toward human receptor in comparison with avian receptor.

Veluraja and Rao (1983, 1984) studied the conformation of monosialogangliosides (GM3, GM2 and GM1), disialogangliosides (GD1A, GD1B and GD3) and higher gangliosides (GT1A and GT1B) and calculated potential energies using semiempirical potential functions. The analysis showed that the binding of monosialogangliosides and disialogangliosides in the binding pockets of CT and tetanus toxin (TT) as well as the binding of higher gangliosides in the binding pocket of TT and sendai virus could explain the biological activity of gangliosides. Later, the different binding specificities of monosialogangliosides (Sharmila and Veluraja, 2004a), disialogangliosides (Sharmila and Veluraja, 2004b) and higher gangliosides (Sharmila and Veluraja, 2006) in the binding site of CT B subunit have been reported using molecular mechanics and MD simulations.

Sharmila and Jino Blessy (2015, 2017) analyzed the conformational behavior of SIA analogues–CT complex in the aqueous environment. SIA analogues showed strong interactions toward the binding pocket of CT through intermolecular hydrogen bond and water-mediated hydrogen bond interactions.

Manna and Mukhopadhyay (2013) studied the atomistic properties of amyloid-β (Aβ)-GM1 (ganglioside) and Aβ–Aβ interactions and the effectiveness of GM1-mediated Aβ oligomerization on the membrane surface using GROMOS87 force field. Shahzadi *et al.* (2018) studied the phase behavior of GM1-containing DMPC–cholesterol monolayer by air–water and air–solid interfaces using Langmuir–Blodgett experiments and scanning electron microscopy

(SEM), respectively. They showed that at liquid expanded–liquid condensed (LE/LC) equilibrium, the GM1 bind cooperatively to the monolayer and MD simulations provided atomistic-level explanations of their experimental results.

Parasuraman *et al.* (2014, 2015) studied the binding specificity of wheat germ agglutinin and Agrocybe cylindracea galectin toward sialylglycans through MD simulations. In comparison with wild-type lectins the double mutation Y59R, N140Q showed the highest binding affinity toward $\alpha(2,3)$–linked sialyldisaccharides. Hence, MD simulations of protein–carbohydrate interactions have a vital role in understanding their functions and lead to the therapeutic design.

12.8. Applications

Carbohydrates are having wide range of applications in biological recognition processes. The adhesion of carbohydrates is an essential property to bind with proteins. The foremost applications of studying protein–carbohydrate interactions are (1) designing of inhibitors to avoid carbohydrate-mediated pathogenic interaction and (2) carbohydrate microarray for cell detection and separation. Further, the recognition process of gangliosides studied with CT, binding sites and HA viruses will be able to recognize the host SIA. Specifically, some of the carbohydrates (e.g., SIA) behave like "biological masks," which prevent the recognition of underlying residues (Varki, 2017).

12.9. Conclusion

Carbohydrates are the major component in biological functions and recognition processes. To understand the structure and conformational models of carbohydrates, the databases are vital. Hence, we discussed different types of carbohydrates and protein–carbohydrate complex databases in this review elaborately. The analysis of binding site residues in protein–carbohydrate complexes and the servers used to predict the carbohydrate-binding sites are detailed out. The 3D structure and conformational analysis of carbohydrates, protein–carbohydrate interactions, stability of proteins and carbohydrates, and their binding affinities are investigated through MD simulations.

Acknowledgments

We thank Department of Biotechnology (DBT), New Delhi, Government of India for partial financial support to KV and MMG (Sanction order No.: BT/PR13410/BID/7/536/2015 and BT/PR12267/BID/7/506/2014). NRS thanks Ministry of Human Resource and Development, India and DST-INSPIRE for fellowship (IF170342).

References

Agarwal, S., Mishra, N.K., Singh, H., *et al*. Identification of mannose interacting residues using local composition. *PLOS ONE*. 2011;e24039.

Agravat, S.B., Saltz, J.H., Cummings, R.D., *et al*. GlycoPattern: A web platform for glycan array mining. *Bioinformatics*. 2014;30:3417–3418.

Ahmed, J., Preissner, S., Dunkel, M., *et al*. SuperSweet — a resource on natural and artificial sweetening agents. *Nucleic Acids Res*. 2010;39: D377–D382.

Alberto, A.J., Facundo, E.A., Aparicio R.M.R., *et al*. Analysis of adaptation mutants in the hemagglutinin of the influenza A (H1N1) pdm09 virus. *PLOS ONE*. 2013;8:e70005.

Asha, K., and Kumar, B. Emerging influenza D virus threat: What we know so far!. *J Clin Med*. 2019;8:192.

Banno, M., Komiyama, Y., Cao, W., *et al*. Development of a sugar-binding residue prediction system from protein sequences using support vector machine. *Comput Biol Chem*. 2017;66:36–43.

Behera, A.K., Chandra, I., and Cherian, S.S. Molecular dynamics simulations of the effects of single (S221P) and double (S221P and K216E) mutations in the hemagglutinin protein of influenza A H5N1 virus: A study on host receptor specificity. *J Biomol Struct Dyn*. 2015;34:2054–2067.

Best, R.B., Zhu, X., Shim, J., *et al*. Optimization of the additive CHARMM all-atom protein force field targeting improved sampling of the backbone phi, psi and side-chain chi(1) and chi(2) dihedral angles. *J. Chem Theory Comput*. 2012;8:3257–3273.

Blessy, J.J., Jawahar, D., and Sharmila, D.J.S. Multivalent interactions of nano-spaced dimers of N-acetylneuraminic acid analogues complex with H5N1 influenza viral neuraminidase and haemagglutinin— A molecular dynamics investigation. *Int J Curr Microbiol App Sci.* 2019;8:1517–1546.

Blessy, J.J., and Sharmila, D.J.S. Molecular simulation of N-acetylneuraminic acid analogs and molecular dynamics studies of cholera toxin-Neu5Gc complex. *J Biomol Struct Dyn.* 2015;33:1126–1139.

Bode, L. Human milk oligosaccharides: Every baby needs a sugar mama. *Glycobiology.* 2012;22:1147–1162.

Böhm, M., Bohne-Lang, A., Frank, M., *et al.* Glycosciences. DB: An annotated data collection linking glycomics and proteomics data. *Nucleic Acids Res.* 2018;47:D1195–D1201.

Bonnardel, F., Mariethoz, J., Salentin, S., *et al.* UniLectin3D, a database of carbohydrate binding proteins with curated information on 3D structures and interacting ligands. *Nucleic Acids Res.* 2018;47: D1236–D1244.

Breschkin, J.L.M. Influenza neuraminidase inhibitors: Antiviral action and mechanisms of resistance. *Influenza Other Respir Viruses.* 2013;7: 25–36.

Brosnan, J.T. Glutamate and glutamine, at the interface between amino acid and carbohydrate metabolism." *J. Nutr.* 2000;130:988–990.

Burley, S.K., Berman, H.M., Bhikadiya, C., *et al.* Protein Data Bank: The single global archive for 3D macromolecular structure data. *Nucleic Acids Res.* 2018;47:D520–D528.

Campbell, M.P., Nguyen-Khuong, T., Hayes, C.A., *et al.* Validation of the curation pipeline of UniCarb DB: Building a global glycan reference MS/MS repository. *Biochim Biophys Acta.* 2014;1844:108–116.

Cantarel, B.L., Coutinho, P.M., Rancurel, C., *et al.* The Carbohydrate-Active EnZymes database (CAZy): An expert resource for glycogenomics. *Nucleic Acids Res.* 2008;37:D233–D238.

Car, R., and Parrinello, M. Unified approach for molecular dynamics and density-functional theory. *Phys Rev Lett.* 1985;55:2471.

Case, D.A., Babin, V., Berryman, J., *et al.* AMBER 14. San Francisco, CA: University of California; 2014.

Chandra, N.R., Kumar, N., Jeyakani, J., *et al.* Lectindb: A plant lectin database. *Glycobiology.* 2006;16:938–946.

Chauhan, J.S., Bhat, A.H., Raghava, G.P., *et al.* GlycoPP: A webserver for prediction of N-and O-glycosites in prokaryotic protein sequences. *PLOS ONE.* 2012;7:e40155.

Chauhan, J.S., Rao, A., and Raghava, G.P. In silico platform for prediction of N-, O- and C-glycosites in eukaryotic protein sequences. *PLOS ONE.* 2013;8:e67008.

Chen, M., Bomble, Y.J., Himmel, M.E., *et al.* Molecular dynamics simulations of the interaction of glucose with imidazole in aqueous solution. *Carbohydr Res.* 2012;349:73–77.

Chen, W., Enck, S., Price, J.L., *et al.* Structural and energetic basis of carbohydrate-aromatic packing interactions in proteins. *J Am Chem Soc.* 2013;135:9877–9884.

Cooper, C.A., Harrison, M.J., Wilkins, M.R., *et al.* GlycoSuiteDB: A new curated relational database of glycoprotein glycan structures and their biological sources. *Nucleic Acids Res.* 2001;29:332–335.

Danne, R., Poojari, C., Martinez-Seara, H., *et al.* doGlycans–Tools for preparing carbohydrate structures for atomistic simulations of glycoproteins, glycolipids, and carbohydrate polymers for GROMACS. *J Chem Inf Model.* 2017;57:2401–2406.

Denning, E.J., Priyakumar, U.D., Nilsson, L., *et al.* Impact of 2´-hydroxyl sampling on the conformational properties of RNA: Update of the CHARMM all-atom additive force field for RNA. *J Comput Chem.* 2011;32:1929–1943.

Dukka, B.K. Structure-based methods for computational protein functional site prediction. *Comput Struct Biotechnol J.* 2013;8:e201308005.

Durrant, J.D., Bush, R.M., and Amaro, R.E. Microsecond molecular dynamics simulations of influenza neuraminidase suggest a mechanism for the increased virulence of stalk-deletion mutants. *J Phys Chem B.* 2016;120:8590–8599.

Ferguson, L., Olivier, A.K., Genova, S., *et al.* Pathogenesis of influenza D virus in cattle. *J Virol.* 2016;90:5636–5642.

Fernández-Alonso, M., Díaz, D., Alvaro Berbis, M., *et al.* Protein-carbohydrate interactions studied by NMR: From molecular recognition to drug design. *Curr Protein Pept Sci.* 2012;13:816–830.

Foni, E., Chiapponi, C., Baioni, L., *et al.* Influenza D in Italy: Towards a better understanding of an emerging viral infection in swine. *Sci Rep.* 2017;7:11660.

Gromiha, M.M., Pujadas, G., Magyar, C., *et al.* Locating the stabilizing residues in (α/β)8 barrel proteins based on hydrophobicity, long-range interactions, and sequence conservation. *Proteins.* 2004;55(2), 316–329.

Gromiha, M.M., Veluraja, K. and Fukui, K. Identification and analysis of binding site residues in proteincarbohydrate complexes using energy based approach. *Protein Peptide Lett.* 2014;21:799–807.

Guan, S., Wang, T., Kuai, Z., *et al.* Exploration of binding and inhibition mechanism of a small molecule inhibitor of influenza virus H1N1 hemagglutinin by molecular dynamics simulations. *Sci Rep.* 2017;7:3786.

Gupta, R., Birch, H., Rapacki, K., *et al.* O-GLYCBASE version 4.0: A revised database of O-glycosylated proteins. *Nucleic Acids Res.* 1999;27: 370–372.

Guvench, O., Greene, S.N., Kamath, G., *et al.* Additive empirical force field for hexopyranose monosaccharide. *J Comput. Chem.* 2008;29: 2543–2564.

Han, N., and Mu Y. Plasticity of 150-loop in influenza neuraminidase explored by Hamiltonian replica exchange molecular dynamics simulations. *PLOS ONE.* 2013;8:e60995.

Hancock, M.K., Haskins, D.J., Su, G., *et al.* Identification of residues essential for carbohydrate recognition by the insulin-like growth factor II/mannose 6-phosphate receptor. *J Biol Chem.* 2002;277:11255–11264.

Harder, E., Damm, W., Maple, J., *et al.* OPLS3: A force field providing broad coverage of drug-like small molecules and proteins. *J Chem Theory Comput.* 2016;12:281–296.

Hart, K., Foloppe, N., Baker, C.M., *et al.* 2012 Optimization of the CHARMM additive force field for DNA: Improved treatment of the BI/BII conformational equilibrium. *J Chem Theory Comput.* 2012;8:348–362.

Hester, S.N., Chen, X., Li, M., *et al.* Human milk oligosaccharides inhibit rotavirus infectivity in vitro and in acutely infected piglets. *Br. J. Nutr.* 2013;110:1233–1242.

Hornak, V., Abel, R., Okur, A., *et al.* Comparison of multiple amber force fields and development of improved protein backbone parameters. *Proteins.* 2006;65:712–725.

Huang, H.L., Lee, H.C., Liou, Y.F., *et al.* Designing predictors of carbohydrate-binding proteins using informative physicochemical properties. *Proc 4th Int Conf Bioinformat Biomed Technol.* 2012;29: 155–160.

Hudson, K.L., Bartlett, G.J., Diehl, R.C., *et al.* Carbohydrate–aromatic interactions in proteins. *J Am Chem Soc.* 2015;137:15152–15160.

Jeyaram, R.A., Priyadarzini, T.R.K., Radha, C.A., *et al.* Molecular dynamics simulations studies on Influenza A virus H5N1 complexed with sialic acid and fluorinated sialic acid. *J Biomol Struct Dyn.* 2019 (in press).

Jo, S., Kim, T., Iyer, V.G., *et al.* Software news and updates-CHARMM-GUI: A web-based graphical user interface for CHARMM. *J Comput Chem.* 2008;29:1859–1865.

Jo, S., Song, K.C., Desaire, H., *et al.* Glycan reader: Automated sugar identification and simulation preparation for carbohydrates and glycoproteins. *J Comput Chem.* 2011;32:3135–3141.

Kairys, V., Baranauskiene, L., Kazlauskiene, M., *et al.* Binding affinity in drug design: Experimental and computational techniques. *Expert Opin Drug Discov.* 2019;1–14.

Kanie, Y., and Kanie, O. Addressing the glycan complexity by using mass spectrometry: In the pursuit of decoding glycologic. *Bio Chem Comp.* 2017;5:3.

Kapoor, M., Thomas, C. J., Bachhawat-Sikder, K., *et al.* Exploring kinetics and mechanism of protein–sugar recognition by surface plasmon resonance. *Method Enzymol.* 2003;362:312–329.

Kasai, K. Frontal affinity chromatography: A unique research tool for biospecific interaction that promotes glycobiology. *Proc Jpn Acad Ser B.* 2014; 90:215–234.

Kim, M., and Cho, A.E. The role of water molecules in stereoselectivity of glucose/galactose-binding protein. *Sci Rep.* 2016;6:36807.

Kirschner, K.N., Yongye, A.B., Tschampel, S.M., *et al.* GLYCAM06: A generalizable biomolecular force field. *J Comput Chem.* 2008;29: 622–655.

Kony, D., Damm, W., Stoll, S., *et al.* An improved OPLS-AA force field for carbohydrates. *J Comput Chem.* 2002;23:1416–1429.

Kulharia, M., Bridgett, S.J., Goody, R.S. *et al.* InCa-SiteFinder: A method for structure-based prediction of inositol and carbohydrate binding sites on proteins. *J Mol Graph Model.* 2009;28:297–303.

Kumar, K.K., Chandra, K.L.P., Sumanthi, J., *et al.* Biological role of lectins: A review. *J Orofac Sci.* 2012;4:20.

Linman, M. J., Taylor, J. D., Yu, H., Surface plasmon resonance study of protein–carbohydrate interactions using biotinylated sialosides. *Anal Chem.* 2008;80:4007–4013.

Liu, H.L., Shu, Y.C., and Wu, Y.H. Molecular dynamics simulations to determine the optimal loop length in the helix-loop-helix motif. *J Biomol Struct Dyn.* 2003;20:741–745.

Liu, T., Lin, Y., Wen, X., *et al.* BindingDB: A web-accessible database of experimentally determined protein–ligand binding affinities. *Nucleic Acids Res.* 2006;35:D198–D201.

Lyubartsev, A.P., and Rabinovich, A.L. Force field development for lipid membrane simulations. *Biochim Biophys Acta, Biomembr.* 2016;1858: 2483–2497.

Malik, A., and Ahmad, S. Sequence and structural features of carbohydrate binding in proteins and assessment of predictability using a neural network. *BMC Struct Biol.* 2007;7:1.

Malik, A., Firoz, A., Jha, V., *et al.* PROCARB: A database of known and modelled carbohydrate-binding protein structures with sequence-based prediction tools. *Adv Bioinformatics.* 2010;436036.

Manna, M., and Mukhopadhyay, C., Binding, conformational transition and dimerization of amyloid-b peptide on GM1-containing ternary membrane: Insights from molecular dynamics simulation. *PLOS ONE.* 2013; 8:e71308.

Mariethoz, J., Khatib, K., Alocci, D., *et al.* SugarBindDB, a resource of glycan-mediated host–pathogen interactions. *Nucleic Acids Res.* 2015; 44:D1243–D1250.

Morrow, A.L., Ruiz-Palacios, G.M., Jiang, X., *et al.* Human-milk glycans that inhibit pathogen binding protect breast-feeding infants against infectious diarrhea. *J Nutr.* 2005;135:1304–1307.

Murrey, H.E., and Hsieh-Wilson, L.C. The chemical neurobiology of carbohydrates. *Chem Rev.* 2008;108:1708–1731.

Murugan, V., Parasuraman, P., Selvin, J.F.A., *et al.* Geometry optimization of carbohydrate binding sites of influenza: A quantum mechanical approach. *J Carbohyd Chem.* 2015;34:409–429.

Murugan, V., Parasuraman, P., Selvin, J.F.A., *et al.* Theoretical investigation on the binding specificity of fluorinated sialyldisaccharides Neu5Acα(2-3) Gal and Neu5Acα(2-6)Gal with influenza hemagglutinin H1—A molecular dynamics study. *J Carbohydr Res.* 2017;36:111–128.

Nakamura-Tsuruta, S., Uchiyama, N., Kominami, J., *et al.* Frontal affinity chromatography: Systematization for quantitative interaction analysis between lectins and glycans. In: Nilsson, C.L., ed. *Lectins.* Amsterdam, the Netherlands: Elsevier; 2007:239–266.

Nassif, H., Al-Ali, H., Khuri, S., *et al.* An inductive logic programming approach to validate hexose binding biochemical knowledge. In *International Conference on Inductive Logic Programming.* Berlin, Heidelberg: Springer; 2009:149–165.

Newburg, D.S., and Neubauer, S.H. Carbohydrates in milks: Analysis, quantities, and significance. In: Jensen, R.G., ed. *Handbook of Milk Composition.* San Diego, CA: Academic press; 1995:273–349.

Pai, P.P., and Mondal, S.MOWGLI: Prediction of protein–MannOse interacting residues With ensemble classifiers usinG evoLutionary Information. *J Biomol Struct Dyn.* 2016;34:2069–2083.

Panchenko, A.R., Kondrashov, F., and Bryant, S. Prediction of functional sites by analysis of sequence and structure conservation. *Protein Sci.* 2004;13:884–892.

Parasuraman, P., Murugan, V., Selvin, J.F.A., *et al.* Insights into the binding specificity of wild type and mutated wheat germ agglutinin towards Neu5Acα(2-3)Gal: A study by in silico mutations and molecular dynamics simulations. *J Mol Recognit.* 2014;27:482–492.

Parasuraman, P., Murugan, V., Selvin, J.F.A., *et al.* Theoretical investigation on the glycan-binding specificity of Agrocybe cylindracea galectin using molecular modeling and molecular dynamics simulations studies. *J Mol Recognit.* 2015;28:528–538.

Parenti, M.D., and Rastelli, G. Advances and applications of binding affinity prediction methods in drug discovery. *Biotechnol Adv.* 2012;30: 244–250.

Pérez, S., Sarkar, A., Rivet, A., *et al.* Glyco3D: A portal for structural glycosciences. In *Glycoinformatics.* New York, NY: Humana Press; 241–258.

Phanich, J., Threeracheep, S., Kungwan, N., *et al.* Glycan binding and specificity of viral influenza neuraminidases by classical molecular dynamics and replica exchange molecular dynamics simulations. *J Biomol Struct Dyn.* 2018;37:3354–3365.

Priyadarzini, T.R.K., Selvin, J.F.A., Gromiha, M.M., *et al.* Theoretical investigation on the binding specificity of sialyldisaccharides with hemagglutinins of influenza A virus by molecular dynamics simulations. *J Biol Chem.* 2012;287:34547–34557.

Qian, X., Nimlos, M.R., Davis, M., *et al.* Ab initio molecular dynamics simulations of β-D-glucose and β-D-xylose degradation mechanisms in acidic aqueous solution. *Carbohydr Res.* 2005;340:2319–2327.

330 *K. Veluraja et al.*

Rapaport, D.C. *The Art of Molecular Dynamics Simulation*. New York, NY: Cambridge University Press; 2004.

Romero, J.M., Trujillo M., Estrin D.A., *et al.* Impact of human galectin-1 binding to saccharide ligands on dimer dissociation kinetics and structure. *Glycobiology.* 2016;26:1317–1327.

Selvin, J.F.A., Priyadarzini, T.R.K., and Veluraja, K. Sialyldisaccharide conformations: A molecular dynamics perspective. *J Comput Aid Mol Des.* 2012;26:375–385.

Seo, M.H., Park J., Kim E., *et al.* Protein conformational dynamics dictate the binding affinity for a ligand. *Nat Commun.* 2014;5:3724.

Shahzadi, Z., Das S., Bala T., *et al.* Phase behavior of GM1 containing DMPC-cholesterol monolayer: Experimental and theoretical study. *Langmuir.* 2018;34:11602–11611.

Shanmugam, N.R.S., Selvin J.F.A., Veluraja K., *et al.* Identification and analysis of key residues involved in folding and binding of protein–carbohydrate complexes. *Protein Pept Lett.* 2018;25:379–389.

Sharma, G., Kumar, S.V., and Wahab, H.A. Molecular docking, synthesis, and biological evaluation of naphthoquinone as potential novel scaffold for H5N1 neuraminidase inhibition. *J Biomol Struct Dyn.* 2016;36:233–242.

Sharmila, D.J.S., and Jino Blessy J., Molecular dynamics of sialic acid analogues complex with cholera toxin and DFT optimization of ethylene glycol-mediated zinc nanocluster conjugation. *J Biomol Struct Dyn.* 2017;35:182–206.

Sharmila, D.J.S., and Veluraja, K. Monosialogangliosides and their interaction with cholera toxin-investigation by molecular modeling and molecular mechanics. *J Biomol Struct Dyn.* 2004a;21:591–613.

Sharmila, D.J.S., and Veluraja, K., Disialogangliosides and their interaction with cholera toxin-investigation by molecular modeling, molecular mechanics and molecular dynamics. *J Biomol Struct Dyn.* 2004b;22:299–313.

Sharmila, D.J.S., and Veluraja, K. Conformations of higher gangliosides and their binding with cholera toxin–Investigation by molecular modeling, molecular mechanics, and molecular dynamics. *J Biomol Struct Dyn.* 2006;23:641–656.

Sharon, N. Carbohydrates as future anti-adhesion drugs for infectious diseases. *Biochim Biophys Acta.* 2006;1760:527–537.

Shionyu-Mitsuyama C., Shirai T., Ishida H., *et al.* An empirical approach for structure-based prediction of carbohydrate-binding sites on proteins. *Protein Eng.* 2003;16:467–478.

Shu, M., Lin Z., Zhang Y., *et al.* Molecular dynamics simulations of oseltamivir resistance in neuraminidase of avian influenza H5N1 virus. *J Mol Model.* 2011;17:587–592.

Singh, A., and Soliman, M.E. Understanding the cross-resistance of oseltamivir to H1N1 and H5N1 influenza A neuraminidase mutations using multidimensional computational analyses. *Drug Des Devel Ther.* 2015;9:4137–4154.

Sly, K.L., and Conboy, J.C. Determination of multivalent protein–ligand binding kinetics by second-harmonic correlation spectroscopy. *Anal Chem.* 2014;86:11045–11054.

Soares, T.A., Hunenberger P.H., Kastenholz M.A., *et al.* An improved nucleic acid parameter set for the GROMOS force field. *J Comput Chem.* 2005;26:725–737.

Sosa, S.M., Martin, M.J., and Hueso, P. The sialylated fraction of milk oligosaccharides is partially responsible for binding to enterotoxigenic and uropathogenic *Escherichia coli* human strains. *J Nutr.* 2002;132: 3067–3072.

Taherzadeh, G., Dehzangi A., Golchin M., *et al.* SPRINT-Gly: Predicting N-and O-linked glycosylation sites of human and mouse proteins by using sequence and predicted structural properties. *Bioinformatics.* 2019 (in press).

Taherzadeh, G., Zhou Y., Liew A.W.C., *et al.* Sequence-based prediction of protein–carbohydrate binding sites using support vector machines. *J Chem Inf Model.* 2016;56:2115–2122.

Taroni, C., Jones, S., and Thornton J.M. Analysis and prediction of carbohydrate binding sites. *Protein Eng.* 2000;13:89–98.

Tiemeyer, M., Aoki K., Paulson J., *et al.* GlyTouCan: An accessible glycan structure repository. *Glycobiology.* 2017;27:915–919.

Toukach, P.V., and Egorova, K.S. Carbohydrate structure database merged from bacterial, archaeal, plant and fungal parts. *Nucleic Acids Res.* 2015;44:D1229–D1236.

Tsai, K.C., Jian, J.W., Yang, E. *et al.* Prediction of carbohydrate binding sites on protein surfaces with 3-dimensional probability density distributions of interacting atoms. *PLOS ONE.* 2012;7:e40846.

Turin, C.G., and Ochoa, T.J. The role of maternal breast milk in preventing infantile diarrhea in the developing world. *Curr Trop Med Rep.* 2014;1: 97–105.

Vanommeslaeghe, K., Hatcher E., Acharya C., *et al.* CHARMM general force field: A force field for drug-like molecules compatible with the

CHARMM all-atom additive biological force fields. *J Comput Chem.* 2010;31:671–690.

Varki, A. Biological roles of glycans. *Glycobiology.* 2017;27: 3–49.

Veluraja, K., and Margulis, C.J. Conformational dynamics of sialyl lewisx in aqueous solution and its interaction with selectine. A study by molecular dynamics. *J Biomol Struct Dyn.* 2005;23:101–111.

Veluraja, K., Selvin J.F.A., Venkateshwari S., *et al.* 3DSDSCAR—a three dimensional structural database for sialic acid-containing carbohydrates through molecular dynamics simulation. *Carbohydr Res.* 2010;345: 2030–2037.

Veluraja, K., and Rao, V.S.R. Theoretical studies on the conformation of monosialogangliosides and disialogangliosides. *Carbohydr Polym.* 1983; 3:175–192.

Veluraja, K., and Rao, V.S.R. Theoretical studies on the conformations of higher gangliosides. *Carbohydr Polym.* 1984;4:357–375.

von Schantz, L., Håkansson M., Logan D. T., *et al.* Structural basis for carbohydrate-binding specificity—A comparative assessment of two engineered carbohydrate-binding modules. *Glycobiology.* 2012;22: 948–961.

Wang, R., Fang X., Lu Y., *et al.* The PDBbind database: Collection of binding affinities for protein–ligand complexes with known three-dimensional structures. *J Med Chem.* 2004;47:2977–2980.

Wimmerová, M., Kozmon S., Nečasová I., *et al.* Stacking interactions between carbohydrate and protein quantified by combination of theoretical and experimental methods. *PLOS ONE.* 2012;7:e46032.

Yokoyama, M., Fujisaki S., Shirakura M., *et al.* Molecular dynamics simulation of the influenza A(H3N2) hemagglutinin trimer reveals the structural basis for adaptive evolution of the recent epidemic clade 3C.2a. *Front Microbiol.* 2017;8:584.

Zhao, H., Taherzadeh G., Zhou Y., *et al.* Computational prediction of carbohydrate-binding proteins and binding sites. *Curr Protoc Protein Sci.* 2018;94:e75.

Zhao, H., Yang, Y., Von Itzstein, M., *et al.* Carbohydrate-binding protein identification by coupling structural similarity searching with binding affinity prediction. *J Comput Chem.* 2014;35:2177–2183.

Chapter 13

Quantitative structure–activity relationship in ligand-based drug design: Concepts and applications

Vishnupriya Kanakaveti[†], P. Anoosha[†], R. Sakthivel,
S.K. Rayala and M. Michael Gromiha*

*Department of Biotechnology, Bhupat and Jyoti Mehta
School of Biosciences, Indian Institute of Technology Madras,
Chennai 600036, Tamil Nadu, India*

Quantitative structure–activity relationship (QSAR), a 50-year-old technology, has gained prominence till date due to its wide applicability in medicinal chemistry, computer-aided drug design (CADD), toxicology, pharmacology and environmental risk assessment. It is one of the major computational tools employed in medicinal chemistry for the development of reliable statistical prediction models that aid drug discovery. This is mainly by depicting the relationships between physicochemical properties of chemical compounds and their biological activities to predict the activities of new chemical structures. Despite many success stories of the technique, challenges run in parallel pertaining to its reliability and applicability. This chapter is devoted to outline the principles, methodology and lessons learned from theory and application of QSAR in identifying potent drug candidates.

*gromiha@iitm.ac.in
[†]Contributed equally to this work.

13.1. Introduction

With the advent of structure-based methods in the drug discovery process, the demand for predictive models has been raised up to a reasonable extent. These predictive models quantify the relationship between the structure and its biological effect (Veldstra, 1953). Among them, quantitative structure–activity relationship (QSAR) occupies a significant stand and is often employed in both academy and industry worldwide. The fundamental principle underlying this technique is that biological activities of chemical compounds are mainly influenced/guided by some of their potential structural or physicochemical properties. The *in silico* modeling of this relationship is termed as SAR or QSAR which include three prime components: (i) input data of chemical structures; (ii) data selected for modeling and prediction; and (iii) method of modeling. The robustness of QSAR models depends on the procedures to mold them into a mathematical form, statistical significance and interpretable features (Chirico and Gramatica, 2011; Dearden *et al.*, 2009; Golbraikh and Tropsha, 2002; Roy *et al.*, 2017).

QSAR models have broad applications for assessing potential impacts of chemicals and their derivatives on human health for treatment of various diseases. For the past few decades, we have witnessed many advances in this field for predicting biological activities of a wide span of chemical compounds and their benefits in screening promising drugs for various treatment strategies (Kanakaveti *et al.*, 2019). From initial models developed by Hansch *et al.* (1962) to advanced machine learning approaches, QSAR has been diversified and well evolved pertaining to its application to a small series of chemicals to large data sets including thousands of chemical compounds (Cronin, 2010). Even though it is a computational technique, QSAR addresses the problems of experimentalists and the end beneficiaries are the medicinal chemists. The perceptions of "developers" and "users" have to be well balanced to guide the synthesis of novel compounds with a better activity (Cherkasov *et al.*, 2014). Also, these models are applied to save the cost of production, reducing animal tests and promoting green chemistry (Cronin, 2010).

In this chapter, we describe the procedures for the development of QSAR models along with molecular descriptors, feature selection and validation of the method. Further, the applications of QSAR in drug design will be discussed with examples.

13.2. Goals of QSAR

QSAR acts as an interface between structure, chemistry and biology of compounds, used to depict the relationships in a meaningful manner, thereby enabling good predictions and interpretations. The purpose of employing QSAR includes

- Predicting the biological activity and key properties
- Rationalizing the drug mechanism of action
- Reducing the cost of production in pharmaceutical, pesticide and nano-industries
- Lessening the requirement of expensive animal testing
- Avoiding pain and discomfort to animals by replacing animal tests
- Endorsing green chemistry by reducing the chemical waste

13.3. Methods based on QSAR

QSAR-based prediction modeling is information driven and it depends on the data set and the method. Over the past five decades several QSAR methods have been proposed by various groups to discern the structure activity relationship and are summarized in Table 13.1.

Among them, Hansch model is the primitive and quite lucid as the model relates lipophilicity and activity using a simple regression technique (Hansch *et al.*, 1962). Several modifications of the model were proposed and the substituent effect on potency of chemical entities was identified (Fujita *et al.*, 1964; Hansch and Fujita, 1964; Jaffé, 1953; Taft, 1956). Finally, all models converged at a point that the biological effect is represented as a function of physicochemical descriptors (Eq. 1):

Biological effect (activity, toxicity) $= f$ *(physicochemical properties)* (1)

Table 13.1.   Different types of QSAR techniques.

QSAR method	Principle	Application
Hansch analysis	Correlation between hydrophobicity and activity	Anti-adrenergic compounds
Free-Wilson	Substituent effect on biological activity	Anti-bacterial compounds
Linear regression analysis	Electronic factors on activity	Plant growth inhibitors
Partial least squares	Generalization of descriptors	Antiviral compounds
Pattern recognition	Knowledge of substituents and core drives modeling	Anti-influenza compounds
Pharmacophores	Knowledge of interactions and key functional groups	Antiviral compounds
3D models, CoMFA	3D properties, molecular field properties on activity	Anticancer compounds, Antiviral compounds
CoMSIA	Similarity indices and molecular field properties drive the activity	Neuraminidase inhibitors
Classification models	Machine learning methods	Mutagenicity of compounds

With the advent of quantum mechanics, the importance of electric and steric determinants was acknowledged in modeling approaches (Cramer *et al.*, 1988). Also, *ab initio* properties and prime determinants of QSAR were identified with the aid of quantum chemistry. The dimensionality of QSAR models was extended from 2D to 3D and as a result, Comparative Molecular Field Analysis (CoMFA) and other 3D approaches have been introduced. Global QSAR models generated using fragment additive approaches for a large data set of compounds were a breakthrough in the medicinal chemistry (Klopman, 1984).

13.4. QSAR based on dimensionality of properties

1D and 2D are classical QSAR methods, which are simpler and faster than 3D-QSAR approaches. 1D-QSAR is correlating the activity of compounds with basic global molecular features such as log P (lipophilicity), pKa (pH of compound at which it is 50% protonated) and so on. 2D-QSAR deals with structural patterns such as connectivity indices and 2D-pharmacophores; 3D-QSAR approaches consider 3D structure of molecules to derive space-related features.

3D-QSAR is a broad term encompassing all those QSAR methods that correlate macroscopic properties including atom-based descriptors derived from the spatial 3D-representation of molecular structures. The methodology has emerged as a natural extension to the classical QSAR approaches pioneered by Hansch and Free-Wilson technique. Advancements in 3D-QSAR lead to 4D-, 5D- and 6D-QSAR techniques, which include induced-fit models and solvation models (Vedani *et al.*, 2005).

13.5. QSAR based on correlation and classification

QSAR methods are classified into two categories depending upon the type of correlation technique utilized to establish a relationship between structural or chemical properties and biological activities of compounds: (i) linear methods including linear regression (LR), multiple linear regression (MLR), partial least squares (PLS) and principal component analysis/regression (PCA/PCR) (Luco and Ferretti, 1997; Yap *et al.*, 2007) and (ii) nonlinear methods consisting of artificial neural networks (ANN) (Roy and Mandal, 2008), k-nearest neighbors (kNN) and Bayesian neural networks (Kauffman and Jurs, 2001).

Unlike traditional QSAR dealing with numerical values for correlating activity and descriptors, machine learning methods such as random forest (Kuz'min *et al.*, 2011), support vector machine (SVM) (Barakat and Bradley, 2010), Bayesian classifiers (Cortes-Ciriano *et al.*, 2014), neural networks (So and Richards, 1992) and logistic regression are being used recently for designing classification models (Martin *et al.*, 1974). These techniques have substantially reduced the time and effort required for the discovery of a new promising drug. The key step in QSAR modeling is to identify the molecular descriptors that represent various properties of analyzed compounds which could predict their biological activities.

13.6. QSAR model development

The methodology mainly consists of two steps: (i) model development and (ii) validation. Choosing an appropriate data set of compounds for the given problem is one of the important steps and first step in QSAR model development. Then, we can derive 2D or 3D physicochemical properties of compounds in the data set. Multiple softwares are available in the literature, which can be used to calculate these descriptors. For example, Dragon, Gaussian (Obrezanova *et al.*, 2007) and PADEL (Yap, 2011) are generally used to derive constitutional, electronic, geometrical, hydrophobic, hydrophilic, solubility and topological features for compounds.

13.6.1. *Molecular descriptors*

Molecular descriptors are the basis for QSAR modeling, which are derived from decoding the chemical information into measurable values. They reflect various levels of chemical structures ranging from molecular formula (1D), two-dimensional structural properties (2D), conformation-dependent properties (3D) and dynamics dependent (4D and higher) (Figure 13.1).

13.6.2. *2D descriptors*

The descriptors depicting the topology and connectivity of atoms in the compounds are known as 2D descriptors. They are quite simple and informative, consistent with dynamics of compounds and highly discriminative in a pool of chemical entities. Mostly, molecular graph is used to calculate these descriptors, which is represented as a function of vertices (atoms), the relationship between pair of vertices and edges (bonds). Generally topological indices and connectivity indices were calculated from the graph theory assumption (García-Domenech *et al.*, 2008).

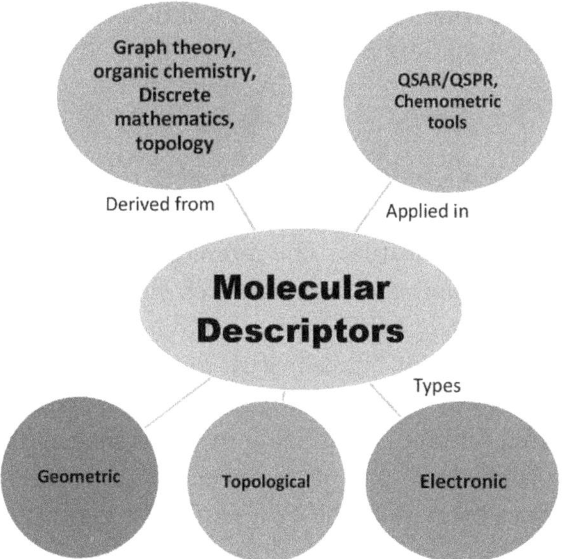

Figure 13.1.   Molecular descriptors: Origin, application and types.

13.6.3. *3D descriptors*

Biological activity is driven by several non-covalent interaction fields in the vicinity of binding interface between ligand and target. However, different compounds have different interaction fields based on their chemical architectures. Thus, the difference in the fields is a plausible metric to predict the biological activity, which was measured with the aid of molecular fields derived from 3D structures. This assumption has led to the development of CoMFA (Cramer *et al.*, 1988).

13.6.4. *Feature selection*

Feature selection is another important step for developing reliable models (Goodarzi *et al.*, 2012). For this purpose, multivariate analysis including MLR and PLS could be employed to identify the best features correlating to observed biological activities of compounds. The following equation represents MLR, where multiple features together have linear relationship with dependent variable, Y:

$$Y = \beta_1 \cdot x_1 + \beta_2 \cdot x_2 + \beta_3 \cdot x_3 + \ldots + \beta_n \cdot x_n + C \qquad (2)$$

where Y is a dependent or response variable representing biological activity of the compound (IC_{50}) and x_i is an independent variable which represents each feature. The intercept C and slopes β_1, β_2,..., β_n are partial regression coefficients, which show the contribution of each feature.

MLR models are not efficient with mutually correlated variables and high-dimensional data. Hence, it is a good practice to remove highly correlated variables by utilizing methods such as "dimensionality reduction coefficient" and also limit the total number of variables to be 5–10 times fewer than the data set. These disadvantages can be overcome by PLS technique, which is a generalization of MLR and can efficiently work with mutually correlated or numerous variables. PLS model can also identify new features which can be estimated and derived from the given set of features for training.

13.6.5. *Validation of QSAR models*

Validation is the process by which the reliability and significance of a method are examined for a specific problem (Veerasamy *et al.*, 2011). The developed models with important features can be further validated internally or externally using different procedures. For internal validation, leave one-out or jack-knife tests are utilized where regression model is developed using n–1 compounds (n: total number of compounds in data set) and tested on the remaining compound to predict its IC_{50} value (activity), and this process is iterated n times to predict IC_{50} of each compound in the data set. Sometimes, n-fold cross-validation is also used for validating the model where entire data set is divided into n equal subsets and single subset is used for validation and n–1 subsets for training. This process is repeated for n times such that each subset is used for testing at least once. The performance of the method is assessed using the measures, correlation and mean absolute error (MAE), which is the absolute difference among experimental predicted activities. The statistical significance is evaluated by various measures mainly using p-value. In addition, the developed model can be validated using external blind data set to evaluate the performance of the model.

13.7. Pitfalls of QSAR modeling

Despite greater scope of QSAR in various fascinating areas, there are certain disadvantages of QSAR pertaining to its reliability and interpretation such as poor statistical methods and failure in following the guidelines of QSAR (Walker *et al.*, 2003). The following are the probable instances related to both descriptors and endpoints, where the QSAR is often failed and overestimated:

   (i)   Heterogeneity of data from experiments and compounds
  (ii)   Errors pertaining to usage of endpoints
 (iii)   Chaotic (duplicate and collinear descriptors), incomprehensible and incorrect descriptors
 (iv)   Replicates of data points, chemicals in the data set
  (v)   Considering end points with minute differences
 (vi)   Overfitting data and improper use of statistical models
(vii)   Inadequacy in the number of data points considered and validation methods.

In nutshell, the eminent QSAR methods are subjected to above-mentioned common errors which can be avoided by proper use of "best practices" for data processing and interpretation.

13.8. Pathway for a successful QSAR model

The path to generate reliable QSAR models is quite straightforward when fundamentals are followed. Few basics are given below:

  (i)   Usage of high-quality data for modeling
 (ii)   Genuine and reliable statistical methods and analysis
(iii)   Mechanistic relevance of model

Organization for economic co-operation and development (OECD) has formulated some principles, namely (i) defined biological activity, (ii) distinct algorithm, (iii) applicability domain and (iv) robust fitting, and are available at http://www.oecd.org/chemical-safety/risk-assessment/37849783.pdf.

13.9. Applications of QSAR modeling

QSAR has been widely used to identify lead compounds. A contest-based approach also revealed the importance of ligand-based QSAR models for identifying hit compounds for cYes kinase (Chiba *et al.*, 2015, 2017). In this section, we discuss a couple of applications on Bcl-2 (B cell lymphoma 2) (Lessene *et al.*, 2008) and mutants of EGFR (epidermal growth factor receptor) (Eck and Yun, 2010) to identify potential inhibitors.

13.9.1. *QSAR modeling of Bcl-2 protein–protein interaction inhibitors*

Bcl-2 family proteins are apoptotic regulators bringing the cytochrome release ultimately leading to mitochondrial membrane damage and cell death. Bcl-2 family proteins interact with each other to its pro-apoptotic partners such as BAX, BAK and BAD (Vogler *et al.*, 2009). Hence, seizing these PPIs can induce the apoptosis. Mimicking the pattern of binding with peptides and small molecule is an attractive approach.

Kanakaveti *et al.* (2017) employed a comprehensive QSAR strategy to design novel scaffolds employing the knowledge of interactions and structure of Bcl-2 protein. The methodology is depicted in Figure 13.2.

They considered 1649 inhibitors belonging to seven diverse scaffolds, namely Apogossypol, Pyrazole, Quinazoline, Quinolone, Benzothiazole, Poly Quinoline and Thiomorpholine, and developed robust MLR models for identifying potential inhibitors. The correlation coefficient (r) of the models ranged from 0.95 to 0.985. The robustness of the models was evaluated using jack-knife

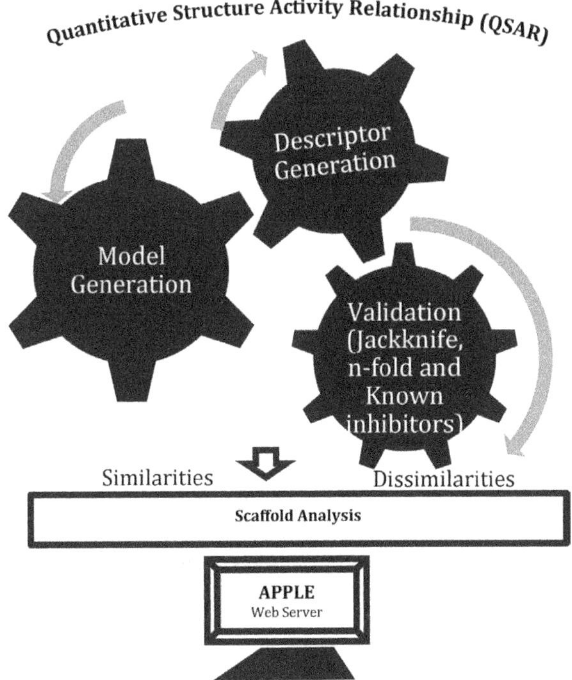

Figure 13.2.   QSAR methodology employed for designing anti-Bcl-2 inhibitors.

validation resulted in correlation coefficient of 0.95–0.98. Figure 13.3 shows the relationship between experimental and predicted IC_{50} values in a set of 1649 compounds belonging to seven scaffolds. Coupling factors emphasizing on both similarities and contribution of descriptors is a novel attempt of getting a mechanistic interpretation of the model. The model has also provided an interaction view point to the SAR equations, which enhanced the prediction to a greater extent and brought meaningful insights to experimentalists as well. Subsequently, they have developed a Web-based tool capable of identifying and categorizing a novel compound as either pan or selective Bcl-2 inhibitor based on MACCS fingerprints and tanimoto similarity coefficient. The APPLE (Applying QSAR for identifying pan or selective Bcl-2 inhibitor) server is a user-friendly tool designed to analyze compounds provided in either smiles or sdf format and it is available at https://www.iitm.ac.in/bioinfo/APPLE/.

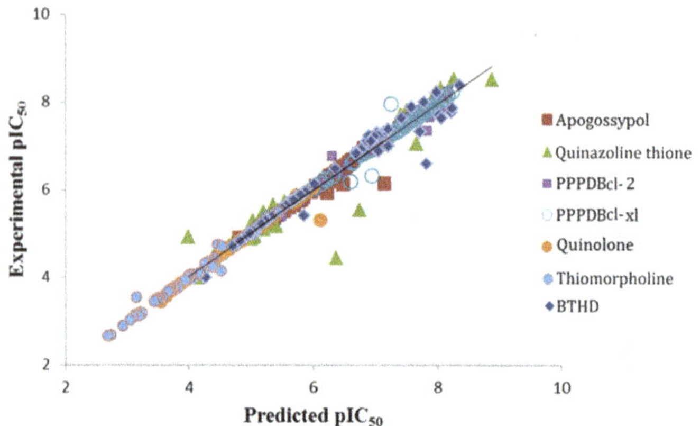

Figure 13.3.   Regression plot of experimental and predicted IC$_{50}$ values of the all the seven families.

QTD: Quinazolinethione derivatives, PPPDBcl-2: Pyrazole pyrimidine phenyl acyl derivatives of Bcl-2, PPPDBcl-xL: Pyrazole pyrimidine phenyl acyl derivatives of Bcl-xL, QS: Quinolone scaffold, BHTD: Benzothiazole hydrazine derivatives, TOAPC: 3-Thiomorpholin-8-oxo-8H-acenaphtho [1; 2-b] pyrrole-9-carbonitrile derivatives.

13.9.2. *QSAR modeling of mutation-specific inhibitors for epidermal growth factor receptor in cancer*

Several studies in the literature show that distinct EGFR mutations have unique inhibitor susceptibilities to different kinase inhibitors (Eck and Yun, 2010). For example, the most well-known tyrosine kinase inhibitor, gefitinib, showed a stronger binding affinity of ~50-fold toward L858R mutant, which is significantly highly sensitive compared to another driver variant G719S (Yun *et al.*, 2007). This revealed that intrinsic differences in inhibitor binding of the aberrant kinases can help to understand the differential sensitivity of variant cell lines and their distinctly activated functional pathways. These important observations trigger the necessity of designing mutation-specific inhibitors for effective treatment of tumors with distinct mutations. Further, the differences in altered drug responses are studied by examining binding patterns of protein-mutant structures to small molecules using computational (Goyal *et al.*, 2015) and biochemical analyses (Singh and Jadhav, 2018; Wang *et al.*, 2014; Yun *et al.*, 2007; Yoshikawa *et al.*, 2013).

Anoosha *et al.* (2017) developed variant-specific QSAR models for eight well-known EGFR driver mutations using a set of anti-cancer compounds with respective biological activities (IC_{50}). The method performed well with self-consistency and jack-knife tests on a set of compounds resulted with a correlation coefficient (r) in the range 0.78–0.92 and 0.72–0.90, respectively, for different mutants. QSAR models were evaluated with blind test data that showed an average correlation (r) of 0.96. Further, the potential features have been systematically analyzed and observed that aromaticity and hydrogen bonding features play vital role in IC_{50} prediction for compounds against EGFR mutants. They have also performed docking analysis on these variants and, interestingly, the analyses have complimented the QSAR predictions and that shows the reliability of prediction models that can be useful in accelerating the development of EGFR mutation-specific drugs.

13.10. Conclusion

QSAR modeling is one of the most popular computer-aided tools employed in medicinal chemistry for drug discovery and lead optimization. Several studies have proven that QSAR techniques make a valuable contribution in discovering new promising drugs and it is especially powerful in the absence of 3D structures of specific drug targets. In the last decades, QSAR has undergone several transformations, ranging from the dimensionality of the molecular descriptors and different methods for finding a correlation between the chemical structures and the biological property. However, there is a need to develop more approaches, which can distinguish the influence of different molecular contexts other than biological activities of chemical compounds. Another important direction is development of online accessible and user-friendly QSAR workflows and tools for analysis and interpretation of QSAR models.

Acknowledgments

We thank the Bioinformatics facility, Department of Biotechnology and Indian Institute of Technology Madras for computational facilities. VK thanks Department of Science and Technology (DST), India, for providing research fellowship. The work is partially supported by the Department of Biotechnology, Government of India, to MMG and SKR (BT/PR15277/BID/7/550/2015).

References

Anoosha, P., Sakthivel, R., and Gromiha, M.M. Investigating mutation-specific biological activities of small molecules using quantitative structure-activity relationship for epidermal growth factor receptor in cancer. *Mutat Res.* 2017;806:19–26.

Barakat, N., and Bradley, A.P. Rule extraction from support vector machines: A review. *Neurocomputing.* 2010;74(1–3):178–190.

Cherkasov, A., Muratov, E.N., Fourches, D., *et al.* QSAR modeling: Where have you been? Where are you going to? *J Med Chem.* 2014;57(12): 4977–5010.

Chiba, S., Ikeda, K., Ishida, T., *et al.* Identification of potential inhibitors based on compound proposal contest: Tyrosine-protein kinase Yes as a target. *Sci Rep.* 2015;5:17209.

Chiba, S., Ishida, T., Ikeda, K., *et al.* An iterative compound screening contest method for identifying target protein inhibitors using the tyrosine-protein kinase Yes. *Sci Rep.* 2017;7(1):12038.

Chirico, N., and Gramatica, P. Real external predictivity of QSAR models: How to evaluate it? Comparison of different validation criteria and proposal of using the concordance correlation coefficient. *J Chem Inf Model.* 2011;51(9):2320–2335.

Cramer, R.D., Patterson, D.E., and Bunce, J.D. Comparative molecular field analysis (CoMFA). 1. Effect of shape on binding of steroids to carrier proteins. *J Am Chem Soc.* 1988;110(18):5959–5967.

Cronin, M.T. Quantitative structure-activity relationships (QSARs)-applications and methodology. In: *Recent Adv. in QSAR Studies.* Dordrecht, the Netherlands: Springer;2010:3–11.

Cortes-Ciriano, I., van Westen, G.J., Lenselink, E.B., *et al.* Proteochemometric modeling in a Bayesian framework. *J Cheminform.* 2014;6(1):35.

Dearden, J.C., Cronin, M.T., and Kaiser, K.L. How not to develop a quantitative structure-activity or structure-property relationship (QSAR/QSPR). *SAR QSAR Environ Res.* 2009;20(3–4):241–266.

Eck, M.J., and Yun, C.H. Structural and mechanistic underpinnings of the differential drug sensitivity of EGFR mutations in non-small cell lung cancer. *Biochim Biophys Acta.* 2010;1804(3):559–566.

Fujita, T., Iwasa, J., and Hansch, C. A new substituent constant, π, derived from partition coefficients. *J Am Chem Soc.* 1964;86(23):5175–5180.

García-Domenech, R., Gálvez, J., de Julián-Ortiz, J.V., *et al.* Some new trends in chemical graph theory. *Chem Rev.* 2008;108(3):1127–1169.

Golbraikh, A., and Tropsha, A. Beware of q2!. *J Mol Graph Model.* 2002;20(4):269–276.

Goodarzi, M., Dejaegher, B., and Heyden, Y. Feature selection methods in QSAR studies. *J AOAC Int.* 2012;95(3):636–651.

Goyal, S., Jamal, S., Shanker, A., *et al.* Structural investigations of T854A mutation in EGFR and identification of novel inhibitors using structure activity relationships. *BMC Genomics.* 2015;16(5):S8.

Hansch, C., Maloney, P.P., Fujita, T., *et al.* (1962). Correlation of biological activity of phenoxyacetic acids with Hammett substituent constants and partition coefficients. *Nature.* 1962;194(4824):178.

Hansch, C., and Fujita, T. p-σ-π Analysis. A method for the correlation of biological activity and chemical structure. *J Am Chem Soc.* 1964;86(8): 1616–1626.

Jaffé, H.H. A reexamination of the Hammett equation. *Chem Rev.* 1953;53(2):191–261.

Kanakaveti, V., Sakthivel, R., Rayala, S.K., *et al.* (2017). Importance of functional groups in predicting the activity of small molecule inhibitors for Bcl-2 and Bcl-xL. *Chem Biol Drug Des.* 2017;90(2):308–316.

Kanakaveti, V., Anoosha, P., Sakthivel, R., *et al.* Influence of amino acid mutations and small molecules on targeted inhibition of proteins involved in cancer. *Curr Top Med Chem.* 2019;19(6):457–466.

Kauffman, G.W., and Jurs, P.C. QSAR and k-nearest neighbor classification analysis of selective cyclooxygenase-2 inhibitors using topologically-based numerical descriptors. *J Chem Inf Comput Sci.* 2001;41(6): 1553–1560.

Klopman, G. Artificial intelligence approach to structure-Activity studies. Computer automated structure evaluation of biological activity of organic molecules. *J Am Chem Soc.* 1984;106(24):7315–7321.

Kuz'min, V.E., Polishchuk, P.G., Artemenko, A.G., *et al.* Interpretation of QSAR models based on random forest methods. *Mol Inf.* 2011;30 (6–7):593–603.

Lessene, G., Czabotar, P.E., and Colman, P.M. BCL-2 family antagonists for cancer therapy. *Nat Rev Drug Discov.* 2008;7(12):989.

Luco, J.M., and Ferretti, F.H. QSAR based on multiple linear regression and PLS methods for the anti-HIV activity of a large group of HEPT derivatives. *J Chem Inf Comput Sci.* 1997;37(2):392–401.

Martin, Y.C., Holland, J.B., Jarboe, C.H., *et al.* Discriminant analysis of the relation between physical properties and the inhibition of monoamine oxidase by aminotetralins and aminoindans. *J Med Chem.* 1974;17(4): 409–413.

Obrezanova, O., Csányi, G., Gola, J.M., *et al.* Gaussian processes: A method for automatic QSAR modeling of ADME properties. *J Chem Inf Model.* 2007;47(5):1847–1857.

Roy, K., Ambure, P., and Aher, R.B. How important is to detect systematic error in predictions and understand statistical applicability domain of QSAR models? *Chemometr Intell Lab Syst,* 2017;162:44–54.

Roy, K., and Mandal, A.S. Development of linear and nonlinear predictive QSAR models and their external validation using molecular similarity principle for anti-HIV indolyl aryl sulfones. *J Enzyme Inhib Med Chem.* 2008;23(6):980–995.

Singh, M., and Jadhav, H.R. Targeting non-small cell lung cancer with small-molecule EGFR tyrosine kinase inhibitors. *Drug Discov Today,* 2018;23(3):745–753.

So, S.S., and Richards, W.G. Application of neural networks: Quantitative structure-activity relationships of the derivatives of 2, 4-diamino-5-(substituted-benzyl) pyrimidines as DHFR inhibitors. *J Med Chem.* 1992;35(17):3201–3207.

Taft, R.W. Separation of polar, steric and resonance effects in reactivity. In: Newman M.S., ed. *Steric Effects in Organic Chemistry,* New York, NY: Wiley; 1956:556–675.

Vedani, A., Dobler, M., and Lill, M.A. Combining protein modeling and 6D-QSAR. Simulating the binding of structurally diverse ligands to the estrogen receptor. *J Med Chem.* 2005;48:3700–3703.

Veerasamy, R., Rajak, H., Jain, A., *et al.* Validation of QSAR models-strategies and importance. *Int J Drug Design Discov.* 2011;2(3):511–519.

Veldstra, H. The relation of chemical structure to biological activity in growth substances. *Annu Rev Plant Physiol.* 1953;4(1):151–198.

Vogler, M., Dinsdale, D., Dyer, M.J., *et al.* Bcl-2 inhibitors: Small molecules with a big impact on cancer therapy. *Cell Death Differ.* 2009;16(3):360.

Wang, B., Shen, W., Yang, H., *et al.* Targeting EGFR mutants with non-cognate kinase inhibitors in non-small cell lung cancer. *Med Chem Res.* 2014;23:4510–4530.

Walker, J.D., Jaworska, J., Comber, M.H., *et al.* (2003). Guidelines for developing and using quantitative structure-activity relationships. *Environ Toxicol Chem.* 2003;22(8):1653–1665.

Yap, C.W., Li, H., Ji, Z.L., *et al.* Regression methods for developing QSAR and QSPR models to predict compounds of specific pharmacodynamic, pharmacokinetic and toxicological properties. *Mini Rev. Med. Chem.* 2007;7(11):1097–1107.

Yap, C.W. PaDEL-descriptor: An open source software to calculate molecular descriptors and fingerprints. *J Comput Chem.* 2011;32(7):1466–1474.

Yun, C.H., Boggon, T.J., Li, Y., *et al.* Structures of lung cancer-derived EGFR mutants and inhibitor complexes: Mechanism of activation and insights into differential inhibitor sensitivity. *Cancer Cell,* 2007;11(3): 217–227.

Yoshikawa, S., Kukimoto-Niino, M., Parker, L., *et al.* Structural basis for the altered drug sensitivities of non-small cell lung cancer-associated mutants of human epidermal growth factor receptor. *Oncogene.* 2013;32(1): 27–38.

Chapter 14

Protein–ligand interactions in molecular modeling and structure-based drug design

Devadasan Velmurugan*,§, Dasararaju Gayathri*,
Chandrasekaran Ramakrishnan† and Atanu Bhattacharjee‡

*Centre of Advanced Study in Crystallography and Biophysics,
University of Madras, Guindy Campus, Chennai, Tamil Nadu, India

†Department of Biotechnology, Bhupat and Jyoti Mehta
School of Biosciences, Indian Institute of Technology Madras,
Chennai 600036, Tamil Nadu, India

‡Department of Biotechnology and Bioinformatics,
North East Hill University (NEHU), Shillong, Meghalaya, India

Radical new opportunities for drug discovery are due to the revolution in biology over the past four decades. Major drug targets in the form of molecular components of disease processes have been developed for use in automated assays. After screening natural compounds, chemical databanks or combinatorial libraries to identify the lead compounds, the target macromolecules are used as a starting point for structure-based approaches. To study complex biological and chemical systems, pharmaceutical research incorporates molecular modeling methods and variety of drug discovery programs. For the identification and development of novel promising compounds,

§shirai2011@gmail.com

the combination of computational and experimental strategies is of great value. In the field of modern drug design, molecular docking methods explore the various ligand conformations adopted within the binding sites of macromolecular targets and select the best conformation. As various docking algorithms are available, an understanding of the advantages and limitations of each method is of great importance. The characterization of protein–ligand binding sites is essential for investigating new functional roles related to major biological research areas like health, food and energy. As the laboratory techniques for drug discovery are very much time-consuming and expensive, computational methods have been developed to assist drug discovery and development, and a number of automated methods have emerged to identify the promising drug candidates.

14.1. Introduction

In living organisms, the basis of all processes is molecular recognition. Here, the macromolecules interact either with each other or with small molecules. Understanding the biology at the molecular level requires the protein–ligand (PL) interactions as these play a major role in cellular metabolism. Docking of proteins with small molecules includes enzyme–ligand binding, protein–saccharide binding, protein–inhibitor binding and protein–peptide interactions. In the binding of small molecules (ligands) with the protein, ligands comprise a very large and structurally diverse group of molecules. The interactions during the binding play a wide variety of physicochemical characteristics.

For a full understanding of PL interactions, the biophysical properties of both the protein and the ligands should be examined independently and fully understood. Particularly, the active site of the binding pocket along with the neighboring amino acids should be thoroughly studied. As far as the ligand is concerned, the Lipinski's rule of five and ADMET properties should be obeyed. The molar dissociation constant K_d (the binding constant $K_a = 1/K_d$) < 5 μM means strong interactions and > 50 μM means weak interactions. In the study of PL interactions, the ultimate goal is to understand the minute molecular details of the function — molecular recognition, structure, kinetics, energetic and dynamics of the system. The analysis of the protein, the ligand and the PL complex involves the experimental variables such as protein–ligand concentration, analogues/homologues of the ligands, protein structural variants, environmental solution condition variables such as temperature, pH, buffers, solvents, salt/ion concentration, denaturants and stabilizers, etc., and also the combinations of the abovementioned experimental variables. But the screening of lakhs of compounds from the databases using the above experimental variables is the never-ending process or is very difficult or very much time-consuming. In silico methods are usually employed in studying the PL interactions and in choosing the best ligand. Here many commercial/free computer programs are available for use. Studying the PL interactions is the main part in structure-based drug design (SBDD) approach. Some of these programs are AuotoDock (MC) [1], MOE-Dock (MC, TS), GOLD (GA) [2, 3], PRO_LEADS (TS) [4], DOCK [5], FlexX [6], Glide [7], Hammerhead [8] and FLOG [9].

In silico methods for docking of PL mostly use high-throughput virtual screening (HTVS) for the effective selection of the best ligand among many others in the fastest way. Here most of the programs assume the receptor site to be rigid. All the docking programs use the crystallographic structures of the protein complexes with the ligands deposited in the Protein Data Bank (PDB) at the start.

14.2. Drug discovery in the past and present

Endogenous ligands were identified in the past long before their bimolecular receptors were isolated and before their chemical structures were established [10]. Using the physiological relevance of these ligands, hypotheses were developed. From the bile acids, steroid hormones were extracted and during 1920 and 1930s several Nobel prizes were awarded for the discoveries of steroids and steroid hormones relating to their biological effects [11]. In 1930s using single crystal X-ray diffraction techniques, the three-dimensional arrangement of the steroid scaffold and the exact positions and stereochemistry of the substituents were known. This allowed relating the positions of the different chemical groups to specific biological activity and in turn helping the development of new medicines [11, 12]. Even though X-ray structures of many co-crystallized ligand–receptors were later available, the effective rational drug design is hindered by the thermodynamics of the ligand–receptor association, which cannot be inferred from the calculation of close contact interactions.

The present day drug discovery is concerned with biological targets, genetic studies, transgenic animal models and technology related to molecular biology, gene technology and protein science. The knowledge of the three-dimensional structure of target macromolecules is of great importance in drug design. The binding affinity of novel molecules with pharmacological activity can be estimated using different computational techniques related to graphics, docking, quantum chemical calculations, molecular simulations and *ab initio* methods. The leads supplied by high-throughput screening are often difficult to be transformed into drugs as the optimization of binding affinity in isolation by traditional medicinal chemistry methods often leads to poor ADMET properties due to many factors. For this reason many researchers have shifted toward SBDD and virtual ligand screening (VLS).

The success of SBDD in the discovery of new drug leads is due to the exploded information of genomic, proteomic and structural information of thousands of macromolecular targets. SBDD has revealed specific, micromolar inhibitors against the HIV-1 RNA target TAR, the Interleukin receptor interaction, the VEGF/VEGF receptor and the breast-cancer-related target Bcl2. Many X-ray structures of pharmaceutical targets co-crystallized with their inhibitors, which have been resolved by the pharma industry, remain secret as they are still unpublished. The missing link that undermines effective SBDD is the lack of reliability in predicting the affinities. The reason for some failures after lead optimization in SBDD is that the program is done in a hurried way within 12–18 months and many properties are not optimized together. In spite of considerable progress achieved in SBDD using VLS and computer-aided drug design (CADD), the practical prediction of binding affinities of ligand remains a big challenge. Although measurements of binding affinity using titration calorimetry and surface plasmon resonance can be done with high accuracy, due to the lack of knowledge of the entropic contribution, which is a measure of the overall system disorder, the binding affinity cannot be inferred reliably. In the X-ray crystallographic target macromolecular structures, there will be conserved water molecules, and in many cases, ligand binding is mediated often by such waters [13]. Attempts to displace them with experimental ligands may not only be difficult to achieve but also have unpredictable unfavorable effects [14–16]. In the PDB, X-ray structures of receptor–ligand complexes for many important families of drug targets like proteases, lipases, nuclear receptors, protein kinases, HIV reverse transcriptase, some G-protein coupled receptors (GPCRs) and RNA are available for SBDD applications. As a result of the structural genomics initiatives globally, the wealth of the macromolecular structural information continues to increase making the SBDD approach in finding lead molecules to treat human ailments feasible. As sequence information available exceeds far the number of target structures available, homology modeling plays a major role in obtaining the macromolecular target structures from the sequence databases. After validation, docking can be proceeded followed by MD simulations to come out with potential lead ligands. The first work describing SBDD was published in 1976 and in the past 40 years, structure-based CADD (SBCADD) has surmounted many hurdles and had played a key role in the developments of several marketed drugs [17, 18].

14.3. Flexibility of macromolecular targets

When a ligand binds at the receptor site of the protein, local rearrangements of side-chains of the amino acids surrounding the binding pocket and sometimes large domain motions are induced and these are called protein conformation changes. The flexibility of proteins is grouped into four methods during docking. In soft docking, a small degree of overlap is allowed between the ligand and the protein by adjusting the van der Waals interaction in docking calculations [19, 20]. Computational efficiency of this method is good and is easier for implementation. The only limitation of this simplest method is that this can account for only small conformational changes. In the second method, side-chain flexibility is focused and the backbones are kept rigid. The various side-chain conformations are sampled. A rotamer library is used and at present, many improved techniques exist to incorporate continuous or discrete side-chain flexibility in ligand docking [21]. In the third method called molecular relaxation, rigid-body docking is used to place the ligand into the binding site and then the protein backbone and nearby side-chain atoms are relaxed. Efficient scoring function is needed as relaxation is given to both backbone and side-chain atoms of the proteins. The fourth approach is the most widely used method and an ensemble of protein structures to represent different possible conformational changes is used [22–24]. Programs like AutoDock and FLEXE use this type of ensemble of protein structures. Wei *et al.* [25] used an algorithm where a protein was decomposed into a rigid part and several flexible paths according to the crystal structures of the protein in the ensemble. The selected local conformers were joined with the rigid part in forming the best-fit conformation. This algorithm is faster than that in FLEXE. In the algorithm developed by Huang and Zou [26, 27], the protein conformation ensemble is treated as an additional dimension to the standard six degrees of freedom, three for translation and three for rotation for the optimization of ligand energy. This algorithm was expanded by Abagyan *et al.*, to four-dimensional docking [28]. To account for the protein flexibility, experimental structures from NMR or single crystal X-ray diffraction and ensembles protein conformation through molecular dynamics simulations, Monte Carlo (MC) simulations and structure prediction are used.

14.4. Sampling of ligands

In PL docking, the most successful area developed is ligand sampling which involves shape matching, systematic search and sarcastic algorithms. Shape matching is also known as shape complementarity. Here, the molecular surface of the ligand should complement the molecular structures of the receptor site. Docking programs like DOCK [4], FRED [29], FLOG [9], EUDOCK [30], LigandFit [31],

SURFLEX [32] and MS-DOCK [33] use shape matching. Its main advantage is computational efficiency. In the systematic search, all possible ligand-binding conformations are generated by exploring all degrees of freedom of the ligand. GLIDE [7] and FRED [29] use this type of sampling. In fragmentation method, the ligand is divided into different rigid parts/fragments. Docking of all fragments is done one by one into the binding site and later the best one is linked covalently. DOCK [4], LUDI [34] and FLEXX [6] are some examples of the docking programs that use fragmentation method. In conformational ensemble methods, an ensemble of pre-generated ligand conformation is rigid docked, the ligand-binding modes are collected and ranked according to binding energy scores. FLOG [9], DOCK3.5 [35], MS-DOCK [33] and Q-DOCK [36] are some of the docking programs that use this method. In stochastic algorithms, MC methods, different search methods like evolutionary algorithms (EA), tabu search methods and swarm optimization methods (SO) are used.

14.4.1. *Scoring functions*

The accuracy of any algorithm is determined by the scoring function in which speed and accuracy are the two important aspects. Although numerous scoring functions have been developed in the past, these can be grouped into three basic categories, namely force field, empirical and knowledge-based scoring function, according to their methods of derivation.

Force-field-based scoring involves the atomic charge and functional forms (bond stretching, angle bending, torsion, improper torsion and non-bonded terms including hydrogen bond interaction). A typical knowledge-based scoring function is based on the experimental data of the PL complexes.

Sometimes consensus scoring includes in finding a correct solution by combining the scoring information from multiple scoring functions. Incorporation of entropy effects is called clustering-based scoring method. As the computing power has rapidly increased, methods are still being developed for better scoring functions and remain one of the major future directions in PL docking. The publicly

available docking benchmarks such as DUD (http://dud.docking.
org/) and CESAR (http://www.cesardock.org/) are extremely valuable for the systematic and consistent evaluation and improvement of new and existing docking algorithms.

PDB contains the macromolecular target structures. CSD and databases like zinc, asinex, drug bank and Chinese database of natural products contain ligands.

14.4.2. *Lipinski's rule of 5*

The rule of 5 plays a major role to design or identify new lead compounds with druggability. This is accomplished due to the basis for this rule that there is a defined range of molecular weight and lipophilicity for most of the orally active drugs [37, 38]. In the context of PL interactions, this rule is satisfied as most of the co-crystallized compounds available in the PDB are small organic compounds and physiologically functional [39]. Hence, this rule is being used widely as the initial filtering criteria in current SBDD and ligand-based drug design protocols [40, 41]. The recent study by Chagas *et al.*, revisiting the rule of 5 concludes that most of the drugs resulted in adverse drug reaction (ADR) due to their MW > 500 Da [42]. However, paclitaxal, everolimus, docetaxel, doxorubicin, goserilin, etc., are exempted [42]. In a nutshell, the rule of 5 defines the boundary for druggability in the chemical space based on the four features such as MW ($\leq$500 Da), number of H-bond acceptors ($\leq$10), number of H-bond donors ($\leq$5) and logP ($\leq$5). Many academic-free and commercial software packages basically use this criteria to filter the large chemical libraries prior to structure-based, pharmacophore-based and ligand-based screening protocols.

14.4.3. *Fragment-based drug design*

Both experimental and computational methods are successful in generating the target-focused fragments and aid in the development of high-affinity compounds with improved target specificity. In the context of PL interactions, the FBDD is an effective method of identification of more promising lead molecules [43]. In principle, the small fragments (<20 heavy atoms) are capable of providing insights into the PL interactions. Moreover, with the small fragments, the chemical space sampling is improved [44]. NMR [45, 46] and X-ray crystallography are robust experimental techniques to validate the FBDD. NMR-based method is proved as a more suitable method for studying protein–protein interactions. The protein crystallography-based screening of fragment library is useful to study the topology of the fragment binding at the active site of the target protein [47]. β-secretase [48], PDE4A [49], HSP90 [50] and bromodomain [51] are some of the targets studied using protein crystallography-based FBDD.

In the process of lead identification using FBDD (Figure 14.1), the fragment library will be filtered based on the "rule of 3" criteria (≤3). Designing a tailored fragment library is the initial step. Chemoinformatics is a key method for building a specific library. Alternately, pharmacophore-based approaches also were used to

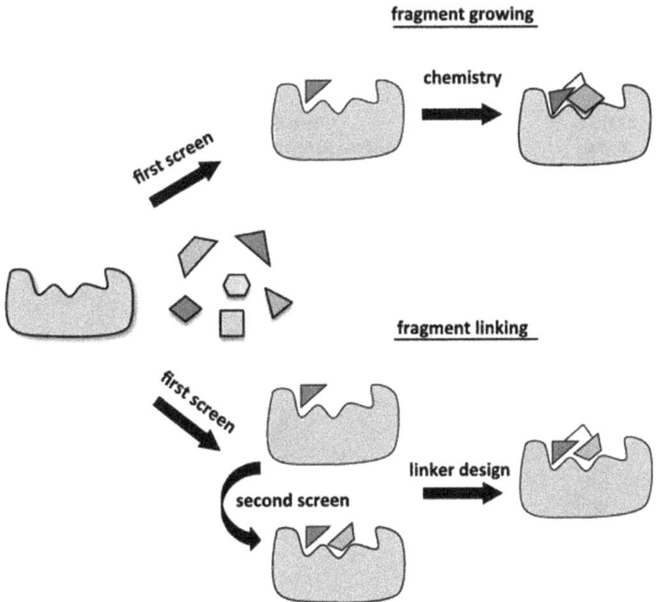

Figure 14.1. Schematic representation of steps involved in FBDD.

generate the target-focused libraries. Then, the active fragments are identified and optimized through: (i) linking of fragments, which involves selection of linker to link non-overlapping fragments to bind adjacent positions and (ii) merging the overlapping fragments at the active site [52].

14.4.4. *Active site prediction and homology modeling*

SBDD is an approach which commits to ensure the appropriate or functional interaction pattern in binding newly identified compounds with the active site. In order to avoid the false positives in the PL interaction patterns which is directly influenced by the energy and scoring schemes, it is important to obtain the relevant conformation of the active site. Figure 14.2 depicts the different active site geometries adopted by the kinases during their functional switch from active to inactive and vice versa.

Moreover, predictive accuracy of the active site region has direct impact on the outcome of the SBDD as well as FBDD, particularly in hit identification. This process is very challenging in the case of

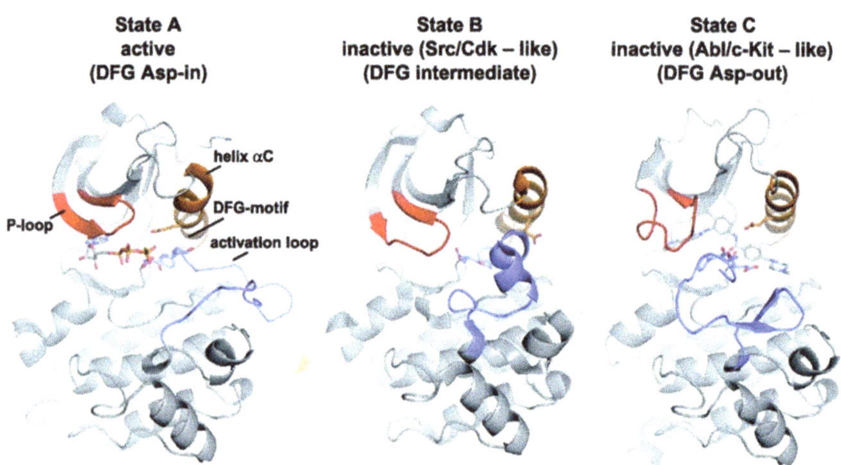

Figure 14.2.   Examples of the active and inactive conformations of kinases.

Source: http://cancerres.aacrjournals.org.

protein–protein complexes, where the interaction modulators necessarily recognize the interface with flat surfaces [53].

Some of the popular active site prediction tools listed in Table 14.1 are with comparable predictive performance, and their algorithms are (i) geometric-based, (ii) energy-based, (iii) docking-based, and (iv) in any combination of these three.

On the other hand, PL binding sites are characterized using comparison of substructure and alignment profile [54].

Homology modeling (in the context of PL interaction) is essential when there is a need for building chimeric proteins or fusion proteins,

Table 14.1. List of some of the popular Web servers to predict the active site/ligand binding site based on the 3D structure of the protein.

Server name	URL
LigSitecsc	http://projects.biotec.tu-dresden.de/cgi-bin/index.php
SURFNET	http://www.biochem.ucl.ac.uk/~roman/
APROPOS	http://www.csb.yale.edu/userguides/datamanip/apropos/manual.html
Binding-response	http://mackerell.umaryland.edu/CADD/CADD_bindingresponse.html
CASTp	http://sts.bioengr.uic.edu/castp/
CAVER	http://loschmidt.chemi.muni.cz/caver/
Fpocket	http://sourceforge.net/projects/fpocket/
GHECOM	http://biunit.naist.jp/ghecom/
McVol	http://www.bisb.uni-bayreuth.de/
PocketDepth	http://proline.physics.iisc.ernet.in/pocketdepth/
PocketPicker	http://gecco.org.chemie.uni-frankfurt.de/pocketpicker
Screen	http://interface.bioc.columbia.edu/screen/
SplitPocket	http://pocket.uchicago.edu/
SURFNET	http://www.biochem.ucl.ac.uk/
VOIDOO	http://xray.bmc.uu.se/usf/voidoo.html
SiteMap	http://www.schrodinger.com/
ICM-PocketFinder	http://www.molsoft.com/
Q-SiteFinder	http://www.modelling.leeds.ac.uk/qsitefinder/

which were not solved using experimental techniques. Here, 3D structure prediction is relying on the fold recognition. I-tasser [55] is one of the popular methods to recognize the fold and to build the structure for complex assemblies. However, the method/theory of the homology modeling is invariable as the cases vary and it can be referred from the literature [56–59].

14.5. Visualization of protein–ligand interactions

PL interactions are mostly non-bonded interactions, broadly, hydrogen bonds and hydrophobic interactions. Salt-bridges, π–π, cation–π, metal co-ordinations, water bridges, halogen bond, etc., are different interaction types that exist in biological macromolecular level (Figure 14.3). The basic information on the interaction potential can be found in the literature. To visualize and analyze these intermolecular interactions, many standalone and Web-based tools

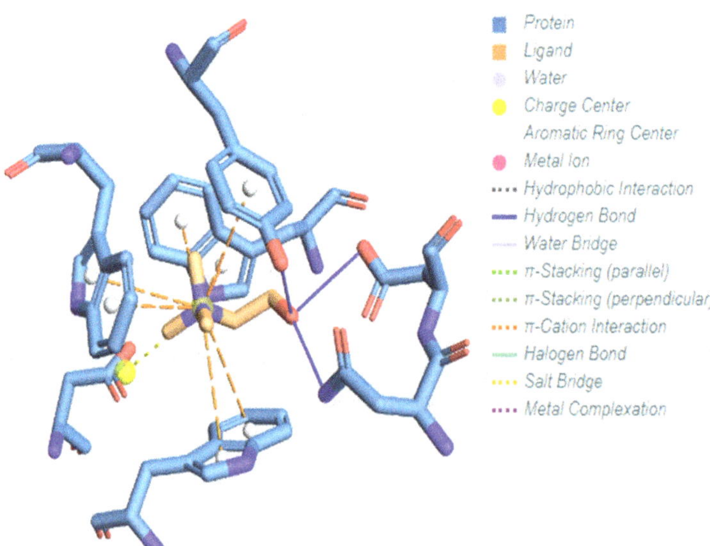

Figure 14.3.   Different types of interactions involved in the PL complexes. Image was generated using PLIP tool for PDB ID: 2REG.

(*Source*: https://projects.biotec.tu-dresden.de/plip-web/plip/index)

available. They accept the complex file from either experimental or computational docking and other molecular modeling methods to visualize the interactions graphically and/or write the summary to the file for further data analyses. Table 14.2 gives the detailed information on these software packages.

Table 14.2. Tools for analyzing and visualizing the PL interactions.

Tool/Server	URL
For analyzing the PL interaction data:	
LPC CSU	http://bip.weizmann.ac.il/oca-bin/lpccsu
MultiBind	http://bioinfo3d.cs.tau.ac.il/MultiBind/
PISA	http://www.ebi.ac.uk/msd-srv/prot_int/pistart.html
PLATINUM	http://model.nmr.ru/platinum/
For visualizing the PL interactions:	
Ligplot	http://www.biochem.ucl.ac.uk/bsm/ligplot/ligplot.html
HBplus	http://www.ebi.ac.uk/thornton-srv/software/HBPLUS/
iview	http://istar.cse.cuhk.edu.hk/iview
PLIP	https://projects.biotec.tu-dresden.de/plip-web/plip/index
PLI	http://bioinformatics.istge.it/pli/
PyPLIF	http://code.google.com/p/pyplif
VMD	https://www.ks.uiuc.edu/Research/vmd/
For comparing the binding sites:	
MultiBind	http://bioinfo3d.cs.tau.ac.il/MultiBind/
PESDserv	http://reccr.chem.rpi.edu/Software/pesdserv/
SiteEngine	http://bioinfo3d.cs.tau.ac.il/SiteEngine/
SuMo	http://sumo-pbil.ibcp.fr/cgi-bin/sumo-welcome
Additionally, below is the list of standalone tools to visualize, analyze and compare the protein structures, complex structures, and interactions.	
Chimera	http://www.cgl.ucsf.edu/chimera/index.html
DeepView	http://expasy.org/spdbv/
Maestro	https://www.schrodinger.com/maestro
Pymol	http://www.pymol.org/
DS	http://www.accelrys.com/products/downloads/ds_visualizer/

Table 14.3.   List of databases available for PLI data obtained from experiments.

Database name	URL
AffinDB	http://pc1664.pharmazie.uni-marburg.de/affinity/
BindingDB	https://www.bindingdb.org/
Chembl	https://www.ebi.ac.uk/chembl/
Credo	http://www-cryst.bioc.cam.ac.uk/credo
Drugbank	https://www.drugbank.ca/
MOAD	http://www.bindingmoad.org/
PDBBind	http://www.pdbbind.org.cn/
PDBeMotif	http://www.ebi.ac.uk/pdbe-site/pdbemotif/
PLID	http://203.199.182.73/gnsmmg/databases/plid/
Protein-Ligand	http://bioinformatics.hsanmartino.it/pli/
Psilo	http://www.chemcomp.com/
Pubchem	https://pubchem.ncbi.nlm.nih.gov/
Relibase	http://relibase.ccdc.cam.ac.uk/

In addition to the interaction data, the experimental and computed binding affinity data associated with the PL complexes help the structural bioinformatics scientists to validate their methods and advance prediction accuracy in SBDD. Some of the widely used such databases are listed in the Table 14.3.

14.6. Conclusion and outlook

PL interaction is a key factor to study the mechanism of binding of the ligands such as small organic compounds (drugs or inhibitors) and macromolecules (proteins/DNA/RNA/polysaccharides, etc.) with the target proteins which are of clinical and industrial importance. Despite the type of ligand that interacts with the target protein, the intermolecular interactions that are acting as stabilizing forces for the complex are common. By paying more attention in the ligands of the small molecular organic compound type, there are many methods such as molecular docking, 3d QSAR and pharmacophore modeling that play a crucial role in the identification of small molecule therapeutics.

In order to study the interactions, quantum mechanics-level calculations are more accurate and they become inaccessible most of the time due to the computational complexity. Hence, this study is relying more on the semi-empirical scoring and other heuristic approaches. In bioinformatics, the knowledge-based approach is employed to study the PL interactions with statistical significance. These approaches are being applied successfully to overcome the time and cost factors that hamper the experimental methods, despite the fact that the phenomena of the non-bonded interactions, biological functions under thermodynamically equilibrium conditions, affinity versus molecular weight, etc., are still unexplored.

X-ray crystallography and NMR play crucial role in analyzing the PL interactions due to advancement in technology to do so even quantitatively. In view of the application of molecular docking to predict the PL interactions, the ligand flexibility is the well-addressed challenge. However, protein flexibility and scoring are still under investigation for better performance.

In summary, the PL interaction is a key factor for successful design of novel therapeutic compounds for any disease based on the target proteins. When more and more human protein structures are available in the future, the design of the novel therapeutic compounds will be more important and meaningful. The experimental and computation methods play an equal and non-redundant role in interaction profiling and make a knowledgebase for further analyses.

Acknowledgments

DV thanks University Grants Commission (UGC) for the award of UGC-BSR Faculty Fellowship and University of Madras for the Honorary Emeritus Professorship. CR thanks Department of Biotechnology, IIT Madras for the postdoctoral fellowship.

References

1. Morris, G.M., Huey, R., Lindstrom, W., *et al.* AutoDock4 and AutoDockTools4: Automated docking with selective receptor flexibility. *J Comput Chem.* 2009;30:2785–2791.

2. Jones, G., Willett, P., and Glen, R.C. Molecular recognition of receptor sites using a genetic algorithm with a description of desolvation. *J Mol Biol.* 1995;245:43–53.

3. Jones, G., Willett, P., Glen, R.C., *et al.* Development and validation of a genetic algorithm for flexible docking. *J Mol Biol.* 1997;267:727–748.

4. Baxter, C.A., Murray, C.W., Clark, D.E., *et al.* Flexible docking using Tabu search and an empirical estimate of binding affinity. *Proteins.* 1998;33:367–82.

5. Kuntz, I.D., Blaney, J.M., Oatley, S.J., *et al.* A geometric approach to macromolecule-ligand interactions. *J Mol Biol.* 1982;161:269–288.

6. Rarey, M., Kramer, B., Lengauer, T., *et al.* A fast flexible docking method using an incremental construction algorithm. *J Mol Biol.* 1996;261:470–489.

7. Friesner, R.A., Banks, J.L., Murphy, R.B., *et al.* Glide: A new approach for rapid, accurate docking and scoring. 1. Method and assessment of docking accuracy. *J Med Chem.* 2004;47:1739–1749.

8. Welch, W., Ruppert, J., and Jain, A.N. Hammerhead: Fast, fully automated docking of flexible ligands to protein binding sites. *Chem Biol.* 1996;3:449–462.

9. Miller, M.D., Kearsley, S.K., Underwood, D.J., *et al.* FLOG: A system to select 'quasi-flexible' ligands complementary to a receptor of known three-dimensional structure. *J Comput Aided Mol Des.* 1994;8:153–174.

10. Eckart, W.U. Science in the third reich. *Minerva.* 2004;42:451–454.

11. Hansen, S.E. The development of adrenal cortical hormones into drugs. *Dan Medicinhist Arbog.* 2008;36:109–132.

12. Ingle, D.J. The biologic properties of cortisone: A review. *J Clin Endocrinol Metab.* 1950;10:1312–1354.

13. Ross, G.A., Morris, G.M., and Biggin, P.C. Rapid and accurate prediction and scoring of water molecules in protein binding sites. *PLOS ONE.* 2012;7:e32036.

14. Kadirvelraj, R., Foley, B.L., Dyekjaer, J.D., *et al.* Involvement of water in carbohydrate-protein binding: Concanavalin A revisited. *J Am Chem Soc.* 2008;130:16933–16942.

15. Li, Z., and Lazaridis, T. Thermodynamics of buried water clusters at a protein-ligand binding interface. *J Phys Chem B.* 2006;110:1464–1475.

16. Li, Z., and Lazaridis, T. The effect of water displacement on binding thermodynamics: Concanavalin A. *J Phys Chem B.* 2005;109: 662–670.

17. Seddon, G., Lounnas, V., McGuire, R., *et al.* Drug design for ever, from hype to hope. *J Comp Aided Mol Des.* 2012;26:137–150.

18. Beddell, C.R., Goodford, P.J., Norrington, F.E., *et al.* Compounds designed to fit a site of known structure in human haemoglobin. *Br J Pharmacol.* 1976;57:201–209.

19. Jiang, F., and Kim, S.H. "Soft docking": Matching of molecular surface cubes. *J Mol Biol.* 1991;219:79–102.

20. Ferrari, A.M., Wei, B.Q., Costantino, L., *et al.* Soft docking and multiple receptor conformations in virtual screening. *J Med Chem.* 2004;47:5076–5084.

21. Nabuurs, S.B., Wagener, M., and de Vlieg, J. A flexible approach to induced fit docking. *J Med Chem.* 2007;50:6507–6518.

22. Carlson, H.A. Protein flexibility is an important component of structure-based drug discovery. *Curr Pharma Des.* 2002;8:1571–1578.

23. Totrov, M., and Abagyan, R. Flexible ligand docking to multiple receptor conformations: A practical alternative. *Curr Opin Struct Biol.* 2008;18:178–184.

24. Knegtel, R.M., Kuntz, I.D., and Oshiro, C.M. Molecular docking to ensembles of protein structures. *J Mol Biol.* 1997;266:424–440.

25. Wei, B.Q., Weaver, L.H., Ferrari, A.M., *et al.* Testing a flexible-receptor docking algorithm in a model binding site. *J Mol Biol.* 2004;337: 1161–1182.

26. Huang, S.Y., and Zou, X. Ensemble docking of multiple protein structures: Considering protein structural variations in molecular docking. *Proteins.* 2007;66:399–421.

27. Huang, S.Y., and Zou, X. Efficient molecular docking of NMR structures: Application to HIV-1 protease. *Protein Sci.* 2007;16:43–51.

28. Bottegoni, G., Kufareva, I., Totrov, M., *et al.* Four-dimensional docking: A fast and accurate account of discrete receptor flexibility in ligand docking. *J Med Chem.* 2009;52:397–406.

29. Claeyssens, F., Harvey, J.N., Manby, F.R., *et al.* High-accuracy computation of reaction barriers in enzymes. *Angew Chem.* 2006;45:6856–6859.

30. Pang, Y.P., Perola, E., Xu, K., *et al.* EUDOC: A computer program for identification of drug interaction sites in macromolecules and drug leads from chemical databases. *J Comput Chem.* 2001;22: 1750–1771.

31. Venkatachalam, C.M., Jiang, X., Oldfield, T., *et al.* LigandFit: A novel method for the shape-directed rapid docking of ligands to protein active sites. *J Mol Graph Model.* 2003;21:289–307.

32. Jain, A.N. Surflex: Fully automatic flexible molecular docking using a molecular similarity-based search engine. *J Med Chem.* 2003;46: 499–511.

33. Sauton, N., Lagorce, D., Villoutreix, B.O., *et al.* MS-DOCK: Accurate multiple conformation generator and rigid docking protocol for multi-step virtual ligand screening. *BMC Bioinformatics.* 2008;9:184.

34. Bohm, H.J. The computer program LUDI: A new method for the de novo design of enzyme inhibitors. *J Comput Aided Mol Des.* 1992;6: 61–78.

35. Shoichet, B.K., and Kuntz, I.D. Protein docking and complementarity. *J Mol Biol.* 1991;221:327–346.

36. Brylinski, M., and Skolnick, J. Q-Dock: Low-resolution flexible ligand docking with pocket-specific threading restraints. *J Comput Chem.* 2008;29:1574–1588.

37. Lipinski, C.A., Lombardo, F., Dominy, B.W., *et al.* Experimental and computational approaches to estimate solubility and permeability in drug discovery and development settings. *Adv Drug Deliv Rev.* 2001;46:3–26.

38. Lipinski, C.A. Lead- and drug-like compounds: The rule-of-five revolution. *Drug Discov Today Technol.* 2004;1:337–341.

39. Michalsky, E., Dunkel, M., Goede, A., *et al.* SuperLigands — a database of ligand structures derived from the Protein Data Bank. *BMC Bioinformatics.* 2005;6:122.

40. Goodwin, R.J., Bunch, J., and McGinnity, D.F. Mass spectrometry imaging in oncology drug discovery. *Adv Cancer Res.* 2017;134: 133–171.

41. Kuo, L.C. How to avoid rediscovering the known. *Methods Enzymol.* 2011;493:159–168.

42. Chagas, C.M., Moss, S., and Alisaraie, L. Drug metabolites and their effects on the development of adverse reactions: Revisiting Lipinski's Rule of Five. *Int J Pharm.* 2018;549:133–149.

43. Erlanson, D.A., Fesik, S.W., Hubbard, R.E., *et al.* Twenty years on: The impact of fragments on drug discovery. *Nat Rev Drug Discov.* 2016;15:605–619.

44. Hall, R.J., Mortenson, P.N., and Murray, C.W. Efficient exploration of chemical space by fragment-based screening. *Prog Biophys Mol Biol.* 2014;116:82–91.

45. Wu, B., Zhang, Z., Noberini, R., *et al.* HTS by NMR of combinatorial libraries: A fragment-based approach to ligand discovery. *Chem Biol.* 2013;20:19–33.

46. Singh, M., Tam, B., and Akabayov, B. NMR-fragment based virtual screening: A brief overview. *Molecules.* 2018;23.

47. Chilingaryan, Z., Yin, Z., and Oakley, A.J. Fragment-based screening by protein crystallography: Successes and pitfalls. *Int J Mol Sci.* 2012;13:12857–12879.

48. Murray, C.W., Callaghan, O., Chessari, G., *et al.* Application of fragment screening by X-ray crystallography to beta-secretase. *J Med Chem.* 2007;50:1116–1123.

49. Recht, M.I., Sridhar, V., Badger, J., *et al.* Fragment-based screening for inhibitors of PDE4A using enthalpy arrays and X-ray crystallography. *J Biom Screen.* 2012;17:469–480.

50. Murray, C.W., Carr, M.G., Callaghan, O., *et al.* Fragment-based drug discovery applied to Hsp90. Discovery of two lead series with high ligand efficiency. *J Med Chem.* 2010;53:5942–5955.

51. Chung, C.W., Dean, A.W., Woolven, J.M., *et al.* Fragment-based discovery of bromodomain inhibitors part 1: Inhibitor binding modes and implications for lead discovery. *J Med Chem.* 2012;55:576–586.

52. Ramakrishnan, C., Joshi, V., Joseph, J.M., *et al.* Identification of novel inhibitors of Daboia russelli phospholipase A2 using the combined pharmacophore modeling approach. *Chem Biol Drug Des.* 2014;84:379–392.

53. Perot, S., Sperandio, O., Miteva, M.A., *et al.* Druggable pockets and binding site centric chemical space: A paradigm shift in drug discovery. *Drug Discov Today.* 2010;15:656–667.

54. Yang, J., Roy, A., and Zhang, Y. Protein-ligand binding site recognition using complementary binding-specific substructure comparison and sequence profile alignment. *Bioinformatics.* 2013;29:2588–2595.

55. Yang, J., and Zhang, Y. Protein structure and function prediction using I-TASSER. *Curr Protoc Bioinformatics.* 2015;52:5.8.1–15.

56. Chothia, C. and Lesk, A.M. The relation between the divergence of sequence and structure in proteins. *EMBO J.* 1986;5:823–826.

57. Marti-Renom, M.A., Stuart, A.C., Fiser, A., *et al.* Comparative protein structure modeling of genes and genomes. *Annu Rev Biophys Biomol Struct.* 2000;29:291–325.

58. Venclovas, C. and Margelevicius, M. Comparative modeling in CASP6 using consensus approach to template selection, sequence-structure alignment, and structure assessment. *Proteins.* 2005;61(supp_7):99–105.

59. Zemla, A. LGA: A method for finding 3D similarities in protein structures. *Nucleic Acids Res.* 2003;31:3370–3374.

Chapter 15

An overview of protein–ligand docking and scoring algorithms

Ruchika Bhat[*,†], Abhilash Jayaraj[†], Anjali Soni[†,§] and B. Jayaram[*,†,‡,¶]

[*]*Department of Chemistry, Indian Institute of Technology, Hauz Khas, New Delhi 110016, India*

[†]*Supercomputing Facility for Bioinformatics and Computational Biology, Indian Institute of Technology, Hauz Khas, New Delhi 110016, India*

[‡]*Kusuma School of Biological Sciences, Indian Institute of Technology, Hauz Khas, New Delhi 110016, India*

[§]*Present address: Biotechnologisches Zentrum, Technische Universität Dresden, Tatzberg 47-51, Dresden 01307, Germany*

An understanding of the rules of receptor interactions with ligand molecules is of utmost importance in the area of computer-aided drug discovery (CADD). Current docking algorithms endeavor toward prediction of biologically relevant ligands, which can bind to a specific cavity/active site of biomolecule(s)/receptor(s) inducing the required upregulation or downregulation. These algorithms aim to predict favorable orientation and conformation of a ligand (small molecule) when bound to a target receptor (protein/DNA) to make a stable complex. The effectiveness of such algorithms largely

---

[¶]bjayaram@chemistry.iitd.ac.in

depends on the adopted mathematical model of scoring, which predicts the binding free energy between the receptor and the ligand. The quality of the available force fields and the extent of conformational sampling make binding free energy estimations challenging. However, incorporation of other approaches, such as machine learning, newer force fields and increased exploration in conformational search space, has made current generation scoring functions more promising. This chapter illustrates the broad classification of various available docking and scoring algorithms, their applications and limitations along with their comparative assessment on PDB-bind core data set of 2018 (same as 2016) release comprising 295 protein–ligand complexes. The advancements in the field of docking and scoring have led to a correlation of ~0.8 with experimental data in generic cases, whereas in specific cases a correlation of over 0.98 has also been reported.

15.1. Introduction

Modern drug discovery is heavily reliant on the discovery of therapeutically relevant small molecules for clinical utilization [1–7] against protein [8–10], miRNA [11, 12] and DNA as targets [13, 14]. This mandates constant development of more efficient means to find such small molecules. One efficacious mode to accelerate this process is to incorporate computer-aided drug design (CADD) approach in the drug discovery pipeline. Docking and scoring serve as cornerstones in this endeavor. The field of protein–ligand docking has been evolving since 1980s and remains an active area of research [15]. There are currently over 200 docking algorithms and over 100 scoring functions available [15–17].

Docking and scoring are accomplished by a combination of a preferred conformational space search algorithm and binding free energy-assessing function. Most of the docking algorithms consider all the possible grid points — to a desired resolution — defining the binding pocket of the receptor and account for translational and rotational degrees of freedom for ligand in this pocket. This would mean that the search space in the complex increases exponentially with increasing resolution. Thus, speed and effectiveness in converging to the relevant configuration have become an important concern [15, 18, 19]. To transcend this concern, docking algorithms are coupled with mathematical models called scoring functions. Role of a scoring function is to favorably rank the true binding pose among the other explored configurations and conformations using a predicted score. The scoring function quantitates all the possible interactions covering a range of thermodynamic quantities to a score. To correctly estimate binding free energy (ΔG) computationally, accurate estimation of various thermodynamic quantities like enthalpy, entropy, solvation effect, electrostatics, stacking, steric clashes, hydrophobicity, and so on is a prerequisite. Numerous studies [20–22] have been published for estimation of these quantities, and a few of these properties can now be calculated accurately.

Hierarchically, docking is a two-stage problem. Typically, the first stage is in the domain of approximate scoring functions, the objective being to screen a library of 10^6–10^9 small molecules and reach an enrichment of 10^2–10^3 small molecules. In the second stage, more rigrous docking methods are deployed based on atomistic level free energy calculations to further optimize these hits to lead molecules. Futher, Boltzmann averaging of binding energy estimations could be done using MMPBSA/MMGBSA [23, 24] techniques that cover the effects of solvation and flexibility implicitly. These calculations require an initial and final conformation of protein–ligand complexes to estimate the change in free energy upon complexation. The dielectric continuum employed by MMPBSA/MMGBSA lacks the molecularity of the solvent, which could play a significant role in binding. To overcome this, the use of free energy simulation methods is deployed, which essentially require multiple intermediate stages apart from initial and final conformations to take care of theoretical rigor needed in capturing the protein–ligand interactions accurately. The limitation of this methodology is the need for extensive computational resources and sampling time for a single protein–ligand complex.

Hence, computational approaches estimating change in binding free energy (ΔG) as a whole can be primarily divided into four categories: docking and scoring functions, MMPBSA/MMGBSA, free energy simulation methods [25] and Quantum Mechanics/Molecular

Mechanics (QM/MM) methods [26, 27]. The latter three are more reliable methodologies for predicting protein–ligand binding energies computationally and assess the contribution of a multitude of factors like enthalpy, entropy, electrostatics, hydrophobicity and effect of solvation, stacking, steric clashes, bridging water, ions, partial charges and other factors toward their binding [28–30]. However, the first one provides a better trade-off between computational time and accuracy by screening milllions of compounds in significantly lesser time.

Statistical mechanically, the states of a protein (P) and ligand (L) as the reactants in aqueous medium and the resultant protein–ligand complex (P*L*) formed in aqueous medium (Figure 15.1) can be described as

$$[P]aq + [L]aq = [P*L*]aq \tag{1}$$

The change in the binding free energy accompanying this process can be estimated using

$$\Delta G^\circ = -RT \ln\left(\frac{Q_{P^*L^*}^{tr}}{Q_P^{tr} * Q_L^{tr}} N_A\right) - RT \ln\left(\frac{Q_{P^*L^*}^{rot}}{Q_P^{rot} * Q_L^{rot}}\right)$$
$$-RT \ln\left(\frac{Z_{P^*L^*.aq}^{int}}{Z_{P.aq}^{int} * Z_{L.aq}^{int}} Q_W\right) + P\Delta V^\circ_{aq} \tag{2}$$

where Qs are molar partition functions for translational and rotational degrees of freedom of protein, ligand and complex, Q_W is the

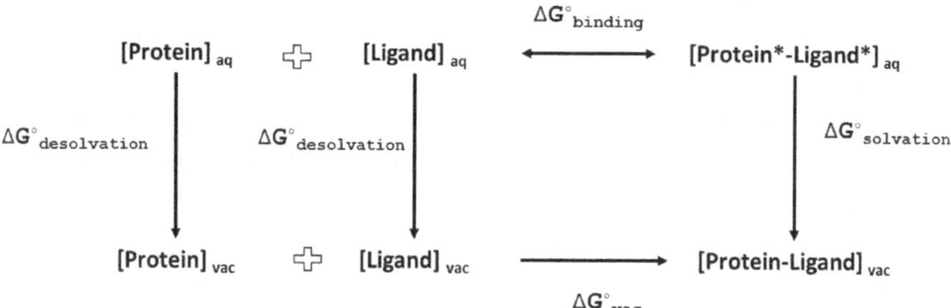

Figure 15.1. A thermodynamic cycle for the protein–ligand binding in aqueous medium.

partition function for pure solvent (water) and $P\Delta V^{\circ}_{aq}$ is the pressure–volume correction to the Helmholtz free energy in the solvent medium, Z^{int} is the configurational partition function.

Equation (2) is an exact expression for noncovalent associations (where the electronic partition function $Q^{el} \approx 1$) in aqueous medium. Equation 2 can be approximated by decoupling the internal/external motions from solvation degrees of freedom as

$$\Delta G^{\circ} = \Delta G^{tr} + \Delta G^{rot} + \Delta G^{int} + \Delta G^{sol} \tag{3}$$

The next approximation is to neglect the solvation effect $P\Delta V^{\circ}_{aq}$ for liquid state work. To make the computations effective in terms of resources and time, certain futher approximations are implemented in Eq. (3), such as freezing of translations and rotations that simplifies Eq. (3) (Figure 15.1) to

$$\Delta G^{\circ} = \Delta G^{\circ}_{vac} + \Delta G^{\circ}_{sol} \tag{4}$$

The translational and rotational contributions in gas phase are computed. Also the entropies are computed through normal mode analysis or its variants such as quasi-harmonic approximations and intramolecular energetics through a force field on the structures generated according to Boltzmann distribution/Newton's laws. Solvation effects are added *post facto* for each structure considered above. Essentially, MMGBSA/MMPBSA fall in this category. The most commonly used is Eq. (4). The simplest is to use static experimental structural information and compute the thermodynamic energies in gas state. To mitigate effect of approximations and stay close to Eq. (2), certain ways to improve them can be adapted. This can be accomplished by generating an ensemble of structures of the reactants and products separately in the solvent medium and then applying Eq. (3). This corresponds to a *post facto* analysis of the molecular dynamics (MD) trajectories. In a sense, the plethora of scoring functions are mimics of Eq. (2), but accomplishing the task computationally expeditiously.

In this chapter, we discuss different types of docking and scoring methods (Figure 15.2). Further, we attempt to provide an overview

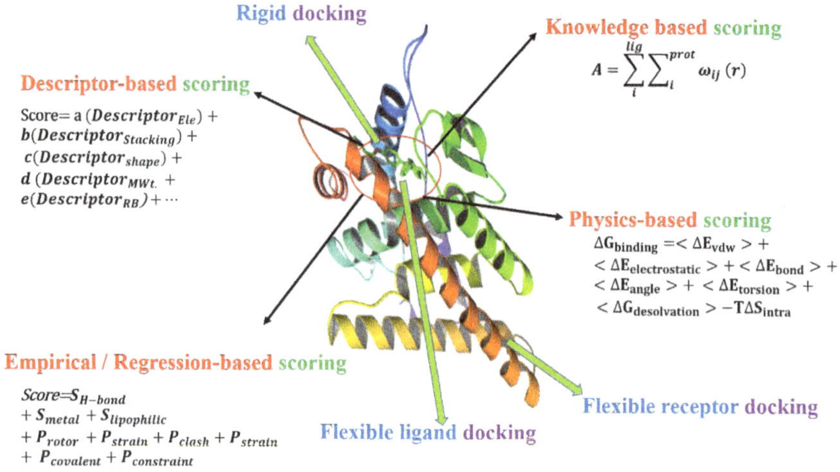

Figure 15.2.   Different types of docking methods and scoring functions.

of the advantages and limitations of these techniques alluding to our own experiences and various published reports from other laboratories.

15.2. Docking algorithms

Docking algorithms intend to find energetically favorable poses of a ligand in the binding pocket of a target biomolecule [17, 31–35]. This enables differentiation of binders from nonbinders from a pool of millions of molecules for any given biomolecular target, in turn, enabling prediction of therapeutically relevant compounds of probable clinical utility. This process has to be both time and cost efficient [8, 10, 36, 37]. The objective of docking algorithms is to generate the conformation of the protein–ligand complex as close as possible to its experimental conformation [38–40]. The accuracy of the algorithm is inversely proportional to the root mean square deviations (RMSDs) between the native and predicted protein–ligand complex. In order to account for the computationally extensive number of degrees of freedom, rigid-body approximations and its variants are usually adapted, which considers six degrees of freedom (rotational and translational) while excluding the overall flexibility of both ligand and protein or considering the protein as rigid and ligand as flexible or considering the protein as partially flexible and ligand as fully flexible [3, 41–44]. Based on this, three basic types of docking algorithms are pursued as discussed below [41].

15.2.1. *Rigid docking*

These algorithms sample different conformations of a ligand in the receptor cavity. They assume receptor rigidity for computational simplicity. Rigid-body docking [7] produces a large number of docked conformations with favorable surface complementarity, followed by a reranking of the conformations using some binding free energy approximation. For example, ParDOCK [45], DOCK [41] and FRED [46] belong to this category. The speed and the relative accuracy of rigid-body docking algorithms make them outstanding especially as a strong filter of hierarchical structure-based virtual screening projects in order to remove nonbinders [47]. Rigid docking followed by MD simulations serves as a dependable alternative to flexible docking. This combined approach is extensively applied nowadays in structure-based drug design protocols. Several cases reporting rigid docking followed by MD simulations have given rise to easy identification of lead molecules in sub micromolar and nanomolar scales [8, 48–51].

15.2.2. *Flexible ligand docking*

For making the ligand flexible, "systematic docking" approach is followed, which covers conformational search methods, fragmentation methods and database methods where all the degrees of freedom of a small molecule are considered. Examples of these methods include LUDI [52], FlexX [53], DOCK [54] and FLOG [55]. Another approach is to generate random conformations of the small molecule. This conformation is accepted or rejected on the basis of certain predefined probability functions. Primarily there are three types of random algorithms, namely Monte Carlo (MC), Genetic Algorithm

(GA) and Tabu search methods. Prominent examples include DockVision [56], QXP [57], MCDOCK [35], GOLD [58], DARWIN [59], AutoDock [60], PRO_LEADS [61] and several others. Also a third approach is used, which can be divided into two types: MD and energy minimization. In energy minimization approach, the steepest descent and conjugate gradient-type approaches are utilized to reach the best possible energetically stable conformation of the molecule. In the MD approach, the conformational sampling is generated based on Newton's second law. Relevant examples include ProDOCK [62], DARWIN [59], DOCK [63] and several others.

15.2.3. *Flexible protein docking*

These approaches assume the flexibility of both ligand and receptor and are expected to provide the best results. However, the complete flexibility assumption tends to exponentially increase the computational cost of screening of a sufficiently large small molecule database. Hence, various restraints are used to keep the computations tractable. Almost all approaches in this class of tools assume flexibility only in the residues of the active site region of receptor and rigidity in the rest of the receptor. The flexible residues are described using rotamer libraries of side chains. Ligand conformations are selected based on energy thresholds of the conformations being sampled. The approach of MD and MC techniques explained under the section of simulation methods of docking also hold true for this method. The primary difference, however, is that the conformational search space increases for the receptor molecules exponentially as compared to small molecules. This method is based on the generation of conformational libraries representing the experimentally observed and preferred rotameric states for each amino acid side chain of the protein. Docking of different conformation structures of protein instead of one rigid structure is viewed as an alternative strategy to encompass protein flexibility. An example could be FlexE [64], which is an extension of FlexX [53]. The solution to efficiently analyze the true conformational space is to utilize a combination of all these methods, which have shown various promising results in the past [28].

15.3. Scoring functions for binding free energy estimations

Scoring functions are commonly used in structure-based drug discovery techniques for evaluating the affinity of protein–ligand complexes. The first scoring function was made available in the early 1990s, and presently, there are over a hundred scoring functions published in literature [65–70]. Scoring functions are not the most accurate methods to find an estimate of binding affinity of a protein–ligand complex as they make various approximations in order to compensate for speed, time and computational resources. These are particularly suitable for some high-throughput tasks, such as library design, virtual screening, molecular docking and so on [18]. Despite having associated error rate, these functions are able to select biologically active molecules with a significantly higher success rate. Such improvement in the rate of finding hits in a database is commonly referred to as enrichment.

Based on how a scoring function is developed, scoring functions can be classified into four types: physics-based methods, knowledge-based potentials, empirical scoring functions and descriptor-based scoring functions [18].

15.3.1. *Physics or force-field-based methods*

"Physics-based" scoring functions utilize force-field parameters. Force fields include bonded and nonbonded parameters. Energy contributions from bonded terms are derived due to bond stretching, bending and variations in angles and torsions. The nonbonded energy terms in a force field include the van der Waals and the electrostatic energies, which implicitly include hydrogen bonding terms [18, 71]. In general, force-field-based energy functions compute potential energy in the gas phase, which can later be augmented by solvation energy terms computed with either Poisson–Boltzmann (PB) or Generalized Born (GB) continuum solvation models [23] to give an improved account of internal and solvation energies. These are also termed as "force-field based" methods. A few examples of such scoring functions are DOCK [2], AutoDock [60], BAPPL [72], BAPPL-Z [73], AADS [37], ParDOCK [45], COMBINE [66], GoldScore [74], MedusaScore [75] and so on.

$$\Delta G_{\text{binding}} = \left\langle \Delta E_{\text{vdw}} \right\rangle + \left\langle \Delta E_{\text{electrostatic}} \right\rangle + \left\langle \Delta E_{\text{bond}} \right\rangle + \left\langle \Delta E_{\text{angle}} \right\rangle$$
$$+ \left\langle \Delta E_{\text{torsion}} \right\rangle + \left\langle \Delta G_{\text{desolvation}} \right\rangle - \text{T} \Delta S_{\text{intra}} \tag{5}$$

where angular brackets denoting ensemble averages collected via MD trajectories could be generated such as in MMBAPPL [72, 32]. However, the usual implementation of Eq. (5) is in the context of a single protein–ligand structure, which means Boltzmann averaging is neglected.

15.3.2. *Knowledge-based or potential of mean force methods*

"Knowledge-based" scoring functions utilize the sum of pairwise statistical potentials between the protein and ligand. SMoG [76], a *de novo* design program, was the first scoring function of this type published in 1996. A decade later, methods based on this type of scoring functions have rapidly evolved such as Muegge's PMF [77], DrugScore [78], IT-Score [79] and KECSA [80]. The basic hypothesis of calculating energy using this type resembles that of the statistical mechanics analysis of liquids, where inverse Boltzmann analysis leads to efficient conversion of a histogram of interatomic distances into potentials of mean force. Due to this reason, the method is also referred to as "potential of mean force" method. However, the limitation of this method is that neither the protein nor the ligand is a randomized assembly of atoms as in liquid; instead they are constrained by covalent bonds in a certain specific order. Studies have pointed out that the occurrence frequency of a certain atom pair in real protein–ligand complex structures could not be assumed to be in a Boltzmann distribution [24]. A few methods combine the potentials with solvation and entropy terms to get an overall view of the estimate of binding affinity.

$$A = \sum_{i}^{\text{lig}} \sum_{i}^{\text{prot}} \omega_{ij}(r) \tag{6}$$

$$\omega_{ij}\left(r\right) = -k_B T \ln\left[g_{ij}\left(r\right)\right] = -k_B T \ln\left[\frac{\rho_{ij}\left(r\right)}{\rho_{ij}*}\right] \tag{7}$$

15.3.3. *Empirical or regression-based methods*

The third type of scoring functions are "empirical" or "regression-based" methods. Böhm in 1994 published [81] first general-purpose empirical scoring function, which is still available presently in the Discovery Studio software. An empirical scoring function utilizes the important energetic factor contributors, such as hydrogen bonding, coordinate bonds with metals, lipophilic contacts, rotation, strain, clash and so on, and estimates the binding by summing up the reward and penalties in terms of their favorability for protein–ligand binding. The examples of this scoring function are ChemScore [82] implemented in the GOLD software [58, 74], X-Score [83], PLP [84], GlideScore [85] and so on. A sample representative scoring equation is as follows [58]:

$$\begin{aligned} \text{Score} = {} & S_{\text{H–bond}} + S_{\text{metal}} + S_{\text{lipophilic}} + P_{\text{rotor}} + P_{\text{strain}} \\ & + P_{\text{clash}} + P_{\text{covalent}} + P_{\text{constraint}} \end{aligned} \tag{8}$$

Although both the physics-based methods and the empirical scoring functions rely on the energy-contributing factors, the empirical methods are more intuitive as they incorporate a variety of energetic contributors. Further, these are amenable to systematic improvements using "regression-based" methods.

15.3.4. *Descriptor or machine learning-based methods*

The fourth type of scoring functions are "descriptor-based" methods, which represent a new trend in this field by the introduction of quantitative structure–activity relationship (QSAR) analyses using various physicochemical, biological and pharmaceutical properties of small molecules and proteins as well as their interaction patterns. One such example of QSAR-based scoring function is RASPD [86]. This QSAR analysis can be applied to derive statistical models that compute protein–ligand binding scores based on a machine learning technique applied to these descriptors of protein and ligand. A few examples of the scoring functions using this method include NNScore [87], RF-Score [88], SFCscore^{RF} [89], BAPPL+ [17], ID-Score [90] and several others. The descriptors accounting for specific interactions such as electrostatic interactions, hydrogen bonds or aromatic stacking, surface and shape properties, molecular weight, number of rotatable single bonds and so on are considered during the learning process. Thus, these are also referred to as "machine learning-based" methods. Different machine learning algorithms, such as random forest, Bayesian classifiers, neural network and support vector machine, are then employed for selecting the descriptors and their weightages necessary to estimate the protein–ligand interaction scores [89]. A sample scoring function equation is as follows:

$$\text{Score} = a(\text{Descriptor}_{\text{Ele}}) + b(\text{Descriptor}_{\text{Stacking}}) + c(\text{Descriptor}_{\text{shape}}) \\ + d(\text{Descriptor}_{\text{MWt.}}) + e(\text{Descriptor}_{\text{RB}}) + \cdots \tag{9}$$

Similar to empirical scoring functions, these methods also need to be trained on a data set of protein–ligand complexes with known structures and binding data to derive their final models and binding energy values. However, more recently "negative" data have also been incorporated to train these scoring functions that have improved the predictability of such approaches to a higher level [91]. Our own analysis in developing BAPPL+ [17], which is a descriptor (machine learning)-based method, has shown significant improvement (Figure 15.2) over other scoring methodologies including our earlier work with physics-based methods such as BAPPL [72] and BAPPL-Z [73]. Thus, these methods have achieved better correlations to protein–ligand binding data than other types of scoring functions.

15.4. CASF, D3R, CASR and other challenges

Computational methods, especially screening and docking, are widely used in different stages of drug discovery. So, it has become important to periodically evaluate the utility and working of all available tools and to identify areas for improvement. This evaluation can be performed at the following three levels:

(i) Screening challenge, that is, identifying new ligands using virtual screening
(ii) Scoring challenge, that is, predicting the binding affinities of related compounds from a known active series
(iii) Ranking challenge, that is, predicting the best binding mode/ pose of a known active ligand

Out of these three challenges, identification of ligand-binding mode/pose in a protein active site has achieved the highest success [33]. Scoring evaluations are done routinely over a specific period of time such as comparative assessment of scoring functions (CASFs) http://www.pdbbind-cn.org/casf.asp/, drug design data resource (D3R) https://drugdesigndata.org, Community Structure-Activity Resource (CSAR) https://www.omg.org/spec/CSAR/1.0 and others. Some special benchmarks have been created by these challenges such as the CASF benchmark [71, 92] and the CSAR exercise [93–95], so that different scoring functions can be compared in a more objective manner.

The result of one such challenge, CASF-2007 (a benchmark data set of 195 diverse protein–ligand complexes), evaluated on a number of conventional scoring functions showed the resultant Pearson correlation coefficients (R) to be in the range of 0.216–0.644 between their binding scores and experimental binding data [92]. During these tests, the "R" values produced by SFCscoreRF [89], RF-Score and ID-Score [90], which are descriptor-based or machine learning-based methods, were 0.779, 0.803 and 0.753, respectively. This demonstrated superior performance by descriptor-based scoring functions as compared to other scoring methods [92].

In the 2010 CSAR challenge, 20 participating research groups were provided with a set of ligands for four proteins (Lpxc, UroKinase, Chk1 and Erk2), which resulted in an overall better ranking of inactive compounds in comparison to experimentally known active compounds [93].

D3R challenge in 2015 attracted 40 research groups, which tried to predict the binding pose and affinity of 180 ligands to HSP90 and 30 ligands for MAP4K4 protein targets. HSP90 has an ATP-binding loop region, which is known to adopt multiple conformations and over 200 different crystal structures of HSP90 are deposited in RCSB databank. The analysis later revealed that the selection of protein crystallized with similar ligand played a vital role in accurate predictions. In real-life situations, the availability of similar ligands crystallized to the target protein is unlikely and this emphasizes that sampling of receptor conformation is vital to success in docking and scoring in real-time situations. The next version of D3R challenge happened from September 2016 to February 2017. The output of this challenge pointed out that while knowledge of experimentally known poses of ligands is a useable guide, it cannot serve as the only guide to docking and scoring. Random forest- and MMGBSA-based methods yielded comparatively better results in this challenge. These challenges highlighted the fact that each docking and scoring methodology has its own areas of applicability and sets of limitations, implying that there is significant room for improvement.

To evaluate the efficiency of ever-evolving docking and scoring methodologies, curated subsets of the structural collection at RCSB have been developed. These subsets are selected based on their reliability and applicability to testing, docking and scoring programs. There are several curated test sets available for comparing docking, scoring and ranking methods, namely PDBbind [96], BindingDB [97], Binding MOAD [98] and CCDC/Astex.

The most accurate methods for predicting free energy of binding in the present time are QM/MM, thermodynamic integration/free energy perturbation, which can calculate the differences in affinities between related molecules to an accuracy of ± 1 kcal/mol [99, 100]. Because these methods demand extensive computation time, they are less useful for screening huge libraries containing millions of compounds. Thus, one has to rely on docking and scoring methods.

A comparison of the scoring ability of different scoring functions on the latest PDBBind core data set of 2018 (same as 2016) release with 295 experimentally known protein–ligand complexes is shown in

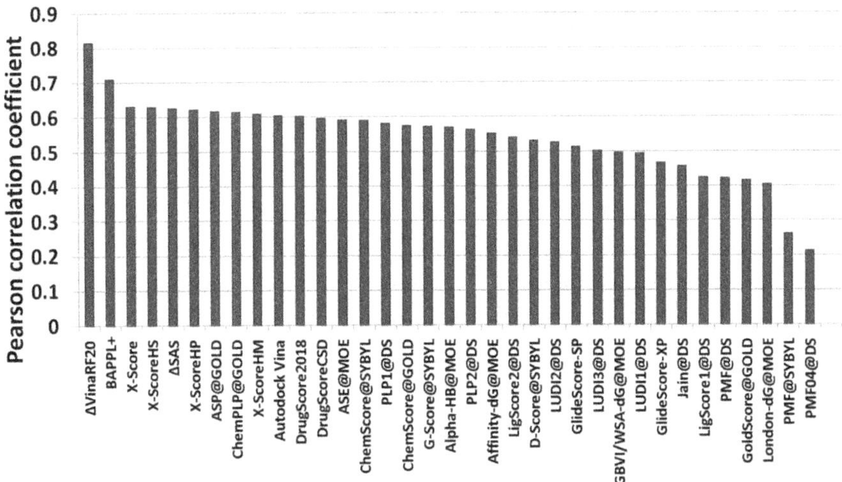

Figure 15.3. Comparison of various software in terms of their scoring efficacy on a core data set of 2016 release of PDBbind [32, 71, 72]. The histogram is showing the Pearson coefficient of correlation and the standard deviation ranges from 1.26 to 2.11. The unit of standard deviation mentioned is in log K. These data are taken from Minyu Su *et al.* (2019) study [71].

Figure 15.3 [32, 71, 72]. As evident from the figure, descriptor-based or machine learning-based methods have been ranked best in terms of their scoring ability in estimating the overall binding energies of protein–ligand complexes.

One way of comparison of experimental versus predicted outcomes is through RMSD. If the RMSD is generated only for the binding pocket overlaps [32, 37] between the actual and predicted poses, it would give a better insight into the accuracy of the methodology adopted. However, most of the challenges consider the RMSD of the two structures for assessing the efficiency of pose generation module of a docking tool.

15.5. Advantages and pitfalls of machine learning methods

In terms of virtual screening, the scoring functions are grappling with efficiently differentiating low- or medium-affinity binders from nonbinders. Screening millions of molecules to select a few binders toward a biomolecular target demands both high-speed and modest accuracy. This challenge has given rise to the ever-evolving current generation of QSAR-based scoring techniques such as RASPD [86], AutoDock Virtual Screening [60], Schrodinger Glide [85] and so on. The outcomes of standard benchmarks suggest that current screening techniques have achieved significant progress in terms of predicting the true positives along with false positives within acceptable limits from a large data set of small molecules. These scoring functions are, however, yet to have comparable performance against experimental values in terms of "docking/scoring/ranking," that is, predicting correct pose and protein–ligand binding affinities.

All the empirical scoring functions, as well as physics-based methods and knowledge-based potentials, are linear functional forms. However, the descriptor-based scoring functions, depending on machine learning techniques are not necessarily linear functions. A descriptor-based scoring function usually consists of a considerably larger number of parameters than any empirical scoring function. One cannot always obtain a converged model, as the input for machine learning is given as different sets of descriptors. An empirical scoring function adopts a theory-inspired hypothesis to calculate the binding score. On the other hand, the descriptor-based scoring function relies on machine learning technique to select the final trained model and thus, is more random but robust. The individual terms in an empirical scoring function normally have an interpretable physical meaning, whereas in the case of a descriptor-based method, the rationale for selecting a combination of descriptors is often ambiguous. In this sense, a descriptor-based scoring function is essentially a "black box" just like many QSAR models. Thus, although machine learning has a wide application due to its accuracy and speed, its limitations originate in the lack of interpretability and the data sets adopted for training and testing.

15.6. Summary and perspectives

The most reliable approaches for estimating binding free energies are QM/MM methods, free energy simulations and MMPBSA/

MMGBSA/MMBAPPL. However, due to their large computational resource and time requirements, they are not ideally suited for multiple calculations. Docking and scoring play an important role to cover the time and resource expense yet maintaining the accuracy to a decent level. For the docking and scoring algorithms to work well, certain parameters need to be taken into account. The location of the binding site is one of them and plays an important role in estimating the protein–ligand binding [66, 71, 78, 93, 101]. Certain cases have been reported where the presence of allosteric sites within the receptor molecules can influence the protein–ligand binding [102–104] to a substantial extent. Also, metal ions and water molecules present in the vicinity of functional sites influence the binding of small molecules to target proteins. There has been a steady increase in the number of crystal structures of the macromolecules with approximately 40,000 protein–ligand structures being available in the PDB library [96] (as of January 2019). These structures also serve as a data set for the type of interactions and conformational inclinations of ligands in the presence of ions and water molecules. These data sets play a vital role in training and testing the docking software.

Additionally, assignment of protonation states and partial charges of ligands plays a crucial role in efficiently quantifying the binding free energies of protein–ligand complexes. The sensitivity of a docking algorithm depends on the correct charge assignment of ligand molecules and active site residues. In the absence of partial charge assignment to the ligand atoms, the scoring function will incorrectly estimate the effect of electrostatic interactions between protein and ligand and the van der Waals interactions will become dominant, thus giving a false estimate of binding free energies. Various methods to assign partial charges of ligands such as AM1-BCC [29], Mulliken [105], Gasteiger [106], Kollman [107], restrained electrostatic potential [108] (RESP) and so on utilize both empirical- and QM-based methods. QM calculations can utilize docking and scoring to address metabolism-related toxitiy issues such as finding sites of metabolism (SOM) [109] of a candidate molecule.

Nevertheless, based on the progress made in the recent years by a few new methods such as ΔVinaRF20 and BAPPL+ [32, 71, 72] as per 2018 PDBbind core data set (same as 2016), we are optimistic that these exciting approaches of scoring will help in increasing the efficacy of prediction of true binders. Industrial and government organizations can contribute toward the growth CADD by providing free access to true and false-positive data sets. These curated data and benchmarking data sets would help in developing advanced docking and scoring methodologies. Although there is room for development in the available docking and scoring techniques, there have been several success stories of low micromolar inhibitors routinely developed using the CADD approaches [8, 51, 110–112]. One such recently published case study utilizing an in-house, *Dhanvantari* Web tool, is shown in Figure 15.4.

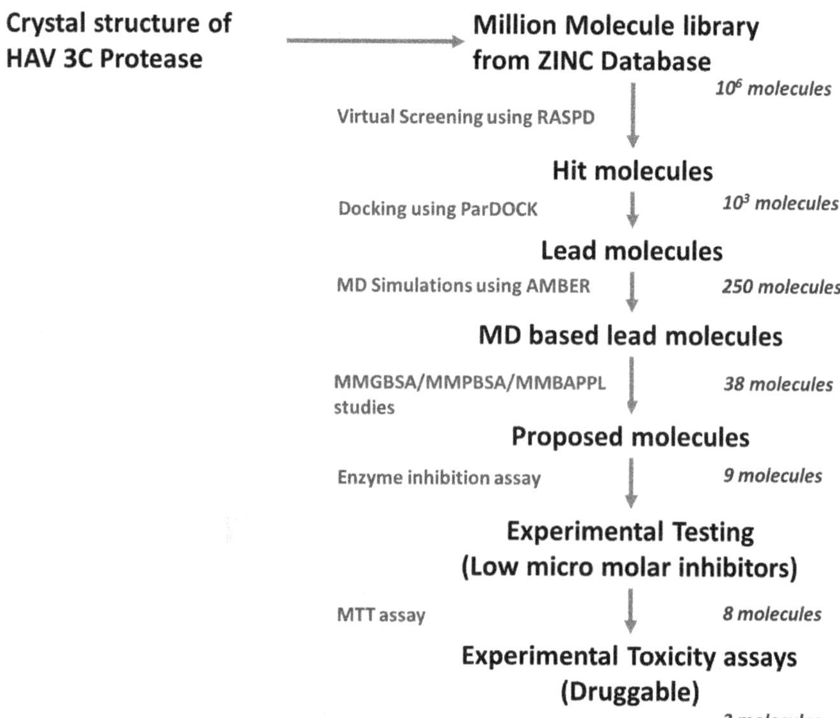

Figure 15.4.   A case study demonstrating the application of structure-based drug discovery process utilizing docking and scoring algorithms [8].

In summary, docking and scoring methlogies have taken a major place in the field of CADD. The challenges of CASF, CSAR and D3R have proven to be efficient in highlighting the scope for improvements. The computational drug discovery does not confine itself to structure-based approach only. One can do reverse docking. In reverse docking, the ligand can be docked to a pool of receptors and to check which receptors bind efficiently to the small molecule to help identify biomolecular targets for bioactive compounds. These can be achieved using software such as BAITOC (http://www.scfbio-iitd. res.in/baitoc), SEA [113], Target hunter [114] and so on. Thus, in combination with advancements in drug design protocols, informatics and experimental validations one can expect some major progresses in drug discovery in the years to come.

Acknowledgments

Funding from the Department of Biotechnology, Government of India, to the Supercomputing Facility for Bioinformatics and Computational Biology, IIT Delhi, is gratefully acknowledged.

References

1. Lionta, E., Spyrou, G., Vassilatis, D., *et al.* Structure-based virtual screening for drug discovery: Principles, applications and recent advances. *Curr Top Med Chem.* 2014;14(16):1923–1938.
2. Meng, E.C., Shoichet, B.K., and Kuntz, I.D. Automated docking with grid-based energy evaluation. *J Comput Chem.* 1992;13(4):505–524.
3. May, A., and Zacharias, M. Accounting for global protein deformability during protein–protein and protein–ligand docking. *Biochim Biophys Acta.* 2005;1754(1–2):225–231.
4. Śledź, P., and Caflisch, A. Protein structure-based drug design: From docking to molecular dynamics. *Curr Opin Struct Biol.* 2018;48: 93–102.
5. Kitchen, D.B., Decornez, H., Furr, J.R., *et al.* Docking and scoring in virtual screening for drug discovery: Methods and applications. *Nat Rev Drug Discov.* 2004;3(11):935–949.

6. Huang, S.Y., Grinter, S.Z., and Zou, X. Scoring functions and their evaluation methods for protein–ligand docking: Recent advances and future directions. *Phys Chem Chem Phys.* 2010;12(40):12899.

7. Pagadala, N.S., Syed, K., and Tuszynski, J. Software for molecular docking: A review. *Biophys Rev.* 2017;9(2):91–102.

8. Banerjee, K., Bhat, R., Rao, V.U.B., *et al.* Toward development of generic inhibitors against the 3C proteases of picornaviruses. *FEBS J.* 2019;286(4):765–787.

9. Jayaram, B., Dhingra, P., Mukherjee, G., *et al. Genomes to hits : The emerging assembly line in silico.* Proceedings of the Ranbaxy Science Foundation 17th Annual Symposium on *"New Frontiers in Drug Design, Discovery and Development"* 2012, Chapter 3, 13–35.

10. Soni, A., Pandey, K., Ray, P., *et al.* Genomes to hits in silico — A country path today, a highway tomorrow: A case study of chikungunya. *Curr Pharm Des.* 2013;19(26):4687–4700.

11. Schmidt, M.F. Drug target miRNAs: Chances and challenges. *Trends Biotechnol.* 2014;32(11):578–585.

12. Zhonghan, L., and Tariq, M.R. Therapeutic targeting of microRNAs: Current status and future challenges. *Nat Rev Drug Discov.* 2014;13(8):622–638.

13. Shaikh, S.A., and Jayaram, B. A swift all-atom energy-based computational protocol to predict DNA–ligand binding affinity and Δ T M. *J Med Chem.* 2007;50(9):2240–2244.

14. Soni, A., Khurana, P., Singh, T., *et al.* A DNA intercalation methodology for an efficient prediction of ligand binding pose and energetics. *Bioinformatics.* 2017;33(10):btx006.

15. Sousa, S.F., Fernandes, P.A., and Ramos, M.J. Protein-ligand docking: Current status and future challenges. *Proteins.* 2006;65(1):15–26.

16. Hughes, J., Rees, S., Kalindjian, S., *et al.* Principles of early drug discovery. *Br J Pharmacol.* 2011;162(6):1239–1249.

17. Soni, A., Bhat, R., and Jayaram, B. Improving the binding free energy estimations for protein-ligand complexes using machine learning approach. 2019, *Manuscript in Submission.*

18. Liu, J., and Wang, R. Classification of current scoring functions. *J Chem Inf Model.* 2015;55(3):475–482.

19. Sliwoski, G., Kothiwale, S., Meiler, J., *et al.* Computational methods in drug discovery. *Pharmacol Rev.* 2014;66(1):334–395.

20. Roy, A., Hua, D.P., Ward, J.M., *et al.* Relative binding enthalpies from molecular dynamics simulations using a direct method. *J Chem Theory Comput.* 2014;10(7):2759–2768.

21. Lai, B., and Oostenbrink, C. Binding free energy, energy and entropy calculations using simple model systems. *Theor Chem Acc.* 2012; 131(10):1272.

22. Du, X., Li, Y., Xia, Y.L., *et al.* Insights into protein–ligand interactions: Mechanisms, models, and methods. *Int J Mol Sci.* 2016;17(2):144.

23. Gohlke, H., and Case, D.A. Converging free energy estimates: MM-PB(GB)SA studies on the protein-protein complex Ras-Raf. *J Comput Chem.* 2004;25(2):238–250.

24. Onufriev, A., Bashford, D., and Case, D.A. Modification of the generalized born model suitable for macromolecules. *J Phys Chem B.* 2000; 104(15):3712–3720.

25. Wu, D. Understanding free-energy perturbation calculations through a model of harmonic oscillators: Theory and implications to improve the sampling efficiency by molecular simulation. *J Chem Phys.* 2010; 133(24):244116.

26. Duarte, F., Amrein, B.A., Blaha-Nelson, D., *et al.* Recent advances in QM/MM free energy calculations using reference potentials. *Biochim Biophys Acta.* 2015;1850(5):954–965.

27. Rod, T.H., and Ryde, U. Accurate QM/MM free energy calculations of enzyme reactions: Methylation by catechol O -methyltransferase. *J Chem Theory Comput.* 2005;1(6):1240–1251.

28. Mukherjee, G., Patra, N., Barua, P., *et al.* A fast empirical GAFF compatible partial atomic charge assignment scheme for modeling interactions of small molecules with biomolecular targets. *J Comput Chem.* 2011;32(5):893–907.

29. Jakalian, A., Jack, D.B., and Bayly, C.I. Fast, efficient generation of high-quality atomic charges. AM1-BCC model: II. Parameterization and validation. *J Comput Chem.* 2002;23(16):1623–1641.

30. Woods, R.J., and Chappelle, R. Restrained electrostatic potential atomic partial charges for condensed-phase simulations of carbohydrates. *Theochem.* 2000;527(1–3):149–156.

31. Berlin, I., Zimmer, R., Thiede, H., *et al.* Comparison of the monoamine oxidase inhibiting properties of two reversible and selective monoamine oxidase-A inhibitors moclobemide and toloxatone, and assessment of their effect on psychometric performance in healthy subjects. *Br J Clin Pharmacol.* 1990;30(6):805–816.

32. Singh, T., Adekoya, O.A., and Jayaram, B. Understanding the binding of inhibitors of matrix metalloproteinases by molecular docking, quantum mechanical calculations, molecular dynamics simulations, and a MMGBSA/MMBappl study. *Mol Biosyst.* 2015;11(4):1041–1051.

33. Yuriev, E., and Ramsland, P.A. Latest developments in molecular docking: 2010–2011 in review. *J Mol Recognit.* 2013;26(5):215–239.

34. Yuriev, E., Holien, J., and Ramsland, P.A. Improvements, trends, and new ideas in molecular docking: 2012–2013 in Review. *J Mol Recognit.* 2015;28(10):581–604.

35. Liu, M., and Wang, S. MCDOCK: A Monte Carlo simulation approach to the molecular docking problem. *J Comput Aided Mol Des.* 1999;13(5):435–451.

36. Jayaram, B. Sanjeevini: A freely accessible web-server for target directed lead molecule discovery. *BMC Bioinformatics.* 2012;13 (suppl 17):S7.

37. Singh, T., Biswas, D., and Jayaram, B. AADS — An automated active site identification, docking, and scoring protocol for protein targets based on physicochemical descriptors. *J Chem Inf Model.* 2011; 51(10):2515–2527.

38. Li, L., Chen, R., and Weng, Z. RDOCK: Refinement of rigid-body protein docking predictions. *Proteins Struct Funct Genet.* 2003;53(3): 693–707.

39. Makeneni, S., Thieker, D.F., and Woods, R.J. Applying pose clustering and MD simulations to eliminate false positives in molecular docking. *J Chem Inf Model.* 2018;58(3):605–614.

40. Awuni, Y., and Mu, Y. Reduction of false positives in structure-based virtual screening when receptor plasticity is considered. *Molecules.* 2015;20(3):5152–5164.

41. Kuntz, I.D., Blaney, J.M., Oatley, S.J., *et al.* A geometric approach to macromolecule-ligand interactions. *J Mol Biol.* 1982;161(2):269–288.

42. Totrov, M., and Abagyan, R. Flexible ligand docking to multiple receptor conformations: A practical alternative. *Curr Opin Struct Biol.* 2008;18(2):178–184.

43. Cavasotto, C.N., Kovacs, J.A., and Abagyan, R.A. Representing receptor flexibility in ligand docking through relevant normal modes. *J Am Chem Soc.* 2005;127(26):9632–9640.

44. Dobbins, S.E., Lesk, V.I., and Sternberg, M.J.E. Insights into protein flexibility: The relationship between normal modes and conformational change upon protein-protein docking. *Proc Natl Acad Sci.* 2008;105(30):10390–10395.

45. Gupta, A., Sharma, P., and Jayaram, B. ParDOCK: An all atom energy based Monte Carlo docking protocol for protein-ligand complexes. *Protein Pept Lett.* 2007;14(7):632–646.

46. McGann, M.R., Almond, H.R., Nicholls, A., *et al.* Gaussian docking functions. *Biopolymers.* 2003;68(1):76–90.

47. Mozziconacci, J.C., Arnoult, E., Bernard, P., *et al.* Optimization and validation of a docking-scoring protocol; application to virtual screening for COX-2 inhibitors. *J Med Chem.* 2005;48(4):1055–1068.

48. Mueller, H., Wildum, S., Luangsay, S., *et al.* A novel orally available small molecule that inhibits Hepatitis B virus expression. *J Hepatol.* 2018;68(3):412–420.

49. Luo, M., Air, G.M., and Brouillette, W.J. Design of aromatic inhibitors of influenza virus neuraminidase. *J Infect Dis.* 1997;176(suppl 1): S62–S65.

50. Hoda, N., Naz, H., Jameel, E., *et al.* Curcumin specifically binds to the human calcium–calmodulin-dependent protein kinase IV: Fluorescence and molecular dynamics simulation studies. *J Biomol Struct Dyn.* 2016;34(3):572–584.

51. Bhatnagar, S., Soni, A., Kaushik, S., *et al.* Nonsteroidal estrogen receptor isoform-selective biphenyls. *Chem Biol Drug Des.* 2018; 91(2):620–630.

52. Bohm, H.J. The computer program LUDI: A new method for the de novo design of enzyme inhibitors. *J Comput Aided Mol Des.* 1992; 6(1):61–78.

53. Rarey, M., Kramer, B., Lengauer, T., *et al.* A fast flexible docking method using an incremental construction algorithm. *J Mol Biol.* 1996;261(3):470–489.

54. Makino, S., and Kuntz, I.D. Automated flexible ligand docking method and its application for database search. *J Comput Chem.* 1997;18(14):1812–1825.

55. Miller, M.D., Kearsley, S.K., Underwood, D.J., *et al.* FLOG: A system to select ?quasi-flexible? Ligands complementary to a receptor of known three-dimensional structure. *J Comput Aided Mol Des.* 1994; 8(2):153–174.

56. Hart, T.N., and Read, R.J. A multiple-start Monte Carlo docking method. 1992;13(3):206–222.

57. McMartin, C., and Bohacek, R.S. QXP: Powerful, rapid computer algorithms for structure-based drug design. *J Comput Aided Mol Des.* 1997;11(4):333–344.

58. Verdonk, M.L., Cole, J.C., Hartshorn, M.J., *et al.* Improved protein-ligand docking using GOLD. *Protein.* 2003;52(4):609–623.

59. Taylor, J.S., and Burnett, R.M. DARWIN: A program for docking flexible molecules. *Proteins.* 2000;41(2):173–191.

60. Morris, G.M., Huey, R., Lindstrom, W., *et al.* AutoDock4 and AutoDockTools4: Automated docking with selective receptor flexibility. *J Comput Chem.* 2009;30(16):2785–2791.

61. Baxter, C.A., Murray, C.W., Clark, D.E., *et al.* Flexible docking using Tabu search and an empirical estimate of binding affinity. *Proteins.* 1998;33(3):367–382.

62. Scheraga, H.A., and Trosset, J.Y. PRODOCK: Software package for protein modeling and docking. *J Comput Chem.* 1999;20(4): 40412.

63. Ewing, T.J.A., and Kuntz, I.D. Critical evaluation of search algorithms for automated molecular docking and database screening. *J Comput Chem.* 1997;18(9):1175–1189.

64. Claußen, H., Buning, C., Rarey, M., *et al.* FlexE: Efficient molecular docking considering protein structure variations. *J Mol Biol.* 2001;308(2):377–395.

65. Jayaram, B., Dhingra, P., Lakhani, B., *et al.* Bhageerath — Targeting the near impossible: Pushing the frontiers of atomic models for protein tertiary structure prediction[#]. *J Chem Sci.* 2012;124(1):83–91.

66. Ortiz, A.R., Pisabarro, M.T., Gago, F., *et al.* Prediction of drug binding affinities by comparative binding energy analysis. *J Med Chem.* 1995;38(14):2681–2691.

67. Brooks, B.R., Bruccoleri, R.E., Olafson, B.D., *et al.* CHARMM: A program for macromolecular energy, minimization, and dynamics calculations. *J Comput Chem.* 1983;4(2):187–217.

68. Wang, J., Wang, W., Kollman, P.A., *et al.* Automatic atom type and bond type perception in molecular mechanical calculations. *J Mol Graph Model.* 2006;25(2):247–260.

69. Weiner, S.J., Kollman, P.A., Nguyen, D.T., *et al.* An all atom force field for simulations of proteins and nucleic acids. *J Comput Chem.* 1986;7(2):230–252.

70. Warren, G.L., Andrews, C.W., Capelli, A.M., *et al.* A critical assessment of docking programs and scoring functions. *J Med Chem.* 2006;49(20):5912–5931.

71. Su, M., Yang, Q., Du, Y., *et al.* Comparative assessment of scoring functions: The CASF-2016 update. *J Chem Inf Model.* 2019;59(2): 895–913.

72. Jain, T., and Jayaram, B. An all atom energy based computational protocol for predicting binding affinities of protein-ligand complexes. *FEBS Lett.* 2005;579(29):6659–6666.

73. Jain, T., and Jayaram, B. Computational protocol for predicting the binding affinities of zinc containing metalloprotein-ligand complexes. *Proteins.* 2007;67(4):1167–1178.

74. Jones, G., Willett, P., Glen, R.C., *et al.* Development and validation of a genetic algorithm for flexible docking. *J Mol Biol.* 1997;267(3): 727–748.

75. Yin, S., Biedermannova, L., Vondrasek, *et al.* MedusaScore: An accurate force field-based scoring function for virtual drug screening. *J Chem Inf Model.* 2008;48(8):1656–1662.

76. DeWitte, R.S., and Shakhnovich, E.I. SMoG: De Novo design method based on simple, fast, and accurate free energy estimates. *J Am Chem Soc.* 1996;118(2):11733–11744.

77. Muegge, I., and Martin, Y.C. A general and fast scoring function for protein–ligand interactions: A simplified potential approach. *J Med Chem.* 1999;42(5): 791–804.

78. Das, A., Kalra, P., Latha, N., *et al.* In silico trends in thermodynamics and kinetics of binding — New tools for de novo drug design. *Recent Trends Chem.* 2002;Chapter-15,218–223.

79. Huang, S.Y., and Zou, X. Inclusion of solvation and entropy in the knowledge-based scoring function for protein–ligand interactions. *J Chem Inf Model.* 2010;50(2):262–273.

80. Zheng, Z., and Merz, K.M. Development of the knowledge-based and empirical combined scoring algorithm (KECSA) To score protein–ligand interactions. *J Chem Inf Model.* 2013;53(5):1073–1083.

81. Bohm, H.J. The development of a simple empirical scoring function to estimate the binding constant for a protein–ligand complex of known three-dimensional structure. *J Comput Aided Mol Des.* 1994; 8(3):243–256.

82. Murray, C.W., Auton, T.R., and Eldridge, M.D. Empirical scoring functions. II. The testing of an empirical scoring function for the prediction of ligand-receptor binding affinities and the use of Bayesian regression to improve the quality of the model. *J Comput Aided Mol Des.* 1998;12(5):503–519.

83. Wang, R., Lai, L., and Wang, S. Further development and validation of empirical scoring functions for structure-based binding affinity prediction. *J Comput Aided Mol Des.* 2002;16(1):11–26.

84. Verkhivker, G., Appelt, K., Freer, S.T., *et al.* Empirical free energy calculations of ligand-protein crystallographic complexes. I. Knowledge-based ligand-protein interaction potentials applied to the

prediction of human immunodeficiency virus 1 protease binding affinity. *Protein Eng.* 1995;8(7):677–691.

85. Friesner, R.A., Murphy, R.B., Repasky, M.P., *et al.* Extra precision glide: Docking and scoring incorporating a model of hydrophobic enclosure for protein–ligand complexes. *J Med Chem.* 2006;49(21): 6177–6196.

86. Mukherjee, G., and Jayaram, B. A rapid identification of hit molecules for target proteins via physico-chemical descriptors. *Phys Chem Chem Phys.* 2013;15(23):9107.

87. Durrant, J.D., and McCammon, J.A. NNScore: A neural-network-based scoring function for the characterization of protein–ligand complexes. *J Chem Inf Model.* 2010;50(10):1865–1871.

88. Ballester, P.J., Schreyer, A., and Blundell, T.L. Does a more precise chemical description of protein–ligand complexes lead to more accurate prediction of binding affinity? *J Chem Inf Model.* 2014;54(3):944–955.

89. Zilian, D., and Sotriffer, C.A. SFCscore RF : A random forest-based scoring function for improved affinity prediction of protein–ligand complexes. *J Chem Inf Model.* 2013;53(8):1923–1933.

90. Li, G.B., Yang, L.L.; Wang, W.J., *et al.* ID-Score: A new empirical scoring function based on a comprehensive set of descriptors related to protein–ligand interactions. *J Chem Inf Model.* 2013;53(3):592–600.

91. Smith, R., Hubbard, R.E., Gschwend, D.A., *et al.* Analysis and optimization of structure-based virtual screening protocols. *J Mol Graph Model.* 2003;22(1):41–53.

92. Cheng, T., Li, X., Li, Y., *et al.* Comparative assessment of scoring functions on a diverse test set. *J Chem Inf Model.* 2009;49(4):1079–1093.

93. Smith, R.D., Dunbar, J.B., Ung, P.M.U., *et al.* CSAR benchmark exercise of 2010: Combined evaluation across all submitted scoring functions. *J Chem Inf Model.* 2011;51(9):2115–2131.

94. Damm-Ganamet, K.L., Smith, R.D., Dunbar, J.B., *et al.* CSAR benchmark exercise 2011–2012: Evaluation of results from docking and relative ranking of blinded congeneric series. *J Chem Inf Model.* 2013; 53(8):1853–1870.

95. Dunbar, J.B., Smith, R.D., Damm-Ganamet, K.L., *et al.* CSAR data set release 2012: Ligands, affinities, complexes, and docking decoys. *J Chem Inf Model.* 2013;53(8):1842–1852.

96. Wang, R., Fang, X., Lu, Y., *et al.* The PDBbind database: Collection of binding affinities for protein–ligand complexes with known three-dimensional structures. *J Med Chem.* 2004;47(12):2977–2980.

97. Chen, X., Lin, Y., and Gilson, M.K. The binding database: Overview and user's guide. *Biopolymers.* 2001;61(2):127–141.

98. Hu, L., Benson, M.L., Smith, R.D., *et al.* Binding MOAD (Mother Of All Databases). *Proteins.* 2005;60(3):333–340.

99. Pearlman, D.A. Evaluating the molecular mechanics Poisson–Boltzmann surface area free energy method using a congeneric series of ligands to p38 MAP kinase. *J Med Chem.* 2005;48(24):7796–7807.

100. Simonson, T., Carlsson, J., and Case, D.A. Proton binding to proteins: P K a calculations with explicit and implicit solvent models. *J Am Chem Soc.* 2004;126(13):4167–4180.

101. Narang, P., Bhushan, K., Bose, S., *et al.* A computational pathway for bracketing native-like structures for small alpha helical globular proteins. *Phys Chem Chem Phys.* 2005;7(11):2364.

102. Yan, Z., Wu, D., Hu, H., *et al.* Direct inhibition of hepatitis B e antigen by core protein allosteric modulator. *Hepatology.* 2019;70(1): 11–14.

103. Zhang, J., Adrián, F.J., Jahnke, W, *et al.* Targeting Bcr–Abl by combining allosteric with ATP-binding-site inhibitors. *Nature.* 2010; 463(7280):501–506.

104. Nwachukwu, J.C., Srinivasan, S., Zheng, Y. *et al.* Predictive features of ligand-specific signaling through the estrogen receptor. *Mol Syst Biol.* 2016;12(4):864.

105. Mulliken, R.S. Electronic population analysis on LCAO–MO molecular wave functions. I. *J Chem Phys.* 1955;23(10):1833–1840.

106. Gasteiger, J. and Marsili, M. Iterative partial equalization of orbital electronegativity — A rapid access to atomic charges. *Tetrahedron.* 1980; 36(22):3219–3228.

107. Singh, U.C. and Kollman, P.A. An approach to computing electrostatic charges for molecules. *J Comput Chem.* 1984;5(2):129–145.

108. Cornell, W.D., Cieplak, P., Bayly, C.I., *et al.* Application of RESP charges to calculate conformational energies, hydrogen bond energies, and free energies of solvation. *J Am Chem Soc.* 1993;115(21): 9620–9631.

109. Mukherjee, G., Lal Gupta, P., and Jayaram, B. Predicting the binding modes and sites of metabolism of xenobiotics. *Mol Biosyst.* 2015; 11(7):1914–1924.

110. Chatterjee, B.K., Jayaraj, A., Kumar, V., *et al.* Stimulation of heat shock protein 90 chaperone function through binding of a Novobiocin analog KU-32. *J Biol Chem.* 2019;294(16):6450–6467.

111. Kumar, J., Gill, A., Shaikh, M., *et al.* Pyrimidine-triazolopyrimidine and pyrimidine-pyridine has potential acetylcholinesterase inhibitors for Alzheimer's disease. *ChemistrySelect.* 2018:3(2):736–747.

112. Nishikawa, J.L., Boeszoermenyi, A., Vale-Silva, L.A., *et al.* Inhibiting fungal multidrug resistance by disrupting an activator–mediator interaction. *Nature.* 2016;530(7591):485–489.

113. Keiser, M.J., Roth, B.L., Armbruster, B.N., *et al.* Relating protein pharmacology by ligand chemistry. *Nat Biotechnol.* 2007;25(2):197–206.

114. Wang, L., Ma, C., Wipf, P., *et al.* TargetHunter: An in silico target identification tool for predicting therapeutic potential of small organic molecules based on chemogenomic database. *AAPS J.* 2013;15(2):395–406.

Index

CPSIA information can be obtained
at www.ICGtesting.com
Printed in the USA
BVHW010826220320
575155BV00008B/1